Nonlinear Optical Materials

ACS SYMPOSIUM SERIES 628

Nonlinear Optical Materials

Theory and Modeling

Shashi P. Karna, EDITOR
U.S. Air Force Phillips Laboratory

Alan T. Yeates, EDITOR
U.S. Air Force Wright Laboratory

Developed from a symposium sponsored
by the Division of Computers in Chemistry
at the 208th National Meeting
of the American Chemical Society,
Washington, DC,
August 21–25, 1994

American Chemical Society, Washington, DC 1996

Library of Congress Cataloging-in-Publication Data

Nonlinear optical materials: theory and modeling: developed from a symposium sponsored by the Division of Computers in Chemistry at the 208th National Meeting of the American Chemical Society, Washington, D.C., August 21–25, 1994 / Shashi P. Karna, Alan T. Yeates.

p. cm.—(ACS symposium series, ISSN 0097–6156; 628)

Includes bibliographical references and indexes.

ISBN 0–8412–3401–9 (alk. paper)

1. Nonlinear optics—Materials—Congresses. 2. Optical materials—Congresses.

I. Karna, Shashi P., 1956– . II. Yeates, Alan T., 1955– III. Series.

QC446.15.N648 1996
621.36′9—dc20

96–956
CIP

This book is printed on acid-free, recycled paper.

PRINTED IN THE UNITED STATES OF AMERICA

Foreword

THE ACS SYMPOSIUM SERIES was first published in 1974 to provide a mechanism for publishing symposia quickly in book form. The purpose of this series is to publish comprehensive books developed from symposia, which are usually "snapshots in time" of the current research being done on a topic, plus some review material on the topic. For this reason, it is necessary that the papers be published as quickly as possible.

Before a symposium-based book is put under contract, the proposed table of contents is reviewed for appropriateness to the topic and for comprehensiveness of the collection. Some papers are excluded at this point, and others are added to round out the scope of the volume. In addition, a draft of each paper is peer-reviewed prior to final acceptance or rejection. This anonymous review process is supervised by the organizer(s) of the symposium, who become the editor(s) of the book. The authors then revise their papers according to the recommendations of both the reviewers and the editors, prepare camera-ready copy, and submit the final papers to the editors, who check that all necessary revisions have been made.

As a rule, only original research papers and original review papers are included in the volumes. Verbatim reproductions of previously published papers are not accepted.

ACS BOOKS DEPARTMENT

Contents

INDEXES

Preface

SINCE THE FIRST THEORETICAL DESCRIPTION of nonlinear optical phenomena was published in 1962, many people have worked to improve our fundamental understanding of the microscopic mechanism and origin of these processes. These efforts extend from the development of simple phenomenological models for explaining experimental observations in terms of a few physical quantities to the development and application of increasingly sophisticated quantum chemical techniques to correctly predict structure–nonlinear optical property relationship information that can be used to design and develop new materials in an efficient and cost-effective manner.

Despite the contributions of many individuals, we felt the field lacked a single central source of information for new researchers to get a good overview of the theoretical aspects of nonlinear optical materials design. The symposium on which this book is based provided a useful exchange of information. We convinced several of the leading authorities in the field to contribute chapters describing the history and state of the art from their individual perspectives in the design and understanding of nonlinear optical materials. This book is the culmination of our efforts.

The interest in quantum mechanical modeling of nonlinear optical materials has, fortunately, coincided with the availability of low-cost computers with fast central processing units, so that it is becoming possible to address more basic and fundamental problems for increasingly complex and realistic model systems. It might even be argued that the increased demand for the computational modeling of nonlinear optical and other materials is largely responsible for the rapid pace of high-performance computer development. In addition, these demands have also made the development of software systems that can be used to model nonlinear optical materials on these machines a profitable and competitive business. With increased access to modeling tools and a spiraling cost of materials and waste disposal, computational modeling probably will acquire a more significant status and make valuable contributions to the development of new nonlinear optical materials. We hope that this book will serve as an important and useful resource for practicing scientists as well as newcomers in the field of computational nonlinear optical materials development. We also hope that experimentally oriented scientists may find this book useful as an aid to understanding the promise, and the limitations, of computational modeling for nonlinear optical design.

The chapters are grouped together by general technique(s) used for the modeling and their applications. After an introduction, graciously contributed by N. Bloembergen, one of the founders of the field, and the overview chapter, the next three chapters (2–4) focus on the ab initio time-dependent Hartree–Fock and post-Hartree–Fock techniques and their application for the calculation of molecular nonlinear optical materials. Then Chapters 5–7 discuss the Hartree–Fock-based semiempirical techniques and their applications to modeling the second- and third-order organic nonlinear optical materials. Chapters 8 and 9 focus on the development of density-functional techniques and their application to predict molecular nonlinear optical coefficients. Chapter 10 discusses a time-dependent perturbation theory for determining nonlinear optical properties of polymers. Chapters 11 and 12, also devoted to the nonlinear optical properties of polymers, discuss model Hamiltonian methods and their applications. We conclude with an experimental chapter devoted to one of the most recent applications of nonlinear optical materials—the resonant nonlinear optical phenomenon, which presents a major challenge to theoretical modeling.

Acknowledgments

We are grateful to the following people whose contributions made this book possible. First we thank all the authors for contributing their time and expertise to create authoritative and stimulating chapters. We also thank John Kester, Robert J. Hildreth, and John Wilkes of the Frank J. Seiler Laboratory, and Bruce Reinhardt and Doug Dudis of Wright Laboratory for their help and valuable discussions throughout the progress of this book. We owe a special thank you to Sylvia Miles at Seiler Laboratory for organizing the chapter contributions and reviews. Finally, we gratefully acknowledge financial support for this research from the Air Force Office of Scientific Research, the Wright Laboratory Materials Directorate, and Biosym, Inc.

SHASHI P. KARNA
Space Electronics Division
U.S. Air Force Phillips Laboratory
3550 Aberdeen Avenue, Southeast
Kirtland Air Force Base, NM 87117–5776

ALAN T. YEATES
Polymer Branch
Wright Laboratory
Wright-Patterson Air Force Base, OH 45433–7750

December 20, 1995

Introduction

THE FIELD OF ORGANIC NONLINEAR OPTICAL materials has been the subject of intense research efforts during the past two decades. Several important applications in nonlinear optical devices appear close to realization. They have been demonstrated in principle, but the economic and engineering considerations require further developments.

It is timely to have a collection of the various theoretical approaches to describe the nonlinear optical properties of large organic molecules and of polymers. The results are compared with the ever-increasing body of experimental results. They include diverse nonlinear optical phenomena such as third-harmonic generation, two-photon absorption, electric field induced second-harmonic generation, intensity-dependent index of refraction, stimulated Raman scattering, and others.

This book provides an up-to-date account and critical discussion that emphasize the issues of theoretical modeling of nonlinear optical properties of organic materials. It is a welcome addition to the ever-growing literature in nonlinear optics.

N. BLOEMBERGEN
Division of Applied Science
Harvard University
Pierce Hall
Cambridge, MA 02138

October 29, 1995

Chapter 1

Nonlinear Optical Materials: Theory and Modeling

Shashi P. Karna[1] and Alan T. Yeates[2]

[1]Space Electronics Division, U.S. Air Force Phillips Laboratory, 3550 Aberdeen Avenue, Southeast, Kirtland Air Force Base, NM 87117–5776
[2]Polymer Branch, Wright Laboratory, Wright-Patterson Air Force Base, OH 45433–7750

With a view to developing new materials for technological applications, the theory and mechanism of nonlinear optical phenomena in various dielectric media are reviewed. The role of theoretical modeling in advancing our understanding of the structure-property relationships of nonlinear optical materials is discussed.

I. INTRODUCTION

Since the discovery by Franken et al[1] of optical second harmonic generation (SHG) in quartz crystal, a number of optical frequency mixing phenomena have been observed in a variety of dielectric media. Detailed descriptions of these processes, collectively known as the nonlinear optical (NLO) processes, the dielectrics in which these are observed, and their practical applications have already appeared in a number of excellent texts[2–6] and symposia proceedings.[7–9] In this chapter, we present a brief account of the theory and modeling of NLO materials with a view of their applications in the current and future technology.

NLO phenomena encompass a broad range of light (electromagnetic radiation) mediated processes from the commonly observed Raman scattering to the less commonly observed two-photon absorption and optical harmonic generation. In this chapter, we will limit our discussion to those NLO phenomena which are principally determined by electronic polarizations alone. It should be noted, however, that even in these cases, there may be some contribution from nuclear motions in the molecule or unit cells.[5,6]

0097–6156/96/0628–0001$15.50/0

The linear and nonlinear polarizations are generally characterized by their respective susceptibilities. The bulk electronic susceptibilities are defined by a power series expansion of the polarization **P** as a function of the applied electric field,[2]

$$\mathbf{P} = \chi^{(1)} \cdot \mathbf{E} + \chi^{(2)} : \mathbf{EE} + \chi^{(3)} \vdots \mathbf{EEE} + \cdots. \tag{1}$$

Here, $\chi^{(n)}$ is the nth-order susceptibility of the medium and **E** represents the total electric-filed experience by the system. The nth order susceptibility is a tensor quantity of rank $(n+1)$ which has $3^{(n+1)}$ elements. Thus, $\chi^{(1)}$ is a second-rank tensor with six elements, $\chi^{(2)}$ is a third-rank tensor with twenty-seven elements, $\chi^{(3)}$ is a fourth-rank tensor with eighty-one elements, and so on. Of course, not all the elements of a susceptibility tensor are linearly independent and in most practical materials much fewer elements are required to describe the tensor.

In general, NLO phenomena are characterized as second-order or third-order depending on whether they are described primarily through the $\chi^{(2)}$ or the $\chi^{(3)}$ terms. The SHG is an example of a second-order phenomenon in which two photons, each of an angular frequency ω, combine to produce a third photon with angular frequency 2ω. Some of the inorganic materials such as dihydrogen phosphate (KDP), lithium niobate, and barium titanate having large $\chi^{(2)}$ value have found applications as frequency doublers for the powerful lasers used in laser fusion. For low power applications, organic dyes with suitable $\chi^{(2)}$ values have been found to be useful for upconverting semiconductor lasers.[5,9,10] These dyes must be incorporated into polymer host for mechanical stability and processibility, which presents the main challenge to the development of SHG materials. Since the dyes are inherently disordered in the host system, there is a center of symmetry which destroys the second-order response. In order to overcome this difficulty, the dye molecules are aligned in an intense static electric field at a temperature above glass transition of the polymer. When the polymer is cooled, the dye molecules are aligned preferentially giving rise to a second-order response. Unfortunately, such systems are unstable with respect to time and temperature, tending to return to their original disordered state. For a material to be commercially viable as a component in an "on-chip" optical device, it must maintain a good second-order response when exposed to temperatures as high as 320 ^{0}C. For military applications it must also maintain 95% of its original response after 10 years at 125 ^{0}C.[11] These are very stringent requirements which will necessitate much further research.

The second-order NLO phenomenon of most interest for military applications is the electrooptic (EO) or Pockels effect. The EO effect, mediated by a $\chi^{(2)}$ term, arises from the change in the index of refraction of a dielectric in the presence of a static or low-frequency optical field. The refractive index of the material varies linearly with the strength of the applied electric field. The EO phenomenon finds applications in the development of active optical interconnects and switches, which will be used in future data processing and communication systems in a wide range of aircraft and satellite systems. Here organic polymers have a great advantage over inorganic materials. While the EO response of organic materials is comparable to or better than the best inorganic materials,[5] the low dielectric constants of the former result in a much *faster* response. This is because the polarization in organics results primarily from electronic motions and have much smaller nuclear (vibrational) component. Thus EO switches can operate at much greater speeds (<ps) and EO modulators can operate at much higher bandwidths (>40 GHz). In addition, organic polymers can be easily processed into films and waveguides, making on-chip applications much easier. However, as with the SHG applications, thermal and temporal stability of the organic materials remains an issue. The optical quality of the materials is also an issue to be taken into account. Optical losses on the order of 1dB/cm or less must be obtained, making the phase stability and purity of the materials a critical issue. Materials with a good combination of properties for second-order on-chip device applications have yet to be developed.

While second-order materials are tantalizingly close to finding application, third-order materials are much further from real applications. However, the possible applications for third-order materials are just as profound. The ultimate goal for the application of third-order materials is the development of "all-optical" (AO) computing and signal processing. This will require the development of materials that have a large *nonresonant* value of $\chi^{(3)}$. Such materials must be able to undergo optically induced changes in the refractive index. So far the largest values of $\chi^{(3)}$ have come from highly conjugated organic polymers.[5-9] While a large number of such compounds have been investigated, none have demonstrated the necessary magnitude of $\chi^{(2)}$ to be of practical, or even laboratory, use in a prototype AO switch. One major problem that exists in the development of linear (or *quasi-linear*) conjugated polymers for AO applications is the interference from low-lying two-photon absorptions.[5] Currently it appears likely that because of high optical losses from such absorptions, long-chain conjugated polymers will probably never be usable for AO switching applications. Alternative materials currently being investigated include two-dimensionally conjugated molecules, which have higher lying two-photon state, and molecules which produce a light induced refractive index change as a result of cascaded second-order optical nonlinearities.[12]

The outlook for the third-order materials is not entirely bleak. Uses have been found recently for materials with a large resonant value of $\chi^{(3)}$. The first is the two-photon upconverted lasing. In this application, a two-photon absorption in the 600 – 900 nm region of the spectrum can be used to produce a population inversion and lasing in the blue region (400 – 480 nm). However, as of this writing, only super-radiance (single-pass amplification) has been demonstrated.[12] It is possible, at least theoretically, to place these materials in a cavity to obtain true lasing. This process, when realized, would provide an alternative to the SHG for the production of coherent blue light. Other potential applications of such materials include higher density optical data storage and high-frequency optical communications. A successful material would have to floresce in the blue, while having a strong two-photon absorption in the near-IR. Since no structure-property relationships have been established for two-photon absorption, computational modeling can have a major influence on the development and successful applications of these materials. These same materials also show promise for another application, namely, optical limiting in the visible and near-IR. Optical limiting is a material property by which low-intensity light is passed and high-intensity light is absorbed. The mechanism is governed again by the two-photon absorption process because, while the fraction of light absorbed is independent of intensity for the linear region (Beer's law), the fraction of light absorbed through two-photon process depends linearly on the intensity of light. Thus intense light will be more strongly absorbed due to the two-photon absorptions leading to a limiting of the passed intensity. As is clear, this property has important commercial and military applications in high-intensity radiation protective devices. The military applications of the optical limiting process center around protection of sensors and eyes from hostile laser threats. Again, theoretical modeling directed toward advancing our understanding of the two-photon absorption process and its relationships with the structural features of materials is of paramount importance for a speedy development of optical limiting materials.

In what follows, a brief survey of the general theory of NLO processes is given in Section II. In Section III, theoretical models used to describe the origin and mechanisms of the second- and the third-order NLO materials are reviewed. The recent observations of SHG in silica glass based materials is discussed in section IV. Finally, the role of theory and modeling in developing new NLO materials is summarized in Section V.

II. QUANTUM MECHANICAL THEORY OF NLO PHENOMENA

The first successful theoretical explanation of NLO processes in an infinite, homogeneous, nonlinear dielectric medium was provided by Armstrong et al.[13] These authors were able to describe various optical frequency mixing phenomena, e.g. SHG and third-harmonic-generation (THG), by introducing a generalized macroscopic nonlinear polarization, $\mathbf{P}^{\mathrm{NL}}$ and solving the Maxwell equation in an infinite, anisotropic dielectric medium. Their theory was subsequently extended by Bloembergen and Pershan[14] to finite boundary conditions which described the optical harmonic generation in experimental situation. According to this theory, in the presence of an external optical electric field $\mathbf{E}$, the total polarization $\mathbf{P}$ of a dielectric medium can be expressed as the sum of a linear polarization $\mathbf{P^L}$ and a nonlinear polarization $\mathbf{P^{NL}}$ defined, respectively, as

$$\mathbf{P^L} = \chi^{(1)} \cdot \mathbf{E} \tag{2}$$

and

$$\mathbf{P^{NL}} = \chi^{(2)} : \mathbf{EE} + \chi^{(3)} \vdots \mathbf{EEE} + \cdots. \tag{3}$$

The susceptibility tensors $\chi^{(n)}$ in the above equations are related to the polarizabilities of the microscopic units of the dielectric. Considering the angular frequency of the incoming light beam to be ω, the linear susceptibility tensor, $\chi^{(1)}$, of the medium can be written as

$$\chi^{(1)}_{a,b}(\omega) = \sum_p \sum_{i,j} N^{(p)}_{a,i}(\omega) N^{(p)}_{b,j}(\omega)\, \alpha^{(p)}_{i,j}(\omega). \tag{4}$$

In the above equation, $\alpha^{(p)}_{i,j}(\omega)$ is the element of the linear polarizability tensor at the pth microscopic site and $N^{(p)}_{b,j}(\omega)$ is related to the local field created by the optical beam of frequency ω at the same site. If we consider two optical beams of angular frequencies ω_1 and ω_2 interacting in a medium to creat a third beam of frequency ω_σ, then the second-order susceptibility tensor $\chi^{(2)}$ is related to the microscopic polarizability tensor, β, as

$$\chi^{(2)}_{a;\,b,c}(\omega_\sigma = \omega_2 + \omega_1) = \sum_p \sum_{i,j,k} N^{(p)}_{a,i}(\omega_\sigma) N^{(p)}_{b,j}(\omega_2) N^{(p)}_{c,k}(\omega_1)\, \beta^{(p)}_{i;\,j,k}(\omega_\sigma = \omega_2 + \omega_1). \tag{5}$$

The third-order susceptibility tensor, $\chi^{(3)}$, corresponding to the generation of a light beam of frequency ω_σ from the mixing of three input beams of frequencies ω_1, ω_2, and ω_3 is related to the microscopic polarizability tensor, γ, as

$$\begin{aligned} &\chi^{(3)}_{a;\,b,c,d}(\omega_\sigma = \omega_3 + \omega_2 + \omega_1) \\ &= \sum_p \sum_{i,j,k,l} N^{(p)}_{a,i}(\omega_\sigma) N^{(p)}_{b,j}(\omega_3) N^{(p)}_{c,k}(\omega_2) N^{(p)}_{d,l}(\omega_1)\, \gamma^{(p)}_{i;\,j,k,l}(\omega_\sigma = \omega_3 + \omega_2 + \omega_1). \end{aligned} \tag{6}$$

In a centrosymmetric medium, the odd-rank tensor (e.g., $\chi^{(2)}$ and β) vanishes, whereas the even-order tensors have nonvanishing values regardless of the symmetry of the medium. It is important to note that the macroscopic susceptibility tensor, $\chi^{(n)}$, has the point symmetry properties of the medium, such as a crystal lattice as a whole, whereas the polarizabilities (α, β, γ, etc.) have the symmetry properties of individual microscopic units, for example an atom on the pth lattice site, a unit cell containing several atoms, or a diatomic bond, that constitute the medium. Both, the macroscopic and the microscopic tensors of a given order n, however, obey some common permutational symmetry. For example, in the second-order case, $\beta_{ijk}(\omega_\sigma = \omega_2 + \omega_1) = \beta_{kij}(\omega_1 = \omega_\sigma - \omega_2) = \beta_{jki}(\omega_2 = \omega_\sigma - \omega_1)$, which is also true for the elements of $\chi^{(2)}$. Similar relation holds for the third-order susceptibility (polarizability) tensor. In other words, the frequency arguments may be permuted at will provided the Cartesian indices i, j, k, etc. are simultaneously permuted so that a given frequency is always associated with the same index. In the limit of low optical frequencies, however, one can freely permute the indices i, j, k, etc., without much loss of generality. This approximation, which considerably reduces the number of independent elements to describe the susceptibility tensor, is known as the Kleinman symmetry.[15]

The susceptibility tensors $\chi^{(2)}$ and $\chi^{(3)}$ in eq. (3) are a measure of the NLO response of a medium. That is, a large value for $\chi^{(n)}$ of a medium is associated with large optical nonlinearity and a fast NLO response. From eqs. (5) and (6), one can easily infer that the large optical nonlinearity of a medium is directly related to the NLO response of its microscopic units. We will see in the next section how this basic concept has played a central role in the development of new NLO materials.

Armstrong et al[13] also derived quantum mechanical expressions to calculate the elements of the microscopic polarizabilities β and γ in terms of the matrix elements of the dipole moment operator and energy of the excited states. A detailed quantum mechanical treatment of microscopic polarizabilities in a nonabsorbing medium was later given by Franken and Ward[16] and Ward[17] which was subsequently generalized by Orr and Ward[18] to include resonance cases. Their expressions for the elements of β and γ, respectively, in a nonresonant case can be written as

$$\beta_{abc}(\omega_\sigma;\omega_1,\omega_2) = \frac{1}{2}K(\omega_\sigma,\omega_1,\omega_2)\cdot\wp(\omega_\sigma,\omega_1,\omega_2)\cdot \left\{\sum_{m,n\neq g}\frac{<g|a|m><m|\bar{b}|n><n|c|g>}{(\Delta E_{mg}-\hbar\omega_\sigma)(\Delta E_{ng}-\hbar\omega_1)}\right\}, \tag{7}$$

$$\gamma_{abcd}(\omega_\sigma;\omega_1,\omega_2,\omega_3) = \frac{1}{6}K(\omega_\sigma,\omega_1,\omega_2,\omega_3)\cdot\wp(\omega_\sigma,\omega_1,\omega_2,\omega_3)\cdot \left\{\sum_{l,m,n\neq g}\frac{<g|a|l><l|\bar{d}|m><m|\bar{c}|n><n|b|g>}{(\Delta E_{lg}-\hbar\omega_\sigma)(\Delta E_{mg}-\hbar\omega_1-\hbar\omega_2)(\Delta E_{ng}-\hbar\omega_1)} - \sum_{m,n\neq g}\frac{<g|a|m><m|d|g><g|c|n><n|b|g>}{(\Delta E_{mg}-\hbar\omega_\sigma)(\Delta E_{ng}-\hbar\omega_1)(\Delta E_{ng}+\hbar\omega_2)}\right\}. \tag{8}$$

In the above equations, $a, b, c, d = x, y, z$; g represents the ground state, l, m, n are the excited states, and ΔE represents the energy difference between two states. The matrix element with barred operator is defined as

$$<m|\bar{r}|n> = <m|r|n> - <g|r|g>. \tag{9}$$

The summations in eqs. (7) and (8) are performed over all terms generated by the permutation of the frequency arguments as indicated by the operator $\wp$. This in eq. (7) leads to six terms and in eq. (8) to twenty-four terms. The values of the numerical coefficients $K(\omega_\sigma;\omega_1,\omega_2)$ and $K(\omega_\sigma;\omega_1,\omega_2,\omega_3)$ depend on particular NLO processes and have been tabulated by Orr and Ward.[18] The Orr-Ward theory of the microscopic nonlinear polarizabilities, also known as hyperpolarizabilities, has played a pivotal role in advancing our understanding of the mechanism and origin of NLO phenomena in atoms and molecules. This theory has also played a crucial role in the development of the "intermediate-state model" which has proved very successful in explaining the mechanism of NLO processes in metals and crystals.[6]

Equations (7) and (8) provide a straight-forward means for a quantitative prediction of the NLO properties of atoms and molecules. All that is required is the matrix elements of the dipole operator and the energies. Although the summation in eqs. (7) and (8) run over the entire excited state space leading up the continuum, in reality there are far fewer matrix elements in the numerator which have sufficient magnitude to be of importance. Therefore, in quantum mechanical calculations, it often suffices to perform the summation over a truncated space comprising of the most important states in the excited-state manifold.

Another quantum mechanical method that has proved to be quite useful in the recent years for theoretical prediction of the linear and NLO polarizabilities is the density matrix approach of Sekino and Bartlett.[19] In Sekino-Bartlett theory, the polarizabilities, which are perturbation enrgies of various order,[20] are calculated as the expectation value of the operator $\mathbf{H}(r,t)$ $(= \mathbf{H}^{(0)} + \mu \cdot \mathbf{E}(r,t))$ as[21]

$$\mathbf{p} = <\Psi(r,t)|\mathbf{H}(r,t)|\Psi(r,t)>, \tag{10}$$

where, $\Psi(r,t)$ is the total wavefunction of the system in the presence of the external optical field, $\mathbf{E}(r,t)$. A Taylor series expansion of the above equation yields the following expressions:[19]

$$\alpha_{ab}(\omega) = -\mathrm{Tr}\ [\mathbf{h}_a^{(1)}\mathbf{D}^{(1)}(\omega)], \tag{11}$$

$$\beta_{abc}(-\omega_\sigma;\omega_1,\omega_2) = -\mathrm{Tr}\ [\mathbf{h}_a^{(1)}\mathbf{D}_{bc}^{(2)}(\omega_1,\omega_2)], \tag{12}$$

$$\gamma_{abcd}(-\omega_\sigma;\omega_1,\omega_2,\omega_3) = -\mathrm{Tr}\ [\mathbf{h}_a^{(1)}\mathbf{D}_{bcd}^{(3)}(\omega_1,\omega_2,\omega_3)], \tag{13}$$

for the elements of the linear and NLO polarizabilities. In the above equations, a, b, c, d $(= x, y, z)$ represent the Cartesian conponents, $\mathbf{h}_a^{(1)}$ is the ath component of the dipole moment matrix, and $\mathbf{D}^{(1)}, \mathbf{D}^{(2)}, \mathbf{D}^{(3)}$ are the ground-sate one-particle perturbed density matrices of order 1, 2, and 3, respectively. Quantum mechanical method to evaluate $D^{(n)}$ by perturbation theory is described in detail in the original paper of Sekino and Bartlett.[19]

The Sekino-Bartlett theory provides a means to calculate (hyper)polarizabilities as a *change in the ground-state charge density of the system in the presence of an external optical field.* Unlike the Orr-Ward theory, where accurate knowledge of the excited state wavefunction is as important as that of the ground-state wavefunction, the Sekino-Bartlett approach requires an accurate description of the ground-state wavefunction alone. Since an accurate determination of the excited state wavefunction, except for a few atomic cases,[22,23] is considerably more difficult than that of the ground-state wavefunction, the Sekino-Bartlett approach offers a better choice than the Orr-Ward theory in the first-principles calculations involving large molecular systems. However, the Sekino-Bartlett theory lacks the advantage of the Orr-Ward theory in its ability to explain NLO processes in terms of the physically useful concept of intermediate states.

The Sekino-Bartlett theory[19] has also served as a catalyst for a number of recent developments related with the first-principles calculation of NLO coefficients. Notable among others are the works of Karna and Dupuis,[24] Rice et al,[25] Årgen et al,[26] Karna,[27] and Sasagane et al[28] to calculate linear and NLO polarizabilities of closed-shell atomic and molecular systems. A recent extension[29] of this theory allows to calculate the linear and the NLO properties in terms of the perturbed-density matrices obtained for individual spins.

As alluded to in the preceding section, the electronic contributions alone cannot account for the observed NLO properties of materials. There is enough experimental evidence[5] to suggest that at sufficiently low optical frequencies the observed NLO processes have sizeable contribution from the nuclear motion of the system as well. In this regard, while the Orr-Ward theory is quite general and can be used to calculate electronic as well as vibrational NLO polarizabilities (in the latter case the wavefunction and the energy difference in eqs. (7) and (8) correspond to *vibrational* states), the Sekino-Bartlett theory does not lend itself to account for such effects. Quantum mechanical theory related to the vibrational NLO properties has been the subject of a number of recent studies.[30–32] It should be, however, noted that unlike the electronic processes, the nuclear process are inherently slow, and although the nuclear NLO phenomena such as the hyper-Raman and hyper-Rayleigh effects[33] are important within their own rights for probing the structural features of materials, at the current time they hold rather little interest for device applications.

III. THEORETICAL MODELING OF NLO MATERIALS

The technological applications of NLO phenomena have been the driving force behind the search for materials with appropriate properties. The SHG and the third harmonic generation (THG) processes are among some of the familiar NLO processes which are routinely used in frequency up-conversion in modern lasers. From two optical beams of frequencies, ω_1 and ω_2, a third beam at frequency ω_3 can be generated via optical parametric up-conversion or down-conversion. Other more ambitious applications of NLO processes are in target recognition and image reconstruction via creation of a time-reversed beam by phase-conjugation, in optical switching through the intensity-dependence of refractive index (IDRI) of materials, in sensors and eye protectors via optical limiting, a phenomenon that too utilizes IDRI, and in optical communications.

Each application of the NLO phenomena requires a certain material characteristic that will enable it to function efficiently according to the demands of a particular device. Among the more practical requirements are the thermal, chemical, optical, and temporal stabilities and processibility for a desired device configuration.[34] While these requirements must be met for practical applications of NLO phenomena, the issue of paramount importance remains the availability of materials with sufficiently large NLO susceptibilities. The number of materials identified to be suitable for frequency-converters, parametric oscillators, and optical inter-connects on the basis of the experimental characterization is rather limited and those already discovered have found appropriate technological applications.[34] The success of the practical realization of other NLO devices rests on the development of appropriate materials. It is in this regard that a clear understanding of the relationships between the NLO properties and the structural features of materials serves as a guide.

Much of our current understanding of the structure-NLO property relationships of materials has emerged from a combination of the experimental observations and theoretical models to explain them. Although, no single theoretical model is capable of explaining the mechanism of NLO phenomena in all materials, most observations can be explained on the basis of the microscopic structure-property relationships in materials.

1. The second-order materials

A number of theoretical models have been proposed to describe the structure-NLO property relationships in second-order, or $\chi^{(2)}$ materials. One of the earliest model is the so called *bond parameter model.*[2] According to this model, the total induced polarization in the system is the sum total of the polarizations induced in individual chemical bonds between two atoms. A slight variation of this model is the *bond charge model.* [35] Collectively known as the *bond additivity model*, since it is based on the assumption that the bond polarization is additive, this model has been quite successful in explaining the structure-NLO property relationships in inorganic crystals and tetrahedrally coordinated semiconductors.[2–4,6] It, however, fails to explain the unusually large NLO susceptibilities in non-σ type crystalline materials or conjugated π-orbital systems.[5,7–9]

Interest in organic crystals as a viable second-order material arose from a number of experimental observations[36–39] in the early 1970's which demonstrated that the NLO susceptibilities of these materials were comparable to or, in some cases, better than the best known inorganic materials.[34] Since organic crystals are generally molecular in nature, i.e. the chemical and physical properties of organic crystals are closely related to the corresponding properties of the molecular units constituting them, we can write from eq. (5)

$$\chi^{(2)} = \sum_p \mathbf{N}^{(p)} \beta^{(p)}, \tag{14}$$

where, now p refers to a molecule as the microscopic unit. We can further simplify the above equation assuming that due to weak intermolecular interaction in organic crystals, the tensor $\mathbf{N}^{(p)}$ is independent of the site p and has a value close to unity. This allows us to write eq. (14) as

$$\chi^{(2)} = \sum_p \beta^{(p)}. \tag{15}$$

Stated otherwise, the macroscopic second-order optical susceptibility $\chi^{(2)}$ of organic crystals is a simple geometrical superposition of its molecular hyperpolarizabilities. Although oversimplified, eq. (15) provides a means to optimize $\chi^{(2)}$ by suitably modifying β of an organic molecule. The question we need to answer now is: How do we optimize (here *maximize*) β for organic molecules? In order to seek an answer to this question, we need to know the origin of microscopic nonlinearity in organic molecules and establish some physical parameters that can be suitably tunned to design and develop materials with large $\chi^{(2)}$.

Much of the progress toward establishing the origin of β and its relationships with the electronic structure and spectrum in conjugated organic molecules was made from the works of Gott,[38] Oudar et al,[40–45] and Levine and Bethea.[46–51] From the powder SHG measurements[52] on a number of aromatic crystals, Gott[38] found that the compounds in which a polarizable and highly inductive substituent group, such as a nitro (-NO_2) group, is coupled to the aromatic ring has the highest NLO susceptibility. Such compounds have high optical polarizability and relatively high dipole moment. The observation by Gott[38] demonstrated an intimate relationship between the inductive effect and microscopic optical nonlinearity, later explored in detail by Oudar et al[40–45] and Levine and Bethea.[46–51] In the case of monosubstituted aromatic compounds, Oudar and Chemla[40] argued that a substituent radical (R) distorts the π electron cloud of the aromatic ring (M) and creates an internal electric field $\mathbf{E}_R$ which induces a dipole moment $\mu_R = \alpha_M \mathbf{E}_R$, where α_M is the linear polarizability of the parent molecule. In the presence of the external optical-field $\mathbf{E}(r,t)$, the induced polarization contains a term which is quadratic in $\mathbf{E}_R$ which leads to second-order polarizability β_{M-R} of the substituted system, given by

$$\beta_{M-R} = 3(\frac{\gamma_M}{\alpha_M})\mu_R, \tag{16}$$

where γ_M represents the third-order or the second-hyperpolarizability of the aromatic molecule. Now, if the molecules are aligned in the crystal such that all the polar (substituent) groups point in the same direction, the above β_{M-R} will lead to a second-order macroscopic susceptibility $\chi^{(2)}$ for the compound. Levine[46] had also arrived at a similar expression for the substituent induced β starting from a purely quantum mechanical arguments. Note that the vector part of β is parallel to μ_R. Equation (16) has a number of far reaching consequences toward modeling π-electron conjugated $\chi^{(2)}$ materials. First, it tells us that $\beta_{M-R} \propto \mu_R$. Since μ_R depends on the strength of the substituent-π-electron interaction, the larger the inductive effect of a substituent radical the greater the value of β_{M-R}. Second, it lets us modulate β in terms of the linear polarizability α and the second-hyperpolarizability γ as $\beta \propto (\gamma/\alpha)$. The second-hyperpolarizability γ of the conjugated systems is known to have unusually large value and increases more quickly than the linear polarizability α with the increase in the π-electron chain-length.[53–55] Therefore, the second-order polarizability β should also have large value. Furthermore, the value of β should increase rapidly with the π-electron chain length. Finally, eq. (16) also lets us determine the sign of β.

The above concept was later extended by Levine[48] and Oudar[44,45] to doubly substituted benzene molecule where one substituent was electron *donor* (D) and the other substituent was electron *acceptor* (A). It was reasoned by Levine[48] that an *intramolecular charge transfer* (ICT) between the donor and the acceptor group that leads to significant spectral shift in charge-transfer complexes[56] could also lead to considerable enhancement of the second-order polarizability β. From the experimental measurements on aniline, nitrobenzene, and nitroaniline with the amino (-NH_2) and the nitro (-NO_2) groups at *ortho*, *meta*, and *para* positions to each other, Levine[48] showed that due to a strong D-A ICT interaction in *para*-nitroaniline (*p*-NA) molecule, its β value was roughly 3 times larger than that of the *ortho*-nitroaniline (*o*-NA) and 5 times larger than that of the *meta*-nitroaniline (*m*-NA). Furthermore, the β value of *p*-NA was an order of magnitude larger than the corresponding value of the monosubstituted aniline and nitrobenzene molecules.

The qualitative model relating the ICT and β value was later given a quantitative treatment by Oudar and Chemla.[43] These authors proposed that in an ICT molecule, such as *p*-NA, the β value can be represented by

$$\beta = \beta_{ct} + \beta_{add}, \tag{17}$$

where, β_{ct} represents the contribution to β from the charge transfer (CT) state and β_{add} is the contribution of the benzene ring-substituent interaction. Then, by making use of the experimental and quantum chemical results of Lutskii and Gorokhova,[57] which showed that the CT state of *p*-NA resulting from the transfer of electronic charge from the nonbonding nitrogen orbital of the amino ($-NH_2$) group to the vacant orbital of the nitro ($-NO_2$) group was the first excited state of the valence manifold, these authors assumed that β_{ct} could be described by the contribution from this (lowest excited) state. Thus from eq. (7), β_{ct} for the SHG of a fundamental beam of angular frequency ω can be written as

$$\beta_{ct} = \frac{3e^2\hbar^2}{2m} \frac{\Delta E_{1g}}{(\Delta E_{1g} - 2\hbar\omega)(\Delta E_{1g} + 2\hbar\omega)\,(\Delta E_{1g} - \hbar\omega)(\Delta E_{1g} + \hbar\omega)} f_{1g}\,\Delta\mu_{1g}, \quad (18)$$

where, f_{1g} and $\Delta\mu_{1g}$ are, respectively, the oscillator strength and the dipole moment difference between the ground state and the first excited state, defined as

$$f_{1g} = \frac{2m}{\hbar^2} \Delta E_{1g} < g|r|\mathbf{1} >, \quad (19)$$

and

$$\Delta\mu_{1g} = < \mathbf{1}|r|\mathbf{1} > - < g|r|g > . \quad (20)$$

In eq. (17), β_{add} is obtained from eq. (8) by summing over the states other than the lowest one (**1**). Oudar and Chemla[43] used the experimental values of ΔE_{1g} and f_{1g} and theoretical values of to estimate β_{ct} for the three isomers (o, m, p) of nitroaniline. They also calculated β_{add} for these molecules from the results of nitrobenzene and aniline using the *additive* model for substituent-ring hyperpolarizability and found the ratio β_{ct}/β_{add} for *o*-NA, *m*-NA, and *p*-NA to be 5.1, 1.29, and 5.6, respectively. They also found that the order in the total measured β for the three isomers of NA was the same as the order in their respective β_{ct}. Furthermore, total β and β_{ct} had the same sign as the latter being the same as $\Delta\mu_{1,g}$.

The above findings of Oudar and Chemla[43] have a number of important implications. (a) First, it shows that the contribution from ICT to β in *p*-NA amounts to more than 80%. (b) Second, even for a non-ICT molecule containing donor-acceptor substituents, such as *o*-NA and *m*-NA, the contribution to β from the single lowest state amounts from 50% to 80% of the total value. (c) When β_{1g} is the dominant contribution, the sign of the β can be deduced from the sign of $\Delta\mu_{1g}$. This leads to the possibility of describing the second-order NLO effects in terms of the excitation and corresponding charge transfer between two states, the ground state (**g**) and the lowest excited state (**1**) of a molecule. Such a possibility has been the foundation of what is popularly known as the 'two-level model'. This oversimplified model has been quite successful in describing the mechanism of second-order NLO effects in the majority of organic solids, despite its obvious failure to account for the dispersion of optical nonlinearity close to a resonance.[51]

It should be noted that while the two-level expression (eq. (18)) does yield a good estimate of the second-order NLO polarizability β, the second term of eq. (17), β_{add}, that incorporates the contribution from all other states in the valence manifold *must* be included for a quantitative prediction. Experience has shown[51] that a two-level model does not only overestimate the magnitude of β but may also yield incorrect sign. Therefore, care must be exercised in associating too much importance to the second-order NLO response of materials based on the results of a two-level model alone.

The second-order NLO effects in organic structures also show a certain dependence on the π-electron conjugation length. An early experimental work of Dulcic et al[58] on a series of ICT molecules in which the D and A groups were separated by different lengths of π-electron conjugated chain, suggested that

$$\beta \propto L^2. \tag{21}$$

Here, L is the conjugation length of the chain and can be considered as proportional to the number of the carbon-carbon double bond (-C=C-) separating the D and the A group. This effect, though central in the modeling of third-order organic materials, has not been fully explored in the $\chi^{(2)}$ materials.

So far our discussion has been concerned with the NLO susceptibility as a function of the microscopic hyperpolarizability of a single molecular structure. As is well known, a molecular structure with large microscopic β does not necessarily lead to a crystal with large $\chi^{(2)}$. The most popular example is the *p*-NA molecule. Despite the large β value that this molecule possesses, it fails to display SHG in the crystalline form because it has a centrosymmetric structure. Therefore, factors other than the microscopic β of a single molecular species also needs to be considered in the modeling and design of NLO materials.

Levine and Bethea[49,50] first recognized that *inter-molecular* interactions and chemical bond-formation also play important roles in determining the second-order NLO susceptibility of organic solids. These authors found that the β value of organic molecules measured in polar solvents were 20% to 30% larger than the corresponding value measured in nonpolar solvents. Furthermore, the β value increased with the concentration of the polar solvent. The concentration-dependence of β in associating liquids, such as methanol and water, also indicated that the hydrogen-bond formation increased the value of β by substantial amount. In other cases, for example the pyridine-I_2 mixture, the large β value could be only explained on the basis of *inter-molecular* charge transfer (IMCT) between I_2 and pyridine molecules. These observations led Levine and Bethea[49,50] to develop a model based on purely electrostatic considerations for the microscopic nonlinearity of an interacting molecule A in the presence of other molecular species B. Using their model, a general expression for the β value of a molecular species A can be written as

$$\beta = \beta_g + \beta_{dip}^{AA} + \beta_{dip}^{AB} + \beta_{hb} + \beta_{IMCT}. \qquad (22)$$

Here, β_g is the first-hyperpolarizability of an isolated (gas-phase) molecule A, β_{dip}^{AA} is the contribution from the dipolar interaction between two A molecules, β_{dip}^{AB} is the contribution from dipolar interaction between molecules A and B ($B \neq A$), β_{hb} is the contribution from hydrogen-bond formation, and β_{IMCT} is the IMCT contribution. Equation (22) provides a useful guide for modeling second-order NLO materials. By modulating the *intra-molecular* as well as the *inter-molecular* terms of eq. (22) it would be possible to design and develop materials with suitably large β values and appropriate bulk structure to yield large $\chi^{(2)}$.

2. The third-order materials

The theory and mechanism of the third-order NLO processes in metals, inorganic crystals, and semiconductors are rather well understood. Most of the observed NLO phenomena in these materials can be described using eqs. (6) and (8) and treating l, m, n, etc. as the *virtual* states of the system.[2–4,6] The situation with organic materials is different, though.[5,34] Despite the numerous experimental observations and theoretical calculations made in the past two decades on the third-order NLO effects involving organic molecules and polymers,[5] the knowledge about their physical mechanism still remains limited.

Interest in organic compounds as potentially useful third-order or $\chi^{(3)}$ materials, as they are popularly known, arose from the experimental observation by Hermann et al[53] of anomalously large values of γ(THG) and $\chi^{(3)}$(THG) for all-trans β-carotene molecule and β-carotene glass, respectively. The large value for $\gamma(\text{THG})=(8 \pm 4) \times 10^{-33}$ esu of β-carotene molecule, which belongs to the polyene family and has 11 carbon-carbon double bonds in its quasi-one-dimensional chain, could not be accounted for on the basis of the bond-additivity schemes. Since, the bond-additivity models successfully describe the observed nonlinearity in σ bonded molecules and solids, Hermann et al.[53] concluded that the origin of the observed NLO effects in the β-carotene molecule as well as the glass lies in the highly delocalized π electron cloud of this system.

It is important to note that while the π-electron charge also plays crucial role in the second-order effects as discussed in the previous section, the mechanism of the two contributions are different. The effect of π electron delocalization on the second-order NLO properties is not *intrinsic*, because there some external perturbation, for example D-A substitution, is required to create appropriate symmetry in the π electron cloud. No such perturbations are needed for third-order effect, where a delocalized π-electron cloud directly affects the NLO response.

The observations by Hermann et al[53] received further support from the theoretical analysis of (hyper)polarizabilities of π-electron conjugated chain by Rustagi and Ducuing.[54] In their study, what is now a classical work, Rustagi and Ducuing considered the π-electron chain of 2N electrons as a one-dimensional box of length 2L. Then, by using the Rayleigh-Schrödinger perturbation theory for this free-electron system, they were able to show that the linear polarizability α and the second-hyperpolarizability γ increased much faster than linearly as the chain-length increased. They derived the relations

$$\alpha = \frac{8L^4}{3a_0\pi^2 N} \tag{23}$$

and

$$\gamma = \frac{256L^{10}}{45a_0^3 e^2 \pi^6 N^5}. \tag{24}$$

between the (hyper)polarizabilities and the length of the π-electron chain. In the above equations a_0 is the Bohr radius and e is the electronic charge. Now, assuming that $N \propto L$, eqs. (23) and (24) can be written as

$$\alpha \propto L^3 \tag{25}$$

and

$$\gamma \propto L^5. \tag{26}$$

The theoretical analysis of Rustagi and Ducuing was experimentally confirmed by Hermann and Ducuing[55] for a number of conjugated polymers. The validity of the Rustagi-Ducuing equation for the π-electron chain-length dependence of γ (eq. (26)) has been extensively investigated in the recent years by increasingly sophisticated quantum mechanical methods.[59–65] Although not all calculations agree on a fifth power dependence of γ on L, this relationship has proved to be one of the most useful guides for modeling third-order NLO active organic polymers.[5,7,8]

Rustagi and Ducuing[54] also introduced an important concept of *bond-alternation* (BA) effect on the third-order NLO susceptibility of conjugated systems. These authors reasoned that electrons in polyenes containing alternate carbon-carbon single-bond and double-bond experience a periodic potential which makes them less mobile (polarizable). Therefore, in such systems the increase of γ with the chain-length will be less rapid than in a free-electron system. Beyond a certain length, the BA in polyenes decreases and eventually becomes homogeneous with the carbon-carbon bond-length somewhere between a single-bond and a double-bond. Thus, in polyenes, the γ value will initially increase with the bond-length but eventually level off and remain constant at its polymeric limit. Although not strictly true, since even polydiacetylene (PA) has a finite BA property, this simple theory has since had a profound effect on the theoretical and experimental research related to the third-order NLO organic polymers.[5,7,8]

Beyond the conjugated π-electron effect and chain-length dependence of γ in organic systems, there has been little new development establishing the structure-NLO property relationships. A coupled anharmonic oscillator (CAO) model recently introduced by Prasad et al[66] correctly accounts for dependence of the band-gap, linear polarizability α, and second-hyperpolarizability γ on the number of repeat units in conjugated organic molecules and polymers. Since the explicit details of the repeat units are not required, this theory can be used for oligomeric chains of polyene or a polymer with large repeat units such as fused aromatic rings. This theory, however, assumes a single resonance frequency corresponding to a two-level model and a single coupling constant to represent the π-electron delocalization throughout the oligomeric series. Thus, for a quantitative treatment, particularly in the short-chain polyenes where a two-level model is not valid, the CAO model may not yield accurate results. However, for polymers and oligomers of large chain-length, the CAO model provides a physically useful and computationally inexpensive method to relate third-order NLO polarizability with the overall structure of the system.

Recently, McWilliams and Soos[67] have proposed an *inter-chain* mechanism for the third-order nonlinearity of conjugated polymers. According to their calculations, the NLO susceptibility for THG in conjugated polymers has considerable contribution from the matrix-elements representing inter-chain interaction. Although a study by Guo and Mazumdar[68] raises questions about the validity of this mechanism, a number of recent studies performed on the clusters of conjugated molecules[69–71] support the view that inter-molecular interaction may have significant effect on the third-order NLO susceptibility. A geometry which resembles the structure of graphite in the sense that the planar-conjugated molecules are *stacked* on each other, has been found to yield very large component of γ in the direction of the stack.[69–71]

The geometry of the repeat units in a polymer has also been found to have significant effect on the $\chi^{(3)}$ values.[72] Theoretical studies on the oligomeric diphenylbenzobisoxazole (PBO) and diphenylbenzobisthiazol (PBT) molecules[72] indicate that γ values in a planar geometry are significantly larger than those in a nonplanar geometry. This effect can be attributed to a free-electron type structure for a nearly planar-conjugated system as opposed to the two-dimensional periodic well type structure in twisted-chain which introduces a break in the free-movement of electrons. The geometrical isomerism (*cis, trans*) has also been identified as one of the important factors influencing the third-order NLO effects of conjugated systems. An isomer that allows greater number of resonance structures (e.g. *trans* isomer of PBO and PBT) and therefore better electron delocalization has considerably enhanced value of γ than that having smaller number of resonance structures (e.g. *cis* isomer of PBO and PBT).

IV. NLO EFFECTS IN SILICA GLASS

Silica glass is a relatively new arrival as an efficient NLO material. The first SHG in a commercial single-mode Ge doped glass fiber was observed by Österberg and Margulis.[73] The relatively high efficiency (3% peak power conversion) of the SHG observed in their experiment was unusual, considering the fact that the structure of glass is known to be a continuous random network (CRN)[74] of tetrahedrally bonded Si and O atoms with a zero net dipole moment. In the experiment of Österberg and Margulis[73] the second-harmonic (SH) signal of the fundamental IR beam was not produced in the beginning. It was only after a steady illumination of the fiber for a certain period of time that the SH signal appeared and grew in intensity with time, eventually reaching a maximum of 3% of the input power.

Stolen and Tom[75] proposed a mechanism of the second-order NLO effect in glass which involved permanent photoinduced changes in the fiber. According to this mechanism, the fundamental beam creates a dipole field which orients the defects in the structure, thus producing a direct, albeit weak, dipole-allowed response. The fundamental and the SH signals then interact via a third-order NLO process to form a strong periodic dc field which produces additional orientation of the defects. During this process the fiber organizes itself to accomplish phase matching, thus yielding efficient SHG. In order to demonstrate the validity of this mechanism, Stolen and Tom[75] used a *seeding* beam of SH frequency and a fundamental to create a strong dipole-field. They observed the SH signals when the seeding beam was removed after 5 minutes of irradiation. The power conversion efficiency of the SH signal was the same (3%) as observed by Österberg and Margulis.[73] However, no SHG was observed after 12 hours of the preparation of the sample and without the seeding light. This indicated that indeed some chemical/physical changes are induced in the structure of the fiber by the combination of the seeding beam and the fundamental beam that leads to efficient SHG.

Since then, SHG of IR light has been observed in poled fused silica glass[76] and in thin-film silica glass wave-guides.[77] A number of mechanisms have been proposed[78,79] to explain the build-up of the periodic dc field in silica glass that leads to a nonvanishing $\chi^{(2)}$. However, none of these theories addresses the microscopic mechanism of second-order NLO process in silica glass. It is generally believed[80] that the intrinsic defects[81] such as the E' center and oxygen-oxygen peroxy bonds (-O-O-) or extrinsic ones created by the dopants and impurities[82] or a combination of both, which create short-range domain structures of non-CRNs in silica glass, may be responsible for the observed SHG. However, the exact nature of these defects and the mechanism of their participation in the second-order NLO processes in silica glass based materials remains to be determined.

Considering the impact of Si and SiO_2 on the electronics industry, it can only be anticipated that with an improved knowledge of the structure-NLO property relationships, amorphous SiO_2 based materials may also prove to be the best yet NLO material for device applications.

VI. SUMMARY

As we have tried to emphasize, the driving force for the interest in NLO materials is their application to emerging technologies. NLO processes are extremely complicated, however, and what prevents the realization of these technologies is a very basic lack in understanding of structure-property relationships in NLO materials. Theory and modeling can provide a powerful means to explore these relationships through isolation of phenomenon, interpretation of experimental data, and development of a microscopic understanding of macroscopic phenomena. The overall goal is to find a method for intelligent materials design. Most modeling efforts to date have concentrated on predicting the magnitude of the microscopic NLO response functions β and/or γ by quantum mechanical methods. It should be emphasized, however, that in some cases, especially those requiring second-order responses, it is not the magnitude of the hyperpolarizabilities alone that determines the applicability of the material. Rather, secondary properties such as thermal and/or mechanical stability, or a combination of properties often determine the suitability of a NLO material for device application. In such cases, theory and modeling can still play a crucial role by providing an enhanced understanding of materials properties. It is important to note that, while the accuracy of quantum chemical techniques used in modeling NLO materials is crucial for the reliability of the predicted results, it is not essential for developing an understanding of the underlying physics in terms of its origin, mechanism, and structural dependence. Often, it is sufficient to obtain a reliable prediction of the qualitative trends among groups of molecules or related structures to isolate potentially important materials. In all cases, however, a close tie between computational modeling and experimental characterizations will lead to a faster and more global understanding of materials NLO and other properties appropriate for technological applications.

Acknowledgment

We would like to thank John Kester, Robert Hildreth, Walter Lauderdale, and John Wilkes of Frank J. Seiler Research Laboratory, John Garth and Kyle Critchfield of Phillips Laboratory, and Doug Dudis and Bruce Reinhardt of Wright Laboratory for fruitful discussions. This work was performed when one of us (SPK) held a National Research Council Senior Associateship.

References

1. P. Franken, A. Hills, C. Peters, G. Weinrich, *Phys. Rev. Lett.* **7**, 118 (1961).
2. N. Bloembergen, *Nonlinear Optics* (Benjamin, New York, 1965).
3. A. Yariv, *Quantum Electronics*, 2nd Ed. (Wiley, New York, 1975).
4. Y. R. Shen, *The Principles of Nonlinear Optics* (Wiley, New York, 1984).
5. P. N. Prasad and D. J. Williams, *Introduction to Nonlinear Optical Effects in Molecules and Polymers* (Wiley Interscience, New York, 1991).
6. R. W. Boyd, *Nonlinear Optics* (Academic Press, New York, 1992).
7. D. J. Williams, Ed. *Nonlinear Optical Properties of Organic and Polymeric Materials* (ACS Symp. Ser. 233, Washington, DC, 1983).
8. D. S. Chemla and J. Zyss, Eds. *Nonlinear Optical Properties of Organic Molecules and Crystals* (Academic Press Inc., New York, 1987).
9. G. A. Lindsay and K. D. Singer, Eds. *Polymers for Second-Order Nonlinear Optics* (ACS Symp. Ser. 601, Washington, DC, 1995).
10. K. D. Singer and J. H. Andrews, in *Quadratic Nonlinear Optics in Poled Polymer Films: From Physics to Device*, J. Zyss, Ed. (Academic Press, Inc., San Diego, CA, 1994).
11. B. A. Reinhardt, R. Kannan, J. C. Bhatt, J. Zieba, and P. N. Prasad, in ref. 9.
12. P. N. Prasad and B. A. Reinhardt, private communication.
13. J. A. Armstrong, N. Bloembergen, J. Ducuing, and P. S. Pershan, *Phys. Rev.* **127**, 1918 (1962).
14. N. Bloembergen and P. S. Pershan, *Phys. Rev.* **128**, 606 (1962).
15. D. A. Kleinman, *Phys. Rev.* **126**, 1977 (1962).
16. P. K. Franken and J. F. Ward, *Rev. Mod. Phys.* **35**, 23 (1963).
17. J. F. Ward, *Rev. Mod. Phys.* **37**, 1 (1965).
18. B. J. Orr and J. F. Ward, *Mol. Phys.* **20**, 513 (1971).
19. H. Sekino and R. J. Bartlett, *J. Chem. Phys.* **85**, 976 (1986).

20. A. D. Buckingham, *Adv. Chem. Phys.* **12**, 107 (1967).

21. P. W. Langhoff, S. T. Epstein, and M. Karplus, *Rev. Mod. Phys.* **44**, 602 (1972).

22. P. Sitz and Y. Yaris, *J. Chem. Phys.* **44**, 3546 (1968).

23. D. M. Bishop, *Rev. Mod. Phys.* **62**, 343 (1990); *Adv. Quantum. Chem.* **25**, 1 (1994).

24. S. P. Karna and M. Dupuis, *J. Comp. Chem.* **12**, 487 (1991).

25. J. E. Rice, R. D. Amos, A. M. Colwell, N. C. Handy, and J. Sanz, *J. Chem. Phys.* **93**, 8828 (1990).

26. H. Årgen, O. Vahatras, J. Koch, P. Jørgensen, and T. Heglaker, *J. Chem. Phys.* **98**, 6417 (1993).

27. S. P. Karna, *Chem. Phys. Lett.*, **214**, 186 (1993).

28. K. Sasagane, F. Aiga, and R. Itoh, *J. Chem. Phys.* **99**, 3738 (1993).

29. S. P. Karna, *J. Chem. Phys.*, to be published.

30. D. M. Bishop and B. Kirtman, *J. Chem. Phys.* **97**, 5255 (1992).

31. D. M. Bishop, B. Kirtman, H. A. Kurtz, and J. E. Rice, *J. Chem. Phys.* **98**, 8024 (1993).

32. G. P. Das, A. T. Yeates, and D. Dudis, *Chem. Phys. Lett.* **212**, 671 (1993).

33. D. A. Long, in *Non-Linear Raman Spectroscopy and Its Chemical Applications*, W. Kiefer and D. A. Long, Eds. (D. Reidel Publ. Co., Amsterdam, 1982).

34. N. Bloembergen, *Int. J. Nonlin. Opt. Phys.* **3**, 439 (1994).

35. B. F. Levine, *Phys. Rev. Lett.* **22**, 787 1969; *ibid.* **25**, 440 (1973); *Phys. Rev.* B **7**, 2600 (1973).

36. B. L. Davydov, L. D. Derkacheva, V. V. Dunina, M. E. Zhabotinski, V. F. Zolin, L. G. Koreneva, and M. A. Samokhina, *JETP Lett.* **12**, 16 (1970); *Opt. Spectrosc.* **30**, 274 (1971).

37. B. L. Davydov, S. G. Kotovshchikov, and V. A. Nefedov, *Sov. J. Quantum Electron.* **7**, 129 (1977).

38. J. R. Gott, *J. Phys. B.* Atom. Mol. Phys. **4**, 116 (1971).

39. P. D. Southgate and D. S. Hall, *J. Appl. Phys.* **43**, 2765 (1971).

40. J. L. Oudar and D. S. Chemla, *Opt. Commun.* **13**, 164 (1975).

41. J. L. Oudar and H. E. LePerson, *Opt Commun.* **15**, 258 (1975); *ibid.* **18**, 410 (1976).

42. D. S. Chemla, J. L. Oudar, and J. Jerphagnon, *Phys. Rev.* B **12**, 4534 (1975).

43. J. L. Oudar and D. S. Chemla, *J. Chem. Phys.* **66**, 2664 (1977).

44. J. L. Oudar, *J. Chem. Phys.* **67**, 446 (1977).

45. J. L. Oudar, D. S. Chemla, and E. Batifold, *J. Chem. Phys.* **67**, 1626 (1977).
46. B. F. Levine *J. Chem. Phys.* **63**, 115 (1975).
47. B. F. Levine and C. G. Bethea, *J. Chem. Phys.* **63**, 2666 (1975).
48. B. F. Levine *Chem. Phys. Lett.* **37**, 516 (1976).
49. B. F. Levine and C. G. Bethea, *J. Chem. Phys.* **65**, 2429 (1976).
50. B. F. Levine and C. G. Bethea, *J. Chem. Phys.* **65**, 2439 (1976).
51. B. F. Levine and C. G. Bethea, *J. Chem. Phys.* **69**, 5240 (1978).
52. S. K. Kurtz and T. T. Perry, *J. Appl. Phys.* **39**, 3798 (1968).
53. J. P. Hermann, D. Ricard, and J. Ducuing, *Appl. Phys. Lett.* **23**, 178 (1973).
54. K. C. Rustagi and J. Ducuing, *Opt. Commun.* **10**, 258 (1974).
55. J. P. Hermann and J. Ducuing, *J. Appl. Phys.* **45**, 5100 (1974).
56. M. Godfrey and J. N. Murrel, *proc. Roy. Soc.* (London) A **278**, 71 (1964).
57. A. E. Lutskii and N. I. Gorokhova, *Opt. Spectrosc.* **27**, 499 (1967).
58. A. Dulcic, C. Flytzanis, C. L. Tang, D. Pépin, M. Fétizon, and Y. Hopilliard, *J. Chem. Phys.* **74**, 1559 (1981).
59. H. F. Hameka, *J. Chem. Phys.* **67**, 2935 (1977).
60. B. Kirtman, W. B. Nilsson, and W. E. Palke, *Solid State Commun.* **46**, 791 (1983).
61. J. R. Heflin, K. Y. Wong, O. Zamani-Khamiri, and A. F. Garito, *Phys. Rev.* B **38**, 1573 (1988).
62. G. J. B. Hurst, M. Dupuis, and E. Clementi, *J. Chem. Phys.* **89**, 385 (1988).
63. B. Kirtman, *Chem. Phys. Lett.* **143**, 81 (1988); *Int. J. Quantum Chem.* **43**, 147 (1992).
64. E. F. Archibong and A. J. Thakkar, *J. Chem. Phys.* **98**, 8324 (1993).
65. B. Kirtman, J. L. Toto, K. A. Robins, and M. Hasan, *J. Chem. Phys.* **102**, 5350 (1995).
66. P. N. Prasad, E. Perrin, and M. Samoc, *J. Chem. Phys.* **91**, 2360 (1989).
67. P. C. M. McWilliams and Z. G. Soos, *J. Chem. Phys.* **95**, 2127 (1991).
68. D. Guo and S. Mazumdar, *J. Chem. Phys.* **97**, (1992).
69. D. R. Kanis, M. A. Ratner, and T. J. Marks, *Chem. Rev.* **94**, 195 (1994).
70. S. P. Karna, unpublished results.
71. H. A. Kurtz, unpublished results.
72. S. P. Karna, V. Keshari, and P. N. Prasad, *Chem. Phys. Lett.* **234**, 390 (1995).
73. U. Österberg and W. Margulis, *Opt. Lett.* **11**, 516 (1986).

74. W. H. Zachareisen, *J. Am. Chem. Soc.* **54**, 3841 (1932).

75. R. H. Stolen and H. W. K. Tom, *Opt. Lett.* **12**, 585 (1987).

76. R. A. Myers, N. Mukherjee, and S. R. J. Brueck, *Opt. Lett.* **16**, 1732 (1991).

77. J. J. Kester, P. W. Wolf, and W. R. White, *Opt. Lett.* **17**, 1779 (1992).

78. V. Dominic and J. Feinberg, *Phys. Rev. Lett.* **71**, 3446 (1993).

79. P. S. Weitzman, J. J. Kester, and U. Österberg, *Electron. Lett.* **30**, 697 (1994).

80. J. J. Kester and S. P. Karna, to be published.

81. D. L. Griscom, *Proc. Mat. Res. Soc.* **61**, 213 (1986).

82. A. H. Edwards, *Proc. Mat. Res. Soc.* **61**, 3 (1986).

RECEIVED December 29, 1995

Chapter 2

Can Quantum Chemistry Provide Reliable Molecular Hyperpolarizabilities?

Rodney J. Bartlett and Hideo Sekino

Quantum Theory Project, Departments of Chemistry and Physics, University of Florida, Gainesville, FL 32611–8435

We analyze the evolution of first principle molecular theory in obtaining reliable values of molecular hyperpolarizabilities. Such quantities place severe demands on the capability of quantum chemistry, as large basis sets, frequency dependence, and high levels of electron correlation are all shown to be essential in obtaining observed, gas phase hyperpolarizabilities to within an error of 10%. *Ab initio* Hartree Fock results are typically in error by nearly a factor of two, while the errors due to the neglect of frequency dependence average ~10–30% depending upon the particular process. It is also shown that for small molecules, standard semi-empirical approaches, like INDO and INDO/S, will often not even give the correct sign for hyperpolarizabilities. It is demonstrated that using state-of-the-art correlated, frequency dependent methods, it is possible to provide results to within 10%. Excluding any one of the essential elements, though, destroys the agreement.

In the late 70's, one of us (RJB) became interested in molecular hyperpolarizabilities as the essential element in non-linear optics (NLO) when Gordon Wepfer at the Air Force Office of Scientific Research called attention to J. F. Ward's results from dc-induced second harmonic generation (dcSHG) experiments and the enormity of their discrepancy with theoretical results. Tables 1 and 2 are extracted from a slightly later, 1979, paper of Ward and Miller (*1*) demonstrating the problem.

Not only was the existing theory results of the time typically in error by a factor of 3 to 5 in magnitude for the electric susceptibility, $\chi_{\|}^{(2)}$, but even the signs were frequently wrong. (The sign is positive if the measured quantity $\mu\chi_{\|}^{(2)}$ is positive, where μ is the permanent dipole moment.) When experimental numbers were obtained by other techniques, which should only differ by different dispersion effects that are typically less than 10%; they, too, had little correspondence with the dcSHG data. Note from Table 1 the -3500 value for NH_3 from refractivity virial data compared to -209±5 for the dcSHG, $\chi_{\|}^{(2)}$.

0097–6156/96/0628–0023$18.75/0

Table 1 $\chi_{\parallel}^{(2)}$ **in units of** 10^{-33} **esu/molecule. Theoretical values from various molecular orbital calculations — semi-empirical (SE), uncoupled Hartree-Fock (HF), and coupled Hartree-Fock (CHF) — and a single other experimental value are included for comparison. The sign of** $\mu\chi_{\parallel}^{(2)}$ **is unambiguously determined by the experiment and is independent of the sense chosen for the molecular** z **axis. This table extracted from J.F. Ward and C.K. Miller. Adapted from ref. 1.**

	μ^a	$\chi_{\parallel}^{(2)}$ *SHG*	*THEORY* Semi-Empirical	*THEORY* Uncoupled HF	*THEORY* Coupled HF	*Other EXP*
C^-O^+	0.112 ±0.005	+129 ±14	-43.5^b $(+95)^d$	$+879^c$ $+420^c$ $+387^c$ -438^e		
N^-O^+	0.15872 ±0.00002	+147 ±17	$+47.7^b$			
$H_2^+S^-$	0.974 ±0.005	-43 ±9				
$N^-H_3^+$	1.474 ±0.009	-209 ±5	56.4^b		-44.4^f -65.1^f -19.0^f -15.6^h -40.8^h	-3500^g
$H_2^+O^-$	1.86 ±0.02	-94 ±4	120^b	90.6	-52.5^f -79.2^f -21.9^f -51.6^h -48.0^h	

[a] Electric dipole moments in Debye units from Landolt-Bernstein, Zahlenwerte und Funktionen, Neue Serie, Vols. II/4 and II/6 (Springer-Verlag, Berlin) and Ref. 18. The sense of the NO moment is suggested by F. P. Billingsley II, J. Chem. Phys. **63** 2267 (1975); **62**, 864 (1975).

[b] N. S. Hush and M. L. Williams, Theoret. Chim. Acta (Berlin) (*25*), 346 (1972). Signs for NH_3 and H_2O are ambiguous.

[c] J. M. O'Hare and R. P. Hurst, J. Chem. Phys. (*46*), 2356 (1967).

[d] $\chi_{zzz}^{(2)}$ (0;0,0) from A. D. McLean and M. Yoshimine, J. Chem. Phys. (*46*), 3682 (1967).

[e] S. P. Liebmann and J. W. Moskowitz, J. Chem. Phys. **54**, 3622 (1971).

[f] P. Lazzeretti and R. Zanasi, Chem. Phys. Lett. (*39*), 323 (1976) *Note added in proof.* Recent reconsideration of the sign of these entries yields the negative signs now shown here — P. Lazzeretti (private communication). Overall consistency is substantially improved by this change.

[g] Refractivity virial data from A. R. Blythe, J. D. Lambert, P. J. Petter and H. Spoel, Proc. R. Soc. (London) A **255**, 427 (1960).

[h] G. P. Arrighini, M. Maestro and R. Moccia, Symp. Farad. Soc. **2**, 48 (1968).

Table 2 $\chi_{\parallel}^{(3)}$ in units of 10^{-39} esu/molecule. Values from other dc-electric field-induced second-harmonic generation (dcSHG), three-wave mixing (TWM), Kerr effect, and third-harmonic generation (THG) experiments, along with theoretical results, are included for comparison. This table extracted from J. F. Ward and C. K. Miller (*1*). Adapted from ref. 1.

	$\chi_{\parallel}^{(3)}$					
	dcSHG Ward & Miller	*dcSHG*[a]	*TWM*[b]	*Kerr*	*THG*[c]	*Theory*
H_2	65.2 ±0.8	79	-	47[d] ±5	80 ±12	34[e]
N_2	86.6 ±1.0	-	104	120[f] ±10	107 ±17	71[g]
O_2	95.3 ±1.6	110	100	-	-	-
CO_2	111.9 ±1.3	-	192	750[f] ±160	156 ±23	
CO	144 ±4	-	138	-	-	-
NO	235 ±7	-	322	-	-	-
H_2S	865 ±22	-	-	-	-	-
NH_3	511 ±9	-	-	-	-	-
H_2O	194 ±10	-	-	-	-	-

[a] Data from G. Mayer, C. R. Acad. Sci. **B276**, 54 (1968) and G. Hauchecorne, F. Kerbervé and G. Mayer, J. Phys. (Paris)**32**, 47 (1971), normalized using the dcSHG coefficient for argon from R. S. Finn and J. F. Ward, Phys. Rev. Lett.**26**, 285 (1971).

[b] Data from W. G. Rado, Phys. Lett.**11**, 123 (1967), normalized using the dcSHG coefficient for argon from R. S. Finn and J. F. Ward, Phys. Rev. Lett. *26*, 285 (1971).

[c] J. F. Ward and G. H. C. New, Phys. Rev.*185*, 579 (1969).

[d] A. D. Buckingham and B. J. Orr, Proc. R. Soc. (London) A**305**, 259 (1968).

[e] χ_{zzzz} (0;0,0,0) from A. D. McLean and M. Yoshimine, J. Chem. Phys. **46**, 3682 (1967).

[f] A. D. Buckingham, M. P. Bogard, D. A. Dunmur, C. P. Hobbs and B. J. Orr, Trans. Fora. Soc.**66**, 1548 (1970).

[g] χ_{zzzz} (0;0,0,0) from R. J. Bartlett and G. D. Purvis (private communication).

In Table 2, the discrepancy among different experiments is even more apparent for $\chi_{\parallel}^{(3)}$ compared to other dcSHG values, three wave mixing (TWM), Kerr effect, and third harmonic generation (THG) experiments. The theory is even off by about a factor of 2 from a static value for H_2. The one reasonable theoretical number they cite is our static, correlated value for N_2, one of our first (unpublished), that is beginning to be in reasonable agreement with $\chi_{\parallel}^{(3)}$.

This chapter, which is intended to be useful to experimentalists who are trying to assess the reliability of the theory, but also to theoreticians, as it provides an overview of the theory which can and has been used, addresses the demands that hyperpolarizabilities place on first-principle *ab initio* electronic structure theory; and the level required for the adequate evaluation of molecular hyperpolarizabilities. After doing some basic elementary perturbation theory that underlies all that is done, and which provides definitions and a framework for discussion, we will explain the various levels of quantum chemical application, illustrated by numerical results we have obtained that emphasize the various approximations employed in electronic structure in its application to hyperpolarizabilities. We will also demonstrate how the theory has necessarily evolved to provide a realistic treatment of such properties.

The order of presentation will follow our own odyssey that led us to introduce electron correlation into molecular hyperpolarizabilities (*2,3*), using, then new, many-body perturbation theory (MBPT) methods (*4,5*); to make a prediction for the FH molecule, that had later ramifications (*6,7*); to use new coupled-cluster (CC) correlated methods including those augmented by triples (*8,9*); to explore vibrational polarizabilities in static electric fields (*10*); and to introduce frequency dependence by developing analytical derivative time-dependent Hartree Fock (TDHF) theory (*11,12*). The latter, which enables a first (decoupled) treatment of both the essential frequency dependent and correlation aspects of the problem (*13*), culminates in a uniform study of ten molecules (*14*). The correlation calculations were assisted by parallel developments in analytical derivative CC/MBPT theory (*15–18*). Finally, we coupled correlation and frequency dependence in the new equation-of-motion (EOM) CC method for hyperpolarizabilities (*19,20*). By taking this evolutionary viewpoint, we hope the current review will complement the other recent excellent reviews (*21–23*) on the topic.

The theory also offers different, conceptual viewpoints on hyperpolarizability evaluation and interpretation, and this degree of flexibility should be better appreciated when comparing theoretical numbers and addressing NLO design criteria. We will conclude with some recommendations for some future developments in the continuing evaluation of predictive quantum chemical methods for NLO material design.

Perturbation Theory of Molecular Hyperpolarizabilities

The quantities of interest are the electric susceptibilities $\chi_{\parallel}^{(2)}$ and $\chi_{\parallel}^{(3)}$. In the gas phase experiments of interest here, there is no particular distinction between macroscopic and microscopic susceptibilities and the hyperpolarizabilities, as they are simply related. Those of particular interest are obtained from electric field (dc) induced second harmonic generation experiments (*1*). Namely, a sample of gas is subjected to a dc field (ε_o) and an optical electric field $\varepsilon_\omega\left(e^{\mathrm{i}\omega t}+e^{-\mathrm{i}\omega t}\right)$ at frequency ω, to induce a dipole moment at frequency 2ω. Allowing for the various manipulations required to relate the molecular quantities to the laboratory fixed observables,

$$\begin{aligned}\vec{\mu}^{2\omega} &= \frac{3}{2}\chi^{e}_{\|}(-2\omega;o,\omega,\omega)\varepsilon_o\varepsilon^2_{\omega o}\\ \chi^{e}_{\|}(-2\omega;o,\omega,\omega) &= \chi^{(3)}_{\|}(-2\omega;o,\omega,\omega)+\frac{\vec{\mu}}{9KT}\chi^{(2)}_{\|}(-2\omega;\omega,\omega)\end{aligned} \tag{1}$$

$\chi^{(2)}_{\|}(-2\omega;\omega,\omega)$ for second harmonic generation (SHG) is separated from $\chi^{(3)}_{\|}(-2\omega;o,\omega,\omega)$, which is dcSHG, by studying $\vec{\mu}^{2\omega}$ as a function of *T*. Numbers relative to a standard, typically He gas, are obtained. Revisions of the He reference value cause some slight rescaling of the experimental numbers (*22*). Furthermore, we relate the susceptibilities to the molecular hyperpolarizabilities via,

$$\begin{aligned}\chi^{(2)}_{\|}(-2\omega;\omega,\omega) &= \beta_{\|}(-2\omega;\omega,\omega)/2\\ &= \frac{1}{2}\left[(\beta_{iij}+\beta_{iji}+\beta_{jji})/5\right]\\ \chi^{(3)}_{\|}(-2\omega;o,\omega,\omega) &= \gamma_{\|}(-2\omega;o,\omega,\omega)/6\\ &= \frac{1}{6}\left[(\gamma_{iijj}+\gamma_{ijji}+\gamma_{ijij})/15\right]\end{aligned} \tag{2}$$

where the Einstein summation convention is employed, meaning all repeated indices are summed, i.e. $\vartheta_{ii}=\vartheta_{xx}+\vartheta_{yy}+\vartheta_{zz}$. The parallel designation ($\|$) means measured parallel to the dc field, while the ($\perp$) component can be similarly obtained. As $\beta_{\|}$ and $\gamma_{\|}$ are properties of molecules, they constitute the objective for first-principle quantum mechanical evaluation; the subject of this chapter.

The basic idea underlying any treatment of a molecular response to an electric field, static or oscillatory, is the solution of the molecular Schrödinger equation in the presence of the field. We will first consider the static case, generalizing the approach for frequency dependence in Section 7. Our perturbation $\vec{\varepsilon}_o\cdot\vec{W}=\vec{\varepsilon}_o\cdot\sum_i e_i\vec{r}_i$ is given by the interaction of an electron e_i at position $\vec{r}_i$ with a static electric field,

$$\vec{\varepsilon}_o=\varepsilon_{xo}\hat{i}+\varepsilon_{yo}\hat{j}+\varepsilon_{zo}\hat{k}, \tag{3}$$

where ε_{xo}, ε_{yo}, and ε_{zo} are the field strengths in the *x*, *y*, and *z* directions of magnitude $|\varepsilon|$. In atomic units, *e* becomes minus unity, so we have a Hamiltonian,

$$\mathcal{H}(\vec{\varepsilon}_o)=\mathcal{H}_o+\vec{\varepsilon}_o\cdot\vec{W}=\mathcal{H}_o-\vec{\varepsilon}_o\cdot\sum_i\vec{r}_i, \tag{4}$$

where H_o consists of the usual kinetic energy, $-\frac{1}{2}\sum_i\nabla_2^2$ and potential energy, $-\sum_{\alpha,i}\frac{Z_\alpha}{r_{\alpha i}}+\frac{1}{2}\sum_{i,j}\frac{1}{r_{ij}}+\frac{1}{2}\sum_{\alpha,\beta}\frac{Z_\alpha Z_\beta}{R_{\alpha\beta}}$, composed of the electron nuclear attraction, the two-electron, electron-electron repulsion operator, and the proton-proton repulsion. Greek letters indicate atoms and i,j electrons. W contains the negative sign. The solution to the time independent Schrödinger equation

$$\mathcal{H}_o\psi_k^o = E_k^o\psi_k^o \tag{5}$$

provides the ground, $\psi_o = \psi_o^o$ and excited states, $\{\psi_k^o\}$, and associated eigenvalues, $\{E_k^o\}$, for the unperturbed molecule. Notice these are the exact, many-particle unperturbed states.

We now seek a solution for the perturbed Hamiltonian in its ground state

$$\mathcal{H}(\vec{\varepsilon}_o)\Psi(\vec{\varepsilon}_o) = E(\vec{\varepsilon}_o)\Psi(\vec{\varepsilon}_o) \tag{6}$$

The natural way to attempt a solution is perturbation theory. Hence, we expand, $\mathcal{H}(\vec{\varepsilon}_o)$, $\Psi(\vec{\varepsilon}_o)$ and $E(\vec{\varepsilon}_o)$ in a perturbation series in $\vec{\varepsilon}_o$. For the time being, we will assume the field lies solely in the z direction, so $\varepsilon_{xo} = \varepsilon_{yo} = 0$ and $\varepsilon_{zo} = \varepsilon_o$. Then,

$$\begin{aligned}(H_o + \varepsilon_o W)\Big(\psi_o + \varepsilon_o\psi^{(1)} + \varepsilon_o^2\psi^{(2)} + \varepsilon_o^3\psi^{(3)} + \ldots\Big) \\ = \Big(E + \varepsilon_o E^{(1)} + \varepsilon_o^2 E^{(2)} + \ldots\Big)\Big(\psi_o + \varepsilon_o\psi^{(1)} + \ldots\Big)\end{aligned} \tag{7}$$

As the equality has to be true for any power of ε_o (i.e. all quantities are linearly independent), gathering terms of a given power of ε_o together, we obtain for ε_o^1;

$$(E_o - \mathcal{H}_o)\psi^{(1)} = \Big(W - E^{(1)}\Big)\psi_o \tag{8}$$

for ε_o^2

$$(E_o - \mathcal{H}_o)\psi^{(2)} = \Big(W - E^{(1)}\Big)\psi^{(1)} - E^{(2)}\psi_o \tag{9}$$

for ε_o^3

$$(E_o - \mathcal{H}_o)\psi^{(3)} = \Big(W - E^{(1)}\Big)\psi^{(2)} - E^{(3)}\psi_o - E^{(2)}\psi^{(1)} \tag{10}$$

and for ε_o^4

$$(E_o - \mathcal{H}_o)\psi^{(4)} = \Big(W - E^{(1)}\Big)\psi^{(3)} - E^{(4)}\psi_o - E^{(3)}\psi^{(1)} - E^{(2)}\psi^{(2)} \tag{11}$$

These are sufficient to take us through the electronic dipole moment, $\underline{\mu}$, the dipole polarizability tensor, $\underline{\alpha}$, and the $\underline{\beta}$ and $\underline{\gamma}$, hyperpolarizability tensors. Multiplying on the left by $\langle\psi_o|$, and using the fact that $\langle\psi_o|(E_o - \mathcal{H}_o) = 0$, we have

$$E^{(1)} = \langle\psi_o|W|\psi_o\rangle \tag{12}$$

Similarly, we obtain

$$\begin{aligned}E^{(2)} &= \langle\psi_o|\Big(W - E^{(1)}\Big)|\psi^{(1)}\rangle = \langle\psi_o|W|\psi^{(1)}\rangle \\ E^{(3)} &= \langle\psi_o|W|\psi^{(2)}\rangle = \langle\psi^{(1)}|W - E^{(1)}|\psi^{(1)}\rangle \\ E^{(4)} &= \langle\psi_o|W|\psi^{(3)}\rangle = \langle\psi^{(2)}|E_o - \mathcal{H}_o|\psi^{(2)}\rangle - E^{(2)}\langle\psi^{(1)}|\psi^{(1)}\rangle\end{aligned} \tag{13}$$

where we have used the intermediate normalization condition, $\langle\psi_o|\psi^{(m)}\rangle = 0$ for $m \neq 0$. The second form comes from manipulations that demonstrate the 2n+1 rule that an nth order wavefunction will determine E^{2n+1} and the analogous 2n rule for even orders.

Instead of straightforward perturbation theory, we can also derive these formulae from explicit differentiation of the expectation value of $\Psi(\varepsilon_o)$ with respect to ε_o. From equation 6, we have the expectation value,

$$E(\varepsilon_o)\langle\Psi(\varepsilon_o)|\Psi(\varepsilon_o)\rangle = \langle\Psi(\varepsilon_o)|\mathcal{H}(\varepsilon_o)|\Psi(\varepsilon_o)\rangle \tag{14}$$

and differentiation gives,

$$\begin{aligned}\left\langle \Psi(\varepsilon_o) \,\middle|\, (\mathcal{H}(\varepsilon_o) - E(\varepsilon_o)) \,\middle|\, \frac{\partial\Psi}{\partial\varepsilon_o}\right\rangle + \left\langle \frac{\partial\Psi}{\partial\varepsilon_o} \,\middle|\, \Big(\mathcal{H}(\varepsilon_o) - E(\varepsilon_o)\Big) \,\middle|\, \Psi(\varepsilon_o)\right\rangle \\ = \left\langle \Psi(\varepsilon_o) \,\middle|\, \frac{\partial H}{\partial\varepsilon_o} - \frac{\partial E}{\partial\varepsilon_o} \,\middle|\, \Psi(\varepsilon_o)\right\rangle\end{aligned} \tag{15}$$

As we want to evaluate this at $\varepsilon_o = 0$, using $(\mathcal{H}_o - E_o)|\psi_o\rangle = 0$, and its complex conjugate (c.c.), gives

$$\left.\frac{\partial E}{\partial\varepsilon_o}\right]_{\varepsilon_o=0} = \left.\left\langle \psi_o \,\middle|\, \frac{\partial\mathcal{H}}{\partial\varepsilon_o} \,\middle|\, \psi_o\right\rangle\right]_{\varepsilon_o=0} = \langle\psi_o \mid W \mid \psi_o\rangle \tag{16}$$

This is simply a statement of the simple Hellman-Feynmann theorem that the derivative of the energy is given as the expectation value of $\frac{\partial\mathcal{H}}{\partial\varepsilon_o}$.

If we proceed to the higher derivatives, we will obtain

$$\left.\frac{\partial^2 E}{\partial\varepsilon_o^2}\right]_{\varepsilon_o=0} = \left\langle \psi_o \,\middle|\, \frac{\partial\mathcal{H}}{\partial\varepsilon_o} - \frac{\partial E}{\partial\varepsilon_o} \,\middle|\, \frac{\partial\psi}{\partial\varepsilon_o}\right\rangle + \text{c.c.} = \left\langle \psi_o \,\middle|\, \frac{\partial\mathcal{H}}{\partial\varepsilon_o} \,\middle|\, \frac{\partial\psi}{\partial\varepsilon_o}\right\rangle + \text{c.c.} \tag{17}$$

or

$$\left.\frac{1}{2}\frac{\partial^2 E}{\partial\varepsilon_o^2}\right]_{\varepsilon_o=0} = \left\langle \psi_o \,\middle|\, \frac{\partial\mathcal{H}}{\partial\varepsilon_o} \,\middle|\, \frac{\partial\psi}{\partial\varepsilon_o}\right\rangle \tag{18}$$

In third and fourth order,

$$\begin{aligned}\left.\frac{1}{3!}\frac{\partial^3 E}{\partial\varepsilon_o^3}\right]_{\varepsilon_o=0} &= \left\langle \psi_o \,\middle|\, \frac{\partial\mathcal{H}}{\partial\varepsilon_o} \,\middle|\, \frac{1}{2!}\frac{\partial^2\psi}{\partial\varepsilon_o^2}\right\rangle = \left\langle \frac{\partial\psi}{\partial\varepsilon_o} \,\middle|\, \frac{\partial\mathcal{H}}{\partial\varepsilon_o} - \frac{\partial E}{\partial\varepsilon_o} \,\middle|\, \frac{\partial\psi}{\partial\varepsilon_o}\right\rangle \\ \left.\frac{1}{4!}\frac{\partial^4 E}{\partial\varepsilon_o^4}\right]_{\varepsilon_o=0} &= \left\langle \psi_o \,\middle|\, \frac{\partial\mathcal{H}}{\partial\varepsilon_o} \,\middle|\, \frac{1}{3!}\frac{\partial^3\psi}{\partial\varepsilon_o^3}\right\rangle\end{aligned} \tag{19}$$

Clearly, this is simply perturbation theory, since up to a numerical factor, $\psi^{(1)} = \frac{\partial\psi}{\partial\varepsilon_o}$, $E^{(1)} = \frac{\partial E}{\partial\varepsilon_o}$, $\psi^{(2)} = \frac{1}{2}\frac{\partial^2\psi}{\partial\varepsilon_o^2}$, $E^{(2)} = \frac{1}{2}\frac{\partial^2 E}{\partial\varepsilon_o^2}$, etc. In other words, we use a McLaurin's series expansion for *E*, $E(\varepsilon_o) = E_o + \frac{\partial E}{\partial\varepsilon_o}\varepsilon_o + \frac{1}{2!}\frac{\partial^2 E}{\partial\varepsilon_o^2}\varepsilon_o^2 + \frac{1}{3!}\frac{\partial^3 E}{\partial\varepsilon_o^3}\varepsilon_o^3 + \ldots$ and Ψ, instead of a straight perturbation series, which introduces the numerical factors that relate the perturbative energies to the derivatives $E^{(n)} = \frac{1}{n!}\frac{\partial^n E}{\partial\varepsilon_o^n}$.

We now define the electric properties as the electric dipole in the *z* direction:

$$\frac{\partial E}{\partial\varepsilon_o} = -\mu_z = \left\langle \psi_o \,\middle|\, \frac{\partial\mathcal{H}}{\partial\varepsilon_o} \,\middle|\, \psi_o\right\rangle = \left\langle \psi_o \,\middle|\, W \,\middle|\, \psi_o\right\rangle = E^{(1)} \tag{20}$$

the dipole polarizability;

$$\frac{\partial^2 E}{\partial \varepsilon_o^2} = -\alpha_{zz} = \left\langle \psi_o \left| \frac{\partial \mathcal{H}}{\partial \varepsilon_o} \right| \frac{\partial \psi}{\partial \varepsilon_o} \right\rangle = 2\left\langle \psi_o \left| W \right| \psi^{(1)} \right\rangle = 2E^{(2)} \quad (21)$$

the first hyperpolarizability;

$$\frac{\partial^3 E}{\partial \varepsilon_o^3} = -\beta_{zzz} = 3\left\langle \psi_o \left| \frac{\partial \mathcal{H}}{\partial \varepsilon_o} \right| \frac{\partial^2 \psi}{\partial \varepsilon_o^2} \right\rangle = 3!\left\langle \psi_o \left| W \right| \psi^{(2)} \right\rangle = 3!E^{(3)} \quad (22)$$

and the second hyperpolarizability;

$$\frac{\partial^4 E}{\partial \varepsilon_o^4} = -\gamma_{zzzz} = 4\left\langle \psi_o \left| \frac{\partial \mathcal{H}}{\partial \varepsilon_o} \right| \frac{\partial^3 \psi}{\partial \varepsilon_o^3} \right\rangle = 4!\left\langle \psi_o \left| W \right| \psi^{(3)} \right\rangle = 4!E^{(4)} \quad (23)$$

That is,

$$E(\varepsilon_o) = E_o - \mu_z \varepsilon_{zo} - \frac{1}{2!}\alpha_{zz}\varepsilon_{zo}^2 - \frac{1}{3!}\beta_{zzz}\varepsilon_{zo}^3 - \frac{1}{4!}\gamma_{zzzz}\varepsilon_{zo}^4 + \ldots \quad (24)$$

which gives the well known series.

More generally, when all components are considered, with the Einstein summation convention, we have

$$E = E - \mu_i \varepsilon_{io} - \frac{1}{2!}\alpha_{ij}\varepsilon_{io}\varepsilon_{jo} - \frac{1}{3!}\beta_{ijk}\varepsilon_{io}\varepsilon_{jo}\varepsilon_{ko} - \frac{1}{4!}\gamma_{ijkl}\varepsilon_{io}\varepsilon_{jo}\varepsilon_{ko}\varepsilon_{lo} - \ldots \quad (25)$$

or for the induced dipole moment,

$$\mu_i(\varepsilon_o) = \mu_i(o) + \alpha_{ij}\varepsilon_{jo} + \frac{1}{2!}\beta_{ijk}\varepsilon_{jo}\varepsilon_{ko} + \frac{1}{3!}\gamma_{ijkl}\varepsilon_{jo}\varepsilon_{ko}\varepsilon_{lo} + \ldots \quad (26)$$

Whether the numerical factors are included determines the choice of conventions for polarizabilities. Ward and co-workers use the perturbative definition, meaning that $E^{(2)} = \underline{\alpha}$, $E^{(3)} = \underline{\beta}$, and $E^{(4)} = \underline{\gamma}$, e.g., while the power series choice directly associates the derivatives with the polarizabilities, obviating the numerical factors. Since this choice has direct correspondence with the energy derivatives, it appeals more to theoreticians.

To evaluate the polarizabilities, we require a knowledge of the perturbed wavefunctions (wavefunction derivatives). It is convenient to introduce the resolvent operator,

$$R_o = (E_o - \mathcal{H}_o)^{-1} Q \quad (27)$$

where Q represents the projector *(Q^2=Q)* of all functions orthogonal to the unperturbed reference, ψ_o. *One* such set consists of all other eigenfunctions of H_o (discrete and continuous), making $Q = \sum_{k \neq o} \left| \psi_k^{(o)} \right\rangle \left\langle \psi_k^{(o)} \right|$. For this particular set,

$R_o|\psi_k^o\rangle = \left(E_o - E_k^{(o)}\right)^{-1}|\psi_k^o\rangle Q = Q\left(E_o - E_k^{(o)}\right)^{-1}|\psi_k^{(o)}\rangle$. By virtue of excluding ψ_o, the inverse R_o operator is well defined. In terms of the resolvent,

$$\begin{aligned} \psi^{(1)} &= R_o\left(W - E^{(1)}\right)\psi_o = R_o W \psi_o \\ \psi^{(2)} &= R_o\left(W - E^{(1)}\right)\psi^{(1)} \\ \psi^{(3)} &= R_o\left(W - E^{(1)}\right)\psi^{(2)} - R_o E^{(2)}\psi^{(1)} \end{aligned} \tag{28}$$

etc., showing that all order wavefunctions are *recursively* computed from the prior ones. Knowing these solutions, we can compute the energy from

$$E^{(n+1)} = \langle\psi_o|W|\psi^{(n)}\rangle \tag{29}$$

or explicitly, in a few cases,

$$\begin{aligned} E^{(2)} &= \langle\psi_o|W R_o W|\psi_o\rangle \\ E^{(3)} &= \langle\psi_o|W R_o\left(W - E^{(1)}\right)R_o W|\psi_o\rangle \\ E^{(4)} &= \langle\psi_o|W R_o\left(W - E^{(1)}\right)R_o\left(W - E^{(1)}\right)R_o W|\psi_o\rangle - E^{(2)}\langle\psi_o|W R_o^2 W|\psi_o\rangle \end{aligned} \tag{30}$$

etc.

The particular choice of expansion of $\psi^{(n)}$ in the set of eigenfunctions of H_o leads to the well known sum-over-state (SOS) formulas for the *z* components,

$$-\mu_z = \langle\psi_o|\hat{z}|\psi_o\rangle \tag{31a}$$

$$-\alpha_{zz} = \sum_{k\neq o} \frac{\langle\psi_o|\hat{z}|\psi_k^{(o)}\rangle\langle\psi_k^{(o)}|\hat{z}|\psi_o\rangle}{E_o - E_k^{(o)}} \tag{31b}$$

$$-\beta_{zzz} = \sum_{k,l\neq o} \frac{\langle\psi_o|\hat{z}|\psi_k^{(o)}\rangle\langle\psi_k^{(o)}|\hat{z} - \langle\psi_o|\hat{z}|\psi_o\rangle|\psi_l^{(o)}\rangle\langle\psi_l^{(o)}|\hat{z}|\psi_o\rangle}{\left(E_o - E_k^{(o)}\right)\left(E_o - E_l^{(o)}\right)} \tag{31c}$$

$$\begin{aligned} -\gamma_{zzzz} = \sum_{k,l,m\neq o} & \frac{\langle\psi_o|\hat{z}|\psi_k^{(o)}\rangle\langle\psi_k^{(o)}|\hat{z} - \langle\psi_o|\hat{z}|\psi_o\rangle|\psi_l^{(o)}\rangle\langle\psi_l^{(o)}|\hat{z} - \langle\psi_o|\hat{z}|\psi_o\rangle|\psi_m^{(o)}\rangle\langle\psi_m^{(o)}|\hat{z}|\psi_o\rangle}{\left(E_o - E_k^{(o)}\right)\left(E_o - E_l^{(o)}\right)\left(E_o - E_m^{(o)}\right)} \\ & + \alpha_{zz}\frac{\langle\psi_o|\hat{z}|\psi_k^{(o)}\rangle\langle\psi_k^{(o)}|\hat{z}|\psi_o\rangle}{\left(E_o - E_k^{(o)}\right)^2} \end{aligned} \tag{31d}$$

The other tensor elements are obtained from exchanging *x* and *y* with *z* in all distinct ways. Henceforth, we will recognize that all practical molecular quantum chemical calculations employ a finite, discrete basis set, so from the beginning we are limited to this choice, and we need not consider any continuum.

Evaluation of Static Hyperpolarizabilities

From the above, it should be abundantly clear that in a basis the derivative approach, equations 20 to 23, and the SOS expressions, equation 31, are simply two equivalent ways of expressing polarizabilities. Furthermore, considering that the choice of excited eigenfunctions of $\mathcal{H}_o$ is just one choice for a complete set representation of the perturbed wavefunctions, $\{\psi^{(n)}\}$, the above SOS forms probably attract more significance in NLO than warranted. For example, since $E_o - E_k^{(o)}$ is the excitation energy, and since $\langle\psi_o|\hat{z}|\psi_k^{(o)}\rangle$ is the z-component of the transition moment, it should be possible to evaluate α_{zz} from equation 31b purely from experiment by knowing *all* electronic excitation energies and transition moments (including those for the continuum in the exact case). The problem is knowing them all—a very large number! Instead, attempts to estimate the SOS by the *few* known excitation energies and transition moments is likely to be very far from the true value (see our contribution later in this volume (*24*)). Note for the ground state, all contributions to α_{zz} from $\psi_k^{(o)}$ have the same sign, so there is no potential cancellation among the neglected contributions. The problem is further compounded for β_{zzz} and γ_{zzzz}, where in addition to transition moments from the ground to excited states, it is necessary to know the transition moments relating two excited states, and that information is hard to obtain. Note that β_{zzz} can have either sign. γ_{zzzz} also can have either sign, since although the lead term corresponds to $\langle\psi_o^{(2)}|E_o - \mathcal{H}_o|\psi_o^{(2)}\rangle$ which must be negative (giving a positive contribution to γ_{zzzz}), the second term is positive, attenuating the value of γ_{zzzz}. Lacking a proof that the magnitude of one must be greater than the other, either sign is possible.

From the above we have two viewpoints on the evaluation of static polarizabilities. We can either evaluate energy derivatives of the Schrödinger equation in the presence of the perturbation, or attempt some approximation to the SOS. Obviously, the former does not require any truncation. (As we will see in Section 8 later, with *proper handling* neither does the finite basis SOS.) The simplest recipe for evaluation of the derivatives is to use what is called the finite-field technique. That simply means solve the Schrödinger equation in the presence of the perturbation by choosing a small finite value for ε_o of 0.001 a.u., e.g. Adding an electric field quantity to H_o gives an unbounded $H(\varepsilon_o)$ operator, and if we obtain its exact solution, the lowest energy state would be the field ionized state, a molecular cation plus an electron; but in practice we must use a finite basis set for its solution which is effectively like putting the molecule into a box, and this gives a valid $E(\varepsilon_o)$. Repeating the procedure at $\varepsilon_o = -0.001$, we could obtain the dipole from the numerical derivative,

$$-\mu_z = \frac{\partial E}{\partial \varepsilon_o} = \lim_{\varepsilon_o \to o} \frac{E(\varepsilon_o) - E(-\varepsilon_o)}{2\varepsilon_o} \tag{32}$$

The accuracy depends upon the size of the finite field strength. If it is too large, the numerical derivative is not very accurate, while if too small, there is not a numerically significant change in $E(\varepsilon_o)$.

To obtain all $\underline{\mu}$, $\underline{\alpha}$, $\underline{\beta}$ and $\underline{\gamma}$, we need several more points, so we obtain expressions like

$$\begin{aligned} \beta_{iii}\varepsilon_{io}^3 &= \frac{1}{2}\Big[E(2\varepsilon_{io}) - E(-2\varepsilon_{io})\Big] + \Big[E(\varepsilon_{io}) - E(-\varepsilon_{io})\Big] + \mathcal{O}\big(\varepsilon_o^5\big) \\ \gamma_{iiii}\varepsilon_{io}^4 &= 4\Big[E(\varepsilon_{io}) + E(-\varepsilon_{io})\Big] - \Big[E(2\varepsilon_{io}) - E(-2\varepsilon_{io})\Big] - 6E(o) + \mathcal{O}\big(\varepsilon_o^6\big) \end{aligned} \tag{33}$$

See (*2*) for others. Note each expression is accurate to the next odd (even) order since only $\underline{\alpha}$, $\underline{\gamma}$, $\underline{\epsilon}$ and $\underline{\mu}$, $\underline{\beta}$, $\underline{\delta}$ are interrelated (*2*). This exclusion of the next higher-order contribution greatly helps the precision of such a calculation. However, note that the energy needs to be accurate to a couple of significant digits better than ε_{io}^3 if we are to get β_{iii}. That is, if ε_{io} =0.001, we require energies to be 10^{-11}. If ε_{io} =0.01, we would require at least 10^{-8}. For γ_{iiii}, we would need 10^{-10} to 10^{-14}. As molecular integrals in quantum chemical calculations are seldom much more accurate than 10^{-12}, not to mention other parts of the calculation, finite field procedures for hyperpolarizabilities can raise serious precision problems.

Another problem lies in the proliferation of tensor elements in $\underline{\beta}$ and $\underline{\gamma}$. *Many* energy calculations involving field strengths in different directions are required to evaluate all the numerical derivatives, and at higher levels of sophistication these are quite expensive calculations.

The solution to the above problem is to *analytically* evaluate the derivatives. The simplest is $-\mu_z = \langle\psi_o|\hat{z}|\psi_o\rangle$, which is just an expectation value. For the others, *analytical* means that while solving the Schrödinger equation for *E*, we also directly obtain all components of the derivatives, $\frac{\partial E}{\partial \vec{\varepsilon}_o}$, $\frac{\partial^2 E}{\partial \vec{\varepsilon}_o \partial \vec{\varepsilon}_o}$, $\frac{\partial^3 E}{\partial \vec{\varepsilon}_o \partial \vec{\varepsilon}_o \partial \vec{\varepsilon}_o}$, etc. in about an equivalent amount of time. This means we differentiate *before* evaluation by using equations 20 to 23 and the explicit solutions for the wavefunctions and derivatives, which are proportional to the perturbed wavefunctions given in equation 28. In a substantial formal and computational achievement of 30 years duration, primarily fueled by the necessity of analytical gradients for atomic displacements in molecules (*15,25,26*), such analytical procedures have been developed in quantum chemistry. Their limitation is that they have not been implemented for all methods. For example, *any-order* analytical higher derivatives with respect to electric field perturbations have been developed for the Hartree-Fock treatment of hyperpolarizabilities by exploiting the recursive nature of perturbation theory (*11,27*), equations 28, 29. For correlated methods, analytical second-order perturbative theory, [MBPT(2)], derivatives are available for $\underline{\alpha}$ and $\underline{\beta}$ (*15*). For other methods, even including highly sophisticated correlated methods like coupled-cluster theory (*28*), the induced dipole $\mu(\vec{\varepsilon}_o)$ can be evaluated analytically, from which numerical derivatives provide $\underline{\alpha}$, $\underline{\beta}$, and $\underline{\gamma}$ (13). In this way, at least one ε_o or two to three orders of magnitude is gained in the precision of $\underline{\beta}$ and $\underline{\gamma}$.

Note that equation 31 represents *analytical* expressions, too, since no finite field is involved in their evaluation. The latter viewpoint leads to the analytical evaluation of the (dynamic) polarizability using the equation-of-motion (EOM) CC method (*20*). Obviously, it would be ideal to be able to analytically evaluate third and fourth derivatives using such powerful CC correlated methods, but the theory and implementation has not yet been developed.

Basis Sets and Hartree-Fock Theory

Now that we know a way to calculate a static hyperpolarizability, we can consider other aspects of the calculation. The first approximation to consider is Hartree-Fock (HF), self-consistent field (SCF) theory. That is, we evaluate the energy and its derivatives for the perturbed Hamiltonian by obtaining the energetically best (lowest) single determinant solution, $\Phi_o = \mathcal{A}(\varphi_1(1), \varphi_2(2)\cdots\varphi_n(n))$, to approximate ψ_o. In the absence of $\varepsilon_o W$, that means $E_{\text{HF}} = \langle\Phi_o|H_o|\Phi_o\rangle$, and further, $\mathcal{H}_o \simeq H_o = \sum_{k=1}^{N} f(i)$ where $f = h + v^{eff}$ and $v^{eff}(1) = \sum_{j=1}^{N} \int \varphi_j(2)\frac{1-P_{12}}{r_{12}}\varphi_j(2)d\tau$. The effective one-particle operator $v^{eff}(1)$ is an average over the two-particle part. The orbitals $\{\varphi_j\}$ are the solutions $f(1)\varphi_j(1) = \epsilon_j\varphi_j(1)$ where ϵ_j is the HF-SCF orbital energy (not electric field ε or $\underline{\epsilon}$ hyperpolarizability!). Self-consistency is required by ensuring that the orbitals $\{\varphi_j\}$ used in $v^{eff}(1)$ in f are self-consistent with the solutions. This model provides the usual molecular orbital (MO) approximation that underlies much of our conceptual understanding of molecules. This averaging procedure introduces the correlation error, which pertains to electrons' instantaneous interaction that keeps them apart. It corresponds to the perturbation $V = \sum_{i,j}\frac{1}{r_{ij}} - \sum_{i} v^{eff}(i)$, which we will consider later in Section 6. The full Hamiltonian of equation 41 is thus $\mathcal{H} = H_o + \varepsilon_o W + V$.

The simplest way to consider doing a HF calculation of a polarizability is the finite-field procedure. That means that we compute the HF approximation to $\mathcal{H} = H_o + V + \varepsilon_o W$ where W is a one-electron operator for a small value of ε_o. Then, we will obtain

$$E_{\text{HF}}(\varepsilon_o) = \langle\Phi_o(\varepsilon_o)|\mathcal{H}(\varepsilon_o)|\Phi_o(\varepsilon_o)\rangle \tag{34}$$

where $\Phi_o(\varepsilon_o) = \mathcal{A}(\varphi_{\varepsilon 1}(1)\varphi_{\varepsilon 2}(2)\ldots\varphi_{\varepsilon n}(n))$ is the HF wavefunction and its component orbitals are all dependent on ε_o. Furthermore,

$$f_\varepsilon(1)\varphi_{j\varepsilon}(1) = \epsilon_j(\varepsilon_o)\varphi_{j\varepsilon}(1), \tag{35a}$$

$$f_\varepsilon(1) = f_o(1) + \varepsilon_o W(1) + v_\varepsilon^{eff}(1) \tag{35b}$$

$$v_\varepsilon^{eff}(1) = \sum_{j=1}^{N}\int \varphi_{j\epsilon}^*(2)\frac{1-P_{12}}{r_{12}}\varphi_{j\epsilon}(2)d\tau_2 \tag{35c}$$

Obtaining $E_{\text{HF}}(\varepsilon_o)$ at various values of ε_o will provide the perturbed energies from which the numerical derivatives may be obtained. This finite-field procedure is frequently called "coupled Hartree-Fock" (CHF) (*29*).

Although it does not change the conceptual content or the numerical values (if done carefully!), the much more computationally convenient analytical equivalent, called coupled *perturbed* Hartree-Fock (CPHF), can be developed by taking the derivatives *before* evaluation by expanding all the equations in perturbation theory and explicitly solving them for perturbed orbitals, $\varphi_{j\varepsilon} = \varphi_{jo} + \varepsilon_o \varphi_j^{(1)} + \cdots$, orbital energies, $\epsilon_j = \epsilon_{jo} + \varepsilon_o \epsilon_j^{(1)} + \cdots$, and using $v_\varepsilon^{eff} = v_o^{eff} + \varepsilon_o v^{eff(1)} + \varepsilon_o^2 v^{eff(2)} + \cdots$ from which $E_{\rm HF}(\varepsilon_o) = E_{\rm HF}^{(0)} + \varepsilon_o E_{\rm HF}^{(1)} + \cdots$ can be obtained.

To avoid too much of a digression, we will not present those equations here. Excellent treatments for the time-independent case are given elsewhere (*27,30,31*). This is also a special case for the time-dependent, TDHF, approach (*11,32*) discussed in Section 7. Suffice it to say that CPHF calculations are preferable to CHF, and several implementations are available.

Even at the HF level, though, we have to pay close attention to the choice of basis set, that is the (usually) contracted Gaussian atomic orbital (AO) basis used to express the MO's. It is apparent that it is essential to have a large, flexible and polarized basis set for hyperpolarizabilities. The basis must correctly describe matrix elements of the long-range operator $\hat{r}$ (i.e. $\hat{x}, \hat{y}, \hat{z}$) while the usual AO basis functions have been selected predominantly to describe the energy, which depends on shorter-range operators like $\frac{1}{r}$ and the kinetic energy operator. A basis like double zeta (DZ) would use 4s and two sets of p functions to describe the energy of a B,C,N,O,F, atom in a molecule, 2s for H. To provide adequate polarization to this basis, particularly to describe the more diffuse and directional part of the charge density, we need at least one or two sets of d-functions and probably more s and p functions. The POL+ basis (*14,33*), optimized to describe polarizabilities has a 5s3p2d distribution for B-F, and 3s2p1d for H, and would be the *minimum* recommended for most hyperpolarizability determinations. The influence of basis set is illustrated in the behavior of the components of $\underline{\alpha}$ and $\underline{\gamma}$ for ethylene, shown in Table III.

Notice that a minimum STO-3G basis underestimates the HF-SCF dipole polarizability by over a factor of 5, with double zeta (DZ) still being in error by ~30%. The essential role of polarization functions is emphasized by the 6–31G+PD, meaning additional diffuse p and d functions on C and H (*34*), and the (5s3p2d/3s2p1d) POL+ basis (*14*).

The basis set effect is amplified dramatically for γ, where *four* products of $\hat{x}$, $\hat{y}$, and $\hat{z}$ are evaluated. Note that in inferior basis sets, including STO-3G and DZ, $\gamma_\parallel^o$, even has the wrong sign, differing from the better converged value by at least *four* orders of magnitude! However, once a few polarization functions are included, even as in the modest 6–31G+PD basis, convergence toward the Hartree-Fock solution is relatively good. The remaining differences between $\gamma_\parallel^o$ and $\gamma_\parallel^{\rm dcSHG}$ and $\gamma_\parallel^{\rm THG}$, which means dc-induced Second Harmonic Generation and Third Harmonic Generation, the potential experimentally observed quantities, lie in the frequency dependence neglected in the static HF calculation. We will discuss that aspect later. Obviously, it, too, is numerically important in providing reliable theoretical predictions.

Semi-empirical Methods

The other numerical values use the INDO (*35*) and spectroscopically parameterized INDO/S methods (*36*), (the latter is also known as ZINDO (*37*)). Like nearly all semi-empirical methods, INDO assumes an underlying minimum Slater-Type-Orbital (STO) basis SCF description, which is close to that of STO-3G. In the INDO case, the parameters are chosen to best reproduce the minimum basis SCF results. Hence, if this were done successfully, results about on the level of STO-3G would be obtained, clearly a level far inferior to that required for hyperpolarizabilities, and this is illustrated by the observed INDO results.

INDO/S would appear to have a little better chance at obtaining reasonable values. Despite the minimum basis set description, which is clearly suspect, the spectroscopic parameterization is chosen to try to describe excitations to the low-lying excited states of molecules and their transition moments within a single excitation configuration interaction (CIS) description. After fitting parameters to known spectra of similar molecules, the method is expected to describe related molecules reliably. If this were true for *all* excited states, from the SOS formula of equation 31b, obviously the dipole polarizability would be well described as it requires only those two pieces of information. However, the low-lying states are only a few of those that contribute to the polarizability. Furthermore, as all terms have the same sign, even though those neglected might have a comparatively high excitation energy, their sum total is significant. That is why there is a large error in $\bar{\alpha}$, particularly the α_{zz} component, compared to the good *ab initio* results. Notice that INDO/S is competitive with the *ab initio* DZ description, but unlike *ab initio* methods, semi-empirical results cannot be systematically improved. You get what you get!

Once you expect to use INDO/S for $\underline{\beta}$ or $\underline{\gamma}$, you now not only require transition moments between ground and excited states but *between excited states* themselves. Such experimental information is very seldom available, much less for the plethora of possible excited states, to offer any help with parameterization.

Table 3 demonstrates the dramatic failure of $\gamma_{\parallel}^{o}$ for ethylene, where the sign is wrong and it is off by 4 to 5 orders of magnitude. The same failure happens for $\underline{\beta}$. For the small molecules, INDO/S and INDO values of $\beta_{\parallel}^{o}$ have the wrong sign for CO and NH_3! (See also (*22*) for other examples.) Consequently, considering the basis set limitations inherent to semi-empirical methods, and their general inability to describe transitions between excited states, only generalizations beyond a minimum basis description, and further and more severe parameterization explicitly for hyperpolarizabilities, should enable such methods to offer any kind of quantitatively reliable results. For design purposes, the hope is to at least reproduce the correct trends among similar molecules, and typically for only one dominant axial component rather than the whole tensor; a simpler problem.

The other widely used semi-empirical methods, like those in MOPAC (*38,39*) (namely MNDO, MINDO, AM1, PM3), share with INDO a minimum basis description, but parameterization is attempted to be made directly to *experimental results*

Table 3 Comparison of *Ab Initio* SCF Hyperpolarizabilities as a Function of Basis Set with Semi-empirical INDO and INDO/S for Ethylene (a.u.)

COMPONENT	*SEMI-EMPIRICAL*		*AB INITIO SCF*				*EXP*
	INDO	*INDO/S*	*STO-3G*	*DZ*	*631+PD*	*POL+*	
α_{xx}	19.9	31.7	11.45	33.6	36.0	36.4	
α_{yy}	15.9	18.1	0.75	18.0	22.9	24.6	
α_{zz}	2.8	3.7	2.84	8.6	19.4	23.1	
$\overline{\alpha}$	**12.9**	**17.8**	**5.01**	**18.4**	**26.1**	**28.0**	**28.7**
γ^{o}_{xxxx}	-155	-2,092	-263	1,961	3,205	3,300	
γ^{o}_{yyyy}	95	194	2	111	2,008	2,800	
γ^{o}_{zzzz}	-6	-13	-23	64	11,303	11,900	
γ^{o}_{xxyy}	81	25	-1	43	1,680	1,600	
γ^{o}_{yyzz}	43	82	9	17	2,344	2,500	
γ^{o}_{xxzz}	98	304	40	231	3,294	3,100	
$\gamma^{o}_{\parallel}$	**76**	**-218**	**-37**	**-241**	**6,230**	**6,500**	
$\gamma^{dcSHG}_{\parallel}$	85	-344	-42	-337	9,251	9,900	9,029 ±203
$\gamma^{THG}_{\parallel}$	96	-811	-49	-538	15,836	17,500	

like dissociation energies and other properties, instead of the minimum basis SCF values. In particular, unlike INDO/S, such approaches have not been developed *to apply* for excited states as recommended by the SOS interpretation. But even from the energy derivative viewpoint, which pertains to INDO without the "S" as well, there is still little reason to believe that such methods have much hope of reliably describing qualities as sensitive to nuances of charge distributions as are hyperpolarizabilities. Several MOPAC examples that demonstrate failures are presented in ref. *(22)*.

Electron Correlation

Up to now, we have only considered HF level methods, *ab initio* or semi-empirical. After ensuring adequate basis sets, there are two particular corrections we need to consider: one is the frequency dependence (discussed in the next section) and the other electron correlation. Both would be essential, including their mutual coupling, to offer the definitive theoretical study for the purely electronic part of molecular hyperpolarizabilities. To initially isolate the effects, we will start with correlation corrections for static hyperpolarizabilities.

As discussed above, HF theory makes the approximation that one electron moves in an average field of *n-1* other electrons, to enable replacing the two-particle operator in the Hamiltonian by the $\sum_i v^{eff}(i)$, *one-particle* Hamiltonian. This ignores that electrons are charged species causing their motions to be *instantaneously* correlated. Clearly, the correlation of electrons bestows an additional degree of stability to the molecule, as the electrons are allowed to avoid each other, and this effect significantly contributes to the molecule's charge distribution and excited states description.

In equation 31, we derived formulas for hyperpolarizabilities based upon knowing the exact solutions to $\mathcal{H}_o$, which is the Hamiltonian in the absence of the electric field perturbation, $\varepsilon_o W$. Such exact solutions properly include all two-electron effects in $\mathcal{H}_o$, meaning their eigenvalues and vectors are *correlated*. This should be contrasted with replacing $\mathcal{H}_o$ by the $H_o = \sum_i f(i)$ operator as employed in the HF theory. These additional effects of correlation, whether used from the energy derivative viewpoint or the SOS, can have a dramatic effect on the observed results.

We can consider various approaches to electron correlation, but those most frequently applied to hyperpolarizabilities are many-body perturbation theory (MBPT) *(4,5,40)*, known in some programs as MP) and coupled-cluster (CC) theory *(7–9,14,41,42)*. In MBPT, in the absence of the electric field, electron correlation corrections could be introduced with the same equations derived in equation 6–11, where $\mathcal{H}_o$ is chosen to be $H_o = \sum_i f(i)$ and the correlation perturbation $V = \sum_{i,j} \frac{1}{r_{ij}} - \sum_i v^{eff(i)}$ replaces *W*. Then $E^{(2)} = \langle \Phi_o | V R_o V | \Phi_o \rangle = \langle \Phi_o | V | \psi^{(1)} \rangle$. This defines MBPT(2). MBPT(2) is the simplest correlated method, consisting of the initial contribution due to double excitations from occupied to unoccupied orbitals (i.e. Φ_{ij}^{ab}). MBPT is usually applied in a given finite order n [MBPT(n)], where n=4 is the highest frequently used.

In the absence of an electric field, proper treatment of correlation in hyperpolarizabilities requires a double perturbation approach (*43*) where all couplings between V and $\varepsilon_o W$ are allowed (*44*), with $\varepsilon_o W$ applied in a given order to describe the particular polarizability, and V is preferably included in *all* orders. A straightforward double perturbation approach is possible, but usually in practice, the requisite coupling between V and W is handled in two other ways. The first way is by correlating states, and then adding the W perturbation; the route taken in equations 28 to 31: This will be the EOM-CC route described in Section 8. Second, we can take the viewpoint that we will first solve the HF problem, and then add correlation to that HF solution straightforwardly, except that H_o, V, Φ_o and $\psi^{(n)}$ are all dependent on ε_o. That is, we evaluate correlation corrections to the electric field perturbed Hamiltonian as $f_\varepsilon = f_o + \varepsilon_o W + v_\varepsilon^{eff}$ as in the CHF method, so

$$\begin{aligned} H_o(\varepsilon_o) &= \sum_i f_\varepsilon(i) = \sum_i \left(f_o + \varepsilon_o W + v_\varepsilon^{eff}\right)(i) \\ V &= \mathcal{H}(\varepsilon_o) - H_o(\varepsilon_o) \\ E(\varepsilon_o) &= E_{\text{CHF}} + E^{(2)}(\varepsilon_o) + E^{(3)}(\varepsilon_o) + \cdots \\ E_{\text{CHF}}(\varepsilon_o) &= \langle \Phi_o(\varepsilon_o)|H(\varepsilon_o)|\Phi_o(\varepsilon_o)\rangle \end{aligned} \tag{36}$$

and all evaluation of correlation corrections pertain to $V(\vec{\varepsilon_o})$; such as $E^{(2)}(\varepsilon_o) = \langle \Phi_o(\varepsilon_o)|V(\varepsilon_o)|\psi^{(1)}(\varepsilon_o)\rangle$, and, similarly, in higher orders of MBPT (*2,3,7,40*). Just as in HF theory, we have the option of doing this analytically or as a finite field.

CC theory offers a natural generalization of MBPT that sums categories of excitations to infinite order. For example, CCSD means all single and double excitations (i.e. $\Psi_{\text{CCSD}} = \exp(T_1 + T_2)\Phi_o$ where Φ_o is the independent particle model and T_1 and T_2 are the single and double excitation operators,

$$T_2\Phi_o = \sum_{\substack{i<j \\ a<b}} t_{ij}^{ab}\Phi_{ij}^{ab} \tag{37}$$

where occupied orbitals i and j are replaced by unoccupied orbitals a and b,

$$T_1\Phi_o = \sum_i t_i^a \Phi_i^a \tag{38}$$

By virtue of the exponential expansion, Ψ_{CCSD} also contains triple excitations, $T_1T_2\Phi_o$, quadruple excitations, $T_2^2\Phi_o$, etc., ensuring a highly correlated solution. CCSD is correct through MBPT(3) with many additional higher-order effects (*9,28*). The CCSD(T) model includes, in addition, a non-iterative (*28*) evaluation of "connected" triples, $T_3\Phi_o = \sum_{\substack{i<j<k \\ a<b<c}} t_{ijk}^{abc}\Phi_{ijk}^{abc}$. This model, correct through MBPT(4), is demonstrably close to the basis set limit or full CI solution—the ultimate result (*28*). We can simply evaluate the CC results, just as described above for

Table 4 Comparison of SCF and Correlated Static Hyperpolarizabilities [c]

	$\chi_{\parallel}^{(2)a}$ *(10^{-32} esu/molecule)*			$\chi_{\parallel}^{(3)b}$ *(10^{-39} esu/molecule)*		
	SCF	*MBPT(2)*	*CCSD(T)*	*SCF*	*MBPT(2)*	*CCSD(T)*
H_2	-	-	-	46	50	51
N_2	-	-	-	61	78	85
CO_2	-	-	-	67	98	97
C_2H_4	-	-	-	546	630	563
CO	9.1	9.7	10.2	85	126	134
HF	-2.3	-3.0	-2.9	27	47	47
H_2O	-4.7	-7.6	-7.8	85	150	151
NH_3	-6.5	-14.0	-14.8	200	340	353
HCl	-1.3	-3.4	-3.3	213	287	295
H_2S	0.6	-4.5	-4.0	470	620	664

[a] $\beta_{\parallel}^{o} = (\beta_{ijj} + \beta_{jij} + \beta_{jji})/5$ with summation convention. $\chi_{\parallel}^{(2)} = \beta_{\parallel}/2$, so conversion to a.u. is 4.3195 x 10^{-33} esu/molecule/a.u.

[b] $\gamma_{\parallel}^{o} = (\gamma_{iijj} + \gamma_{ijij} + \gamma_{ijji})/15$ with summation convention. $\chi_{\parallel}^{(3)} = \gamma_{\parallel}/6$, so conversion to a.u. is 8.395 x 10^{-41} esu/molecule/a.u.

[c] Basic sets listed in Tables 7 and 8.

MBPT, by letting ψ_{CC} to be ε_o dependent, $[(\psi_{CC} = \exp(T(\varepsilon_o))\Phi_o(\varepsilon_o))$, along with $\mathcal{H}(\varepsilon_o) = H_o + V(\varepsilon_o) + \varepsilon_o W$ and $E_{CC}(\varepsilon_o) = \langle\Phi_o(\varepsilon_o)|\mathcal{H}(\varepsilon_o)e^{T(\varepsilon_o)}|\Phi_o(\varepsilon_o)\rangle]$.

Even better, because of the advantage of analytical evaluation even for the induced dipole, we might describe how that is done in CC theory. MBPT follows as a special case (*18*). Because $\psi_{CC}(\varepsilon_o) = \exp(T(\varepsilon_o))\Phi_o(\varepsilon_o)$ is an infinite series, we do *not* evaluate the induced dipole $\vec{\mu}(\varepsilon_o)$, as

$$\vec{\mu}(\varepsilon_o) = \langle\psi_{CC}(\varepsilon_o)|W|\psi_{CC}(\varepsilon_o)\rangle / \langle\psi_{CC}(\varepsilon_o)|\psi_{CC}(\varepsilon_o)\rangle \tag{39}$$

because the expression would have to be truncated. Instead, we can derive the form (*18–28*),

$$\begin{aligned}\vec{\mu}(\varepsilon_o) &= \langle\Phi_o(\varepsilon_o)|(1+\Lambda(\varepsilon_o))e^{-T(\varepsilon_o)}We^{-T(\varepsilon_o)}|\Phi_o(\varepsilon_o)\rangle \\ &= \sum_{pq} \vec{r}_{pq}\gamma_{pq}\end{aligned} \tag{40}$$

where Λ is a de-excitation operator, complementary to the excitation operator T and γ_{pq} is the element of the one-particle "relaxed density" matrix (*28*). Both Λ and T have to be determined from the CC equations. Subsequent finite-field differentiation relative to $\vec{\varepsilon}_o$ provides the various polarizabilities like $\frac{\partial\vec{\mu}(\varepsilon_o)}{\partial\vec{\varepsilon}_o} = \underline{\alpha}$, $\frac{\partial^2\mu(\varepsilon_o)}{\partial\vec{\varepsilon}_o\partial\vec{\varepsilon}_o} = 2!\underline{\beta}$, etc. As mentioned, this greatly improves precision and diminishes the number of distinct $\vec{\varepsilon}_o$ values required.

Table 4 shows the effect of correlation on static $\chi_{\parallel}^{(2)}(0;0,0)$ and $\chi_{\parallel}^{(3)}(0;0,0,0)$. Note the factor of 2 change for $\chi_{\parallel}^{(2)}$ of NH_3, and the order of magnitude and sign change for H_2S. Similarly, $\chi_{\parallel}^{(3)}$ changes substantially. This large correlation effect has been observed since the initial *correlated* studies of molecular hyperpolarizabilities (*2,3*), and makes it apparent that "predictive" *ab initio* methods for hyperpolarizabilities must include electron correlation.

Frequency Dependence

All experiments are frequency dependent, and frequency dependence introduces many different processes that become the same in the static limit.

Instead of the static field perturbation, $\vec{\varepsilon}_o \cdot \vec{W}$, consider the expansion of the induced dipole analogous to that in equation 24, for a time-dependent, oscillatory field, $\varepsilon = \varepsilon_o + \varepsilon_\omega \cos \omega t$,

$$\begin{aligned}\mu_i &= \langle\Psi(\varepsilon,t)|r_i|\Psi(\varepsilon,t)\rangle \\ &= \mu_i(o) + \alpha_{ij}(o,o)\varepsilon_{oj} + \alpha_{ij}(-\omega;\omega)\varepsilon_{oj} \cos \omega t \\ &+ \frac{1}{2!}\beta_{ijk}(o;o,o)\varepsilon_{oj}\varepsilon_{ok} + \frac{1}{4}\beta_{ijk}(o;\omega,-\omega)\varepsilon_{\omega j}\varepsilon_{\omega k} \\ &+ \beta_{ijk}(-\omega;\omega,o)\varepsilon_{\omega j}\varepsilon_{ok} \cos \omega t + \frac{1}{4}\beta_{ijk}(-2\omega;\omega,\omega)\varepsilon_{\omega j}\varepsilon_{\omega k} \cos \omega t \\ &+ \cdots\end{aligned} \tag{41}$$

Now, in addition to the static terms we have previously considered, we obtain a number of terms that correspond to different incoming and outgoing frequencies.

Table 5 Representative Non-Linear Optical Processes, with Corresponding Resultant (ω_σ) and Incident (ω_1,ω_2 . . .) Frequencies

Non-Linear Optical Process	$-\omega_\sigma$	$-\omega_1$	$-\omega_2$	$-\omega_3$
First Static Hyperpolarizability	0	0	0	-
Second Harmonic Generation (SHG)	-2ω	ω	ω	-
Electro-Optics Pockels Effect (EOPE)	$-\omega$	0	ω	-
Optical Rectification (OR)	0	ω	$-\omega$	-
Two-Wave Mixing	$-(\omega_1+\omega_2)$	ω_1	ω_2	-
Second Static Hyperpolarizability	0	0	0	0
Third Harmonic Generation (THG)	-3ω	ω	ω	ω
Intensity-Dependent Refactive Index (IDRI)	$-\omega$	ω	ω	$-\omega$
Optical Kerr Effect (OKE)	$-\omega_1$	ω_1	ω_2	$-\omega_2$
D.C.-Induced Optical Rectification (DCOR)	0	0	ω	$-\omega$
D.C.-Induced SHG (DC-SHG)	-2ω	0	ω	ω
Electro-Optic Kerr Effect (EOKE	$-\omega$	ω	0	0
Three-Wave Mixing	$-\omega_\sigma$	ω_1	ω_1	ω_2
D.C.-Induced Two-Wave Mixing	$-(\omega_1+\omega_2)$	0	ω_1	ω_2

For example, $\beta_{ijk}(-2\omega;\omega,\omega)$ corresponds to second harmonic generation (SHG) with incoming frequencies $\omega+\omega$ resulting in an outgoing frequency, 2ω. Similarly, $\alpha_{ij}(-\omega,\omega)$ corresponds to the dynamic polarizability, and $\beta_{ijk}(o;\omega,-\omega)$ is called optical rectification (OR), and $\beta_{ijk}(-\omega;\omega,o)$ corresponds to the electro optic Pockels effect (EOPE). If we allowed the frequencies to be different, $\beta(-\omega_1-\omega_2;\omega_1,\omega_2)$ would correspond to two-wave mixing. Similarly, many such components occur in γ. A summary of them is shown in Table 5. All become the same in the static limit; hence, *without inclusion of frequency dependence in the quantum mechanical method, we cannot distinguish between the different processes.* Obviously, we can obtain each of these quantities from an appropriate derivative,

$$\frac{\partial^2\mu_i(\varepsilon)}{\partial\varepsilon_{\omega j}\partial\varepsilon_{\omega k}} = \frac{1}{2!}\beta_{ijk}(-o;\omega,-\omega) \tag{42}$$

etc. Just as in the static case, that derivative can be further related to a quasi-energy derivatives *(45,46)* from the Frenkel variational principle *(11)*.

The first way to augment a static calculation with some measure of frequency dependence is to recognize that it may be rigorously shown *(47)* that for low frequencies,

$$\beta(-\omega_\sigma;\omega,\omega) = \beta(o;o,o)\left(1 + A\omega_L^2 + B\omega_L^4 + \cdots\right) \tag{43}$$

where $\omega_L = \omega_\sigma^2 + \omega_1^2 + \omega_2^2$ and *A* and *B* are unknown constants. Without an independent evaluation of *A* and *B*, the most we can conclude are the ratios of the various dispersion effects.

For example, neglecting the smaller quartic term, the SHG and EOPE values are

$$\begin{aligned} \beta(-2\omega;\omega,\omega) &= \beta(o;o,o)\left(1+6A\omega^2\right) \\ \beta(\omega;\omega,o) &= \beta(o;o,o)\left(1+2A\omega^2\right) \end{aligned} \tag{44}$$

showing that their ratio,

$$\frac{\beta(-2\omega;\omega,\omega)}{\beta(-\omega;\omega,o)} = \frac{1+6A\omega^2}{1+2A\omega^2} \simeq 1 + 4A\omega^2 - 8A^2\omega^4 + \cdots \tag{45}$$

Obviously, SHG has a much greater dispersion effect than EOPE. Similarly, we have for γ,

$$\gamma(-\omega_\sigma;\omega_1,\omega_2,\omega_3) = \gamma(0;0,0,0)\left(1 + A'\omega_L^2 + B'\omega_L^4 + \cdots\right) \tag{46}$$

The components of γ, OKE, IDRI, DCSHG and THG correspond to the values 1 plus $2A'\omega^2$, $4A'\omega^2$, $6A'\omega^2$ and $12A'\omega^2$, respectively, providing similar ratios. Obviously, the degree of dispersion follows the order THG > DCSHG > IDRI > OKE. However, we must at least know the constants to relate the various processes, particularly, to static quantities. It appears the only way at present to obtain any quantitative relationship is to evaluate the frequency dependent quantum chemical results.

Table 6 Comparison of Static and Dynamic[a] Hartree Fock Hyperpolarizabilities [b]

	$\chi_{\parallel}^{(2)}$ *(10⁻³² esu/molecule)*		$\chi_{\parallel}^{(3)}$ *(10⁻³⁹ esu/molecule)*	
	SCF	*TDHF*	*SCF*	*TDHF*
H_2	-	-	46	53.6
N_2	-	-	61	69.0
CO_2	-	-	67	76.6
C_2H_4	-	-	546	832
CO	9.1	10.5	85	102
HF	-2.3	-2.5	27	30
HCl	-1.3	-1.6	213	270
H_2O	-4.7	-5.4	85	100
NH_3	-6.5	-9.3	200	280
H_2S	0.58	0.64	470	690

[a] Values by TDHF for SHG and dcSHG at 694.3 nm.

[b] Basis sets listed in Tables 7 and 8.

The first such viable approach is the time-dependent Hartree-Fock (TDHF) method. Just as HF theory provides the simplest approach for static polarizabilities, TDHF provides the simplest *ab initio* solution to the time-dependent Schrödinger equation $\mathcal{H}(t)\Psi(t) = i\frac{\partial\Psi(t)}{\partial t}$ and, thereby, frequency dependent processes. Just as in the static case, we can impose a single determinant approximation for $\Psi(t) \approx \Phi_o(t)$, and by applying the time-dependent (Frenkel) variational/principle obtain the TDHF (or RPA) equations. The RPA equations date from long ago (*48*); however, this only pertains to excitation energies and $\underline{\alpha}$. In 1986, we made the generalization to $\underline{\beta}$, $\underline{\gamma}$, $\underline{\delta}$, $\underline{\epsilon}$, etc. and showed that *analytically* we can evaluate the frequency dependent hyperpolarizabilities from the derivatives in equation 36 in *any order* (*11*) giving ready TDHF access to hyperpolarizabilities. Others have since implemented equivalent TDHF approaches for $\underline{\beta}$ (*45*) and for $\underline{\beta}$ and $\underline{\gamma}$ (*32*). Recently, the unrestricted Hartree-Fock (open shell) generalization has been made (*49*). Figs. 1 and 2 graphically illustrate the dispersion behavior for various processes, which reflect the relative numerical proportions.

Table 6 shows the effect of dispersion in HF-level calculations of frequency dependent polarizabilities to SHG and dcSHG processes. For $\chi_{\parallel}^{(2)}$ the average change is 20% and $\chi_{\parallel}^{(3)}$ is 26%. Obviously, from the static values the other dynamic processes can show greater and lesser effects. Just as usual, we can take the derivative or SOS viewpoint in TDHF. In the latter case, the excited TDHF states are the usual RPA ones (*24*).

At this point, we have a procedure based upon TDHF to evaluate the dispersion, and a procedure to add electron correlation to static hyperpolarizabilities. Clearly, both are critical in obtaining predictive values. Hence, guided by the fact that equations 37 and 40 are exact in the low-frequency limit, it makes sense to use a percentage TDHF dispersion correction (*13*) to augment static, correlated hyperpolarizabilities; namely,

$$\begin{aligned}\beta(\omega_\sigma;\omega_1,\omega_2) &= \beta(0;0,0) \times \frac{\beta_{\mathrm{TDHF}}(\omega_\sigma;\omega_1,\omega_2)}{\beta_{\mathrm{HF}}(0;0,0)} \\ \gamma(\omega_\sigma;\omega,\omega_1,\omega_2) &= \gamma(0;0,0,0) \times \frac{\gamma_{\mathrm{TDHF}}(\omega_\sigma;\omega,\omega_1,\omega_2)}{\gamma_{\mathrm{HF}}(0;0,0,0)}\end{aligned} \tag{47}$$

By equating the percentage correction to $A\omega_L^2$ at a particular frequency, a value of A could be extracted, as well; or fitting to several different processes, A and B. In the few cases where frequency dependent *correlated* results have been obtained (*20,50,51*), the percentage TDHF dispersion estimate has been well supported. However, it is clear that if the TDHF=RPA result for the excitation energies are poor, then the slope of the curve in Figs. 1 and 2 will have to eventually change to be able to approach the different asymptotic values of the excitation energy.

Using this decoupled TDHF dispersion, static correlation procedure, we obtain the results shown in Tables 7 and 8. All fall within 10% error of the experimental result except for FH, whose errors are 28% for $\chi_{\parallel}^{(2)}$ and 24% for $\chi_{\parallel}^{(3)}$; for C_2H_4,

Table 7 Theoretical[a] vs. Experimental[b] Hyperpolarizabilities of Molecules

	$\chi_{\parallel}^{(2)} = \frac{1}{2}\beta_{\parallel}^{\mathrm{SHG}}$				
	TDHF	*MBPT(2)*	*CCSD*	*CCSD(T)*	*EXPERIMENT*
CO	10.5	11.2	11.4	11.7	12.9 ±1.4
HF	-2.5	-3.3	-3.2	-3.4	-4.70 ±0.41
HCl	-1.6	-4.0	-3.3	-3.8	-4.22 ±0.50
H_2O	-5.4	-8.8	-8.2	-9.1	-9.4 ±0.4
NH_3	-9.3	-20.1	-18.4	-21.2	-20.9 ±0.5
H_2S	0.6	-5.0	-3.4	-4.5	-4.3 ±0.9

[a]Value corrected for the dispersion effect at 694.3 nm using the formula,

$$\chi_{\parallel}^{(n)}(\omega) = \chi_{\parallel\mathrm{corr}}^{(n)}(\omega = 0) \times \frac{\mathrm{TDHF}(\omega)}{\mathrm{TDHF}(\omega = 0)}$$

The calculations are performed with basis sets [5s3p2d] for C, N, O and F; [7s5p2d] for S; and [3s2p1d] for H. Lone-pair functions are added for HF, H_2O and NH_3. For HCl, basis is [8s6p3d1f] for Cl and [3s2p1d] for H. All molecules at experimental geometries and there is no estimate of vibrational corrections. H_2S values in lone-pair augmented basis are 1.0, -4.4, -2.8 and -3.8, respectively (*14*). The lone-pair basis HCl values are -1.0, -3.1, -2.6 and -3.0.

[b]Value obtained by dc-induced Second Harmonic Generation.

Table 8 Theoretical[a] vs. Experimental[b] Hyperpolarizabilities of Molecules

	$\chi_{\parallel}^{(3)} = \frac{1}{6}\gamma_{\parallel}^{\mathrm{dcSHG}}$				
	TDHF	*MBPT(2)*	*CCSD*	*CCSD(T)*	*EXP*
H_2	54	58.2	59.3	59.3	60.5[c]
N_2	69	88.7	90.6	96	86.6 ±1.0
CO_2	77	111.5	107.9	110	111.9 ±1.3
C_2H_4	822	960	820	860	758 ±17
CO	102	151	149	160	144 ±4
HF	30	52	49	53	70 ±10
HCl	270	364	352	374	347 ±15
H_2O	100	180	170	180	194 ±10
NH_3	280	460	430	470	511 ±9
H_2S	690	910	870	930	865 ±22

[a]Value corrected for the dispersion effect at 694.3 nm using the formula,

$$\chi_{\parallel}^{(n)}(\omega) = \chi_{\parallel \mathrm{corr}}^{(n)}(\omega = 0) \times \frac{\mathrm{TDHF}(\omega)}{\mathrm{TDHF}(\omega = 0)}$$

The calculations are performed with basis sets [5s3p2d] for C, N, O and F; [7s5p2d] for S; and [3s2p1d] for H. The lone-pair functions are added for HF, H_2O and NH_3. For HCl, basis is [8s6p3d1f] for Cl and [3s2p1d] for H. All molecules at experimental geometries and there is no estimate of vibrational corrections. The lone-pair augmented H_2S basis values are 731, 970, 930 and 980, respectively. The lone-pair basis HCl values are 274, 369, 356 and 378.

[b]Value obtained by dc-induced Second Harmonic Generation.

[c]Exact electronic value [D. M. Bishop, J. Pipin and S. M. Cybulski, Phys. Rev. A**43**, 4845 (1991)]. The experimental value of 67.2 includes a significant vibrational effect.

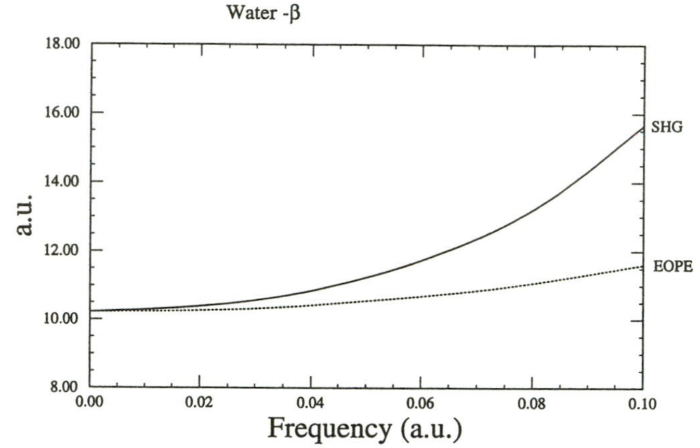

Figure 1. Water-β: NLO Processes As a Function of Frequency (a.u.)

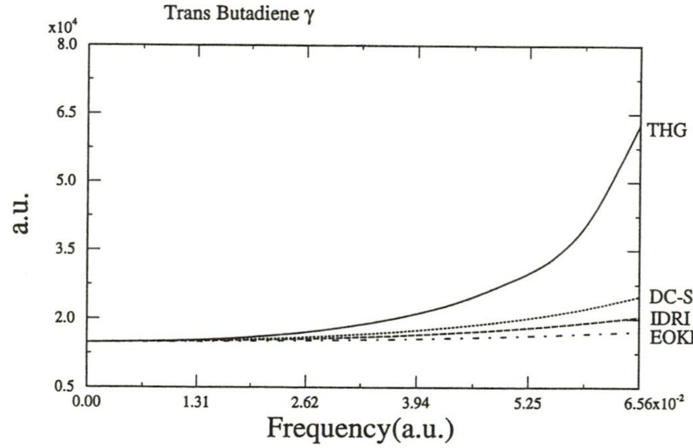

Figure 2. Trans Butadiene γ: NLO Processes As a Function of Frequency (a.u.)

whose $\chi_{\|}^{(3)}$ is too large by 13.4% and N_2, and CO, whose error is a modest 11%. We have discussed FH at length elsewhere (*7,14*), so we will not repeat that here, except to say that we dispute the experimental values, expecting an error in its determination. Extensive studies of the convergence of FH's hyperpolarizabilities to basis set, correlation, frequency dependence and vibrational corrections (*14*), show no convergence to the experimental values. Instead, we propose that the correct value for $\chi_{\|}^{(2)}$ is -3.6 ± 0.3x10^{-32} esu/molecule and $\chi_{\|}^{(3)}$ is 55 ± 5x10^{-39} esu/molecule. For the similar HCl molecule, our CCSD(T) results fall within our goal of a 10% error.

For C_2H_4 and N_2, we can consider more recent or rescaled experimental values (*22*). These are 76.6 ± 17 and 88.8 ± 1, respectively. This does reduce the C_2H_4 error to 12.3% and that for N_2 to 7.9%. Using the experimental results in the tables, the average error for CCSD(T) $\chi_{\|}^{(3)}$ is 9.4%. Excluding FH, it becomes 7.3%. Similarly, the average error for $\chi_{\|}^{(2)}$ is 9.4%, or 5.7% without the FH example, falling within our 10% error target objective. The various contributions are shown in Table 9. Clearly, correlation and frequency dependence are critical, and CCSD(T) is better than CCSD. However, it is gratifying that MBPT(2), which is a lot cheaper than CCSD(T), maintains about a 10% error. This level of correlation can be applied to much larger molecules (*52*) than can CCSD(T).

Correlated Frequency Dependent Polarizabilities

Despite the success of the decoupled correlated/TDHF results shown above, the most rigorous method would include the full coupling between correlation and dispersion. There have only been three attempts of this type for hyperpolarizabilities: a second-order [MBPT(2)=MP(2)] level method (*50*); a multi-configurational (MCSCF) linear response approach (*51*); and our EOM-CC method (*20*). The first two are developed from the energy (or quasi-energy) derivative viewpoint, while the latter refers, conceptually, to the SOS expressions in equation 31. In other words, EOM-CC provides excited states, $\{\psi_k^{(o)}\}$, their excitation energies, $\left(E_o - E_k^{(o)}\right)$, and generalized transition moments, $\langle\psi_k^{(o)}|r_i|\psi_o\rangle$, from which the SOS expressions could be formally constructed. That form is particularly convenient, since frequency dependence can be trivially added to such an SOS expression. For example, for some frequency ω,

$$-\alpha_{ij}(-\omega;\omega) = \left\{\sum_{k\neq o} \frac{\langle\psi_o|r_i|\psi_k^{(o)}\rangle\langle\psi_k^{(o)}|r_j|\psi_o\rangle}{\omega_k+\omega} + \frac{\langle\psi_o|r_j|\psi_k^{(o)}\rangle\langle\psi_k^{(o)}|r_i|\psi_o\rangle}{\omega_k-\omega}\right\} \quad (48)$$

When $\omega = E_o - E_k^{(o)} = \omega_k$, we have a pole, whose residue is the dipole strength $\langle\psi_o|r_j|\psi_k^{(o)}\rangle\langle\psi_k^{(o)}|r_i|\psi_o\rangle$.

The basic idea of EOM is very simple (53). Consider the solutions to the Schrödinger equation for an excited state, ψ_k^o and for the ground state, ψ_o,

$$\mathcal{H}_o\psi_o = E_o\psi_o \quad (49a)$$

$$\mathcal{H}_o\psi_k^{(o)} = E_k^{(o)}\psi_k^{(o)} \tag{49b}$$

Now we will choose to write $\psi_k^{(o)} = \mathcal{R}_k\psi_o$, where $\mathcal{R}_k$ is an operator that creates excitations from ψ_o. If we limit ourselves to single and double excitations, we will define an EOM-CCSD approximation. Inserting the excited state expression into equation 49b, multiplying equation 49a by $\mathcal{R}_k$ from the left and subtracting, we obtain

$$(\mathcal{H}_o\mathcal{R}_k - \mathcal{R}_k\mathcal{H}_o)\psi_o = \left(E_k^{(o)} - E_o\right)\mathcal{R}_k\psi_o \tag{50a}$$

$$[\mathcal{H}_o, \mathcal{R}_k]\psi_o = \omega_k\mathcal{R}_k\psi_o \tag{50b}$$

with the commutator between $\mathcal{H}_o$ and $\mathcal{R}_k$, [i.e. $(\mathcal{H}_o, \mathcal{R}_k)$], expression gives the "equation-of-motion," from the obvious connection with Heisenberg's form. To introduce CC theory, we simply choose for the correlated ground state, $\psi_o = \psi_{\mathrm{CC}} = \exp{(T_1 + T_2)}\Phi_o$. As $\mathcal{R}_k$ and T_n are all excitation operators, $[\mathcal{R}_k, T_n] = 0$, and we can commute the operators to give,

$$[e^{-T}\mathcal{H}_o e^T, \mathcal{R}_k]\Phi_o = [\bar{\mathcal{H}}_o, \mathcal{R}_k]\Phi_o = \omega_k\mathcal{R}_k\Phi_o \tag{51}$$

recognizing that

$$\hat{\mathcal{R}}_k\Phi_o = r_o + \sum_{i,a} r_i^a\Phi_i^a + \sum_{\substack{i<j\\a<b}} r_{ij}^{ab}\Phi_{ij}^{ab} \tag{52}$$

the coefficients $\{r_i^a, r_{ij}^{ab}\}$ are to be determined by the matrix equation,

$$\bar{\mathbf{H}}\mathbf{r}_k = \mathbf{r}_k\omega_k \tag{53}$$

Since $\bar{\mathcal{H}}_o$ is not Hermitian, we also have left-side eigenvectors, $\Phi_o\mathcal{L}_k$, which unlike $\mathcal{R}_k\Phi_o$, correspond to de-excitation processes. The $\mathcal{L}_k$ and $\mathcal{R}_l$ states are biorthogonal, $\langle\Phi_o|\mathcal{L}_k\mathcal{R}_l|\Phi_o\rangle = \delta_{kl}$. Their matrix equation is, $\mathbf{l}_k\bar{\mathbf{H}}_o = \omega_k\mathbf{l}_k$. The corresponding generalized transition moments are obtained from $\langle\Phi_o|\mathcal{L}e^{-T}r_ie^T\mathcal{R}_l|\Phi_o\rangle = \langle\Phi_o|\mathcal{L}_k\bar{r}_i\mathcal{R}_l|\Phi_o\rangle$. Notice $\langle\Phi_o|\mathcal{L}_k\bar{r}_i\mathcal{R}_l|\Phi_o\rangle \neq \langle\Phi_o|\mathcal{L}_l\bar{r}_i\mathcal{R}_k|\Phi_o\rangle$ since the operator $\bar{r}_i$ is not Hermitian. The observable quantity, however, is the dipole strength (the product), *not* the transition moment itself.

Now if we return to equation 48, in terms of EOM-CC solutions, we have

$$\begin{aligned}-\alpha_{ij}(-\omega;\omega) = \sum_{k\neq o}\Bigg\{&\frac{\langle\Phi_o|(1+\Lambda)\bar{r}_i\hat{\mathcal{R}}_k|\Phi_o\rangle\langle\Phi_o|\mathcal{L}_k\bar{r}_j|\Phi_o\rangle}{\omega_k+\omega}\\ &+\frac{\langle\Phi_o|(1+\Lambda)\bar{r}_j\mathcal{R}_k|\Phi_o\rangle\langle\Phi_o|\mathcal{L}_k\bar{r}_i|\Phi_o\rangle}{\omega_k-\omega}\Bigg\}\end{aligned} \tag{54}$$

Notice, we used the fact that $\langle\Phi_o|\mathcal{L}_o = \langle\Phi_o|(1+\Lambda)$, which is the same Λ operator introduced in equation 40, and that $\mathcal{R}_o = 1$.

Though informative in this form, we would still have to truncate the SOS. To avoid any such truncation, we need to recognize that all EOM-CC states are ultimately represented in terms of their underlying single and double excitations, i.e. the expression in equation 54. If we collectively represent all $\{\Phi_i^a;\Phi_{ij}^{ab}\}$ as $|\mathbf{h}\rangle$, it may be shown (*19*) that equation 48 may be written as

$$\begin{aligned}-\alpha_{ij}(-\omega;\omega) = (&\langle\Phi_o|(1+\Lambda)(\bar{r_i}-\langle\bar{r_i}\rangle)T_\pm^{(j)}|\Phi_o\rangle \\ &+\langle\Phi_o|(1+\Lambda)(\bar{r_j}-\langle\bar{r_j}\rangle)T_\pm^{(i)}|\Phi_o\rangle)\end{aligned} \tag{55}$$

where $\langle\bar{r_i}\rangle$ is the generalized expectation value like that in equation 40, without ε_o dependence). Analogous to ordinary perturbation theory [equations 28 to 30], the first-order perturbed correlated wavefunction is given in terms of the resolvent operator matrix $\mathbf{R}_o$,

$$\langle\mathbf{h}|T_\pm^{(j)}|\Phi_o\rangle = \langle\mathbf{h}|E_{\mathrm{CC}}\pm\omega-\bar{\mathbf{H}}_o|\mathbf{h}\rangle^{-1}\langle\mathbf{h}|\bar{r}_j|\Phi_o\rangle = \mathbf{R}_o\langle\mathbf{h}|\bar{r}_j|\Phi_o\rangle \tag{56}$$

In practice, instead of inversion, we solve the *very large* linear equation,

$$\langle\mathbf{h}|E_{\mathrm{CC}}\pm\omega-\mathcal{H}_o|\mathbf{h}\rangle\langle\mathbf{h}|T_\pm^{(j)}|\Phi_o\rangle = \langle\mathbf{h}|\bar{r}_j|\Phi_o\rangle \tag{57}$$

at a given value of ω for the $T_\pm^{(1)}$ coefficients. Hence, we can evaluate the SOS dynamic polarizability *without any truncation* (*19,20*)!

From the above evaluation of $\alpha(-\omega;\omega)$, we can obtain the hyperpolarizabilities for the optical Kerr effect (OKE) as follows. $\beta(-\omega;\omega,o)$, sometimes also called the EOPE, is the second-order hyperpolarizability obtained from a Kerr effect experiment, while the EOKE corresponds to $\gamma(-\omega;\omega,o,o)$. Because of the static fields in both processes, we are able to obtain $\beta(-\omega;\omega,o)$ and $\gamma(-\omega;\omega,o,o)$ from finite-field differentiation of $\alpha(-\omega;\omega)$. That is, we evaluate $\alpha_{ij}(-\omega;\omega)$ analytically, using equations 56 and 57, where we use the perturbation $\vec{\varepsilon_o}\cdot\vec{r}$ instead of just $\vec{r}$. Then we obtain $\alpha(-\omega;\omega_o,\varepsilon_o)$, from which

$$\beta^k(-\omega;\omega,o) = \frac{\partial\alpha(-\omega;\omega;\varepsilon_o)}{\partial\varepsilon_o} \tag{58a}$$

$$\gamma^k(-\omega;\omega,o,o) = \frac{\partial^2\alpha(\omega;\omega;\varepsilon_o)}{\partial\varepsilon_o^2} \tag{58b}$$

Results are shown in Table 10 for NH_3 and trans-butadiene as a function of frequency, for EOM-CC and TDHF. The usual very large effect of correlation accounts for the much larger magnitude for the EOM-CC values, while the comparative dispersion values are indicated as the percent dispersion in parentheses. For NH_3, there is about 10% greater dispersion as measured by EOM-CC compared to TDHF at the high frequencies (0.1 a.u.), but not at the low frequency (0.0656 a.u.) value used in the dcSHG experiments we previously described. For butadiene, the percentage dispersion is close, but the slightly smaller value helps to reduce our calculated values

Table 9 Percent Error of Hyperpolarizabilities at Various Levels Compared to Experiment

	$\chi_{\|}^{(2)}$	$\chi_{\|}^{(3)}$
HF (ω=0)	57	46
TDHF (ω)	50	31
MBPT(2)	12 (9)*	9 (7)*
CCSD	19 (16)*	8 (6)*
CCSD(T)	9 (6)*	9 (7)*
*Without FH example.		

Table 10 Kerr Effect Tensors for NH_3 and C_4H_6 Calculated at Different Frequencies (in au) [a]

Frequency (au/nm)	*0*	*0.043/1060*	*0.0656/694.3*	*0.1/455.6*
NH_3[a] (POL+)				
β^k [b]	-35.9 -14.7	-	-41.5 (15.7%) -16.6 (12.7%)	-51.5 (43.5%) -19.7 (33.8%)
γ^k [c]	4136.63 2404.99	-	4703.00 (13.7%) 2646.36 (10.0%)	5711.56 (38.1%) 3039.95 (26.4%)
C_4H_6[d] (631G+PD)				
$\gamma_{xxxx}(-\omega;\omega,0,0)$	41200 23514	44000 (6.9%) 25571 (8.7%)	48100 (16.8%) 28733 (22.2%)	-
γ^k	20700 14812	21900 (5.7%) 15794 (6.6%)	23700 (14.1%) 17277 (16.7%)	-

[a] The numbers in upper and lower rows are evaluated by EOM-CC and TDHF, respectively.x is theC_3 molecular axis.

[b] $\beta^k = \frac{3}{10}(3\beta_{ixi} - \beta_{iix})$

[c] $\gamma^k = \frac{1}{10}(3\gamma_{ijij} - \gamma_{iijj})$

[d] The numbers in upper and lower rows are evaluated by EOM-CC and TDHF, respectively. The$x$ component corresponds to the longitudinal molecular axis and the z component is perpendicular to the molecular plane.

to be in somewhat better agreement with experiment. Unlike most other molecules we have studied, TDHF results for ethylene and butadiene are fortuitously close to experiment, while correlation hurts the agreement. The origin of this is not yet clear, but the fact that the correlated dispersion is smaller than that for TDHF could be ascribed to the TDHF=RPA excitation energies being too low, causing the curve in Fig. 2 to rise too quickly to approach the wrong asymptotic values. For multi-bonded molecules like ethylene and butadiene, the restricted HF result is not triplet stable, although this only prohibits RPA from correctly describing triplet excited states, the RPA singlet excitations tend to be lower than experiment; contrary to that for most systems. This may partially account for TDHF giving higher percentage dispersion corrections.

We have not discussed vibrational contributions to predictive studies of hyperpolarizabilities, but these can sometimes be important (*21*). In equation 48, e.g., we could have contributions from all vibronic states indexed by *k*, instead of just the electronic ones. For an optical frequency, ω, which is much greater than a vibrational energy, such slight changes in $E_k^{(o)}$ would have negligible numerical value. However, for NLO processes that involve static fields, the SOS formulas will have some denominators without an additional large ω value, causing the vibronic changes in $E_k^{(o)}$ to be more significant to the final result. The vibrational energy levels can be substantially perturbed by such an electric field as we have shown numerically for FH and H_2 (*10*). Because of the static fields in OKE, this is an example where attention needs to be paid to such effects. Assisted by a determination of $\beta_k^v$and γ_k^v (*54*), we can extract from our calculations predictions for $\beta^k = \beta_k^{el} + \beta_k^v$ and $\gamma^k = \gamma_k^{el} + \gamma_k^v$ that could be compared with experiment. For NH_3, MBPT(2) values for β_k^v and γ_k^v are 3.8 and 135 a.u. at 0.07 a.u. This suggests that $\beta^k \approx -36$, $\gamma^k \approx 4800$ at ω=0.0656. For butadiene, there is a larger (SCF), γ^v=1395 to 1762 a.u., [55] giving $\gamma^k \approx$25000.

Future Extensions

There are a couple of fairly obvious extensions that should be made in future theoretical work for NLO material design. In the short term, we obviously need to generalize analytical frequency dependent EOM-CC for all the components of $\underline{\beta}$ and $\underline{\gamma}$. Also, considering the good accuracy of MBPT(2) level correlation, purely analytical, frequency dependent versions for $\underline{\beta}$ and $\underline{\gamma}$ are strongly recommended, that should remove the current constraints (*50*). Other routes to partitioned EOM-CC approximations that are operationally second-order, can be envisioned (*56*) and should be pursued.

Obviously, it would also be nice to be able to treat hyperpolarizabilities for molecules in solution. Several such solvation methods, ranging from continuous reaction fields to more detailed solvation models are becoming available (*57*). These should help in sorting out the large discrepancies among results from solvation experiments (EFISH) (*58*).

In the longer term, we need theoretical methods comparable to that presented for small molecules that are applicable to extended, polymeric systems (*59*). The first such approaches should employ periodicity, with future extensions directed at the inclusion of impurity effects.

Today, it is not possible to use analytical gradient techniques with correlation to move atoms around in polymers, as it is for molecules. Nor, are there the quality *ab initio* methods for band gaps, and excited states, and polarizabilities as there are for molecules. Clearly, developing the tools for rational NLO polymer design should have a high priority.

As current high-level *ab initio* methods will eventually encounter limitations, even for periodic systems, simplified techniques should be pursued, simultaneously. The questionable reliability of semi-empirical MO theory suggests that a better "semi-empirical" approach is likely to be offered by modern density functional theory (DFT). Although DFT has a rigorous base, in application it is semi-empirical. Such methods are well known for extended systems, and have decided computational advantages, compared to *ab initio* correlated methods, but they have not yet been demonstrated to provide comparable results to those presented in this chapter. In fact, one paper says that Kohn-Sham DFT does not work for molecular hyperpolarizabilities (*60*). We have considered other DFT variants, however, and find that competitive results can be obtained (*61*). The significant computational advantages of DFT make this a profitable area for study. Frequency dependent approaches need to be developed, however. Also, the conventional wisdom is that DFT does not admit treatments for excited states. Exploiting the equivalent derivative viewpoint should avoid any such formal restrictions for polarizabilities. Also, the ultimate limitation of applied DFT methods, like semi-empirical MO methods, is that there is no way to systematically converge to the exact result. New methods that combine elements of *ab initio* correlated theory with DFT methods will be forthcoming and might alleviate this failing.

Acknowledgments

Our work on molecular hyperpolarizabilities has been supported by the U.S. Air Force Office of Scientific Research since 1978, most recently under Grant No. AFOSR-F49620–93–I-0118. We are greatly indebted to their long-term commitment to this project. Their enlightened support made possible the fundamental developments in the basic theory that, today, pays dividends in a wealth of numerical results that were not possible even a year ago. Only with equal commitment to basic and applied research can many problems, like this one, be intelligently addressed.

Literature Cited

1. Ward, J. F. ; Miller ,C. K, *Phys. Rev. A* **1979**, 19, 826.
2. Bartlett, R. J.; Purvis III, G. D. *Phys. Rev.* **1979**,20, 1313.
3. Purvis, G. D.; Bartlett, R. J. *Phys. Rev. A* **1981**, 23, 1594.
4. Kelly, H. P. *Adv. Chem. Phys.* **1963**14, 12.

5. Bartlett, R. J.; Silver, D. M., *Int. J. Quantum Chem. Symp.* **1974**, 8, 271; *Phys. Rev. A* **1974**, 10, 1927; *J. Chem. Phys.* **1975**, 62, 3258.
6. Dudley, J. W. ; Ward, J. F. *J. Chem. Phys.* **1985**82, 4673.
7. Sekino, H. ; Bartlett, R. J. *J. Chem. Phys.* **1986**,84, 2726.
8. Cizek, J. *Adv. Chem. Phys.* **1969**, 14, 35.
9. Bartlett, R. J.; Purvis III, G. D. *Int. J. Quantum Chem.* **1978**, 14, 561; Purvis III, G. D. ; Bartlett, R. J. *J. Chem. Phys.* **1982**, 76, 1910.
10. Adamowicz, L; Bartlett, R. J. *J. Chem. Phys.* 1986, 84, 4988; (E) **1987**, 86, 7250.
11. Sekino, H.; Bartlett, R. J. *J. Chem. Phys.* **1986**, 85, 976.
12. Sekino, H.; Bartlett, R. J. *Int. J. Quantum Chem.* 1992, 43, 199.
13. Sekino, H. ; Bartlett, R. J. *J. Chem. Phys.* **1991**, 94, 3665.
14. Sekino, H. Sekino; Bartlett, R. J. J. Chem. Phys. **1993**, 98, 3022 .
15. For reviews, see Amos, R. D. in *Ab Initio Methods in Quantum Chemistry, Part I.* Wiley, Chichester, U.K., 1987; *Adv. Chem. Phys.* **1987**, 67, 99; Bartlett, R. J.; Stanton, J. F. ; Watts, J. D. *Advances in Molecular Vibrations and Collision Dynamics*, ed. J. Bowman,Vol. 1B, JAI Press, Inc., Greenwich, CT, pp. 139–167 (1991).
16. Bartlett R.J., In *Geometrical Derivatives of Energy Surfaces and Molecular Properties*, Editors. Jørgensen, P.; Simons J., Reidel: Dordrecht, The Netherlands, 1986, p. 35.
17. Handy, N. C.; Amos, R. D.; Gaw, J. F.; Rice, J. E.; Simandrias, E. D.; Lee, T. J.; Harrison, R.J.; Laidig, W.D.; Fitzgerlad, G.B.; Bartlett, R.J. In *Geometrical Derivatives of Energy Surfaces and Molecular Properties*; Editors, Jørgensen, P.; Simons J.; Reidel: Dordrecht, The Netherlands, 1986, p. 179.
18. Salter, E. A.; Trucks, G. W.; Bartlett, R. J. *J. Chem. Phys* **1989**, 90, 1752.
19. Stanton, J. F.; Bartlett, R. J. *J. Chem. Phys.* **1993**, 99, 5178.
20. Sekino,H.; Bartlett, R.J. *Chem. Phys. Lett.* **1995**, 234, 87.
21. Bishop, D. M. Revs. *Mod. Phys.* **1990**, 62, 343; *Adv. Quantum Chem.* **1994**, 25, 1.
22. Shelton, D. P; Rice, J. E. *Chem. Rev.* **1993**, 29, 3.
23. Ratner, M. A. *Int. J. Quantum Chem.* **1992**, 43, 5, and this special issue.
24. Sekino, H.; Bartlett, R. J. in *Theoretical and Computational Modeling of NLO and Electronic Material*, eds. Karna, S.; Yeates, T.A.; American Chemical Society Publications.
25. Gerratt, J.; Mills, I. *J. Chem. Phys.* **1968,** 49, 1719.
26. Pulay, P. *Mol. Phys.* **1969**, 17, 197; *Adv. Chem. Phys.* **1987**, 69, 241.
27. Dykstra, C. E.; Jasien, P. G. *Chem. Phys. Lett.* **1984**, 109, 388 ; Dykstra, C. E. *J. Chem. Phys.*, **1985**, 82, 4120.
28. Bartlett, R.J.; Stanton, J.F. In *Reviews in Computational Chemistry;* Editors, Boyd, D.; Lipkowitz, K; VCH Publishers: New York, NY, 1994; Vol 5; pp 65.

29. Cohen, H. D.; Roothaan, C. C. J. *J. Chem. Phys.* **1965**, 43, 534 .
30. Lipscomb, W. N. In *Advances in Magnetic Resonance*; Editor, Waugh, J. S.; Academic Press: New York, 1987, Vol. 2, pp. 137.
31. Nee, T.-S.; Parr, R. G.; Bartlett, R. J. *J. Chem. Phys.* **1976**, 64, 2216.
32. Karna, S.; Dupuis, M. *J. Comp. Chem.* **1991**, 12, 487; Karna, S: Dupuis, M. *Chem. Phys. Lett.* **1990**, 171, 201.
33. Sadlej, A. J. *Coll. Czech. Chem. Commun.* **1988**, 53, 1995; *Theoret. Chim. Acta.* **1991**, 79, 123.
34. Hurstm G. J. B.; Dupuis, M.; Clementi, E. *J. Chem. Phys.* **1988**, 89, 385.
35. Pople, J. A.; Beveridge, D.; Dobosh, P. A. *J. Chem. Phys.* **1967**, 47, 2026.
36. Del Bene, J.; Jaffe, H. H. *J. Chem. Phys.* **1968**, 48, 1807.
37. Zerner, M. C. *Rev. Comput. Chem.* **1991**, 2, 313; Kanis, D. R.; Ratner, M. A.; Marks, T. J.; Zerner, M. C. *Chem. Mater.* **1991**, 3, 19.
38. Dewar, M. J. S.; Stewart, J. J. P. *Chem. Phys. Lett* **1984**. 111, 416.
39. Kurtz, H. A.; Stewart, J. J. P.; Dieter, K. M. *J. Comput. Chem.* 1990, 11, 82.
40. Maroulis, G.; Thakkar, A. J. *J. Chem. Phys.* **1988**, 88, 7623; **1989**, 90, 366; 1990, 93, 4164.
41. Chong, D. P. ; Langhoff, S. R. *J. Chem. Phys.* **1990**, 93, 570 .
42. Taylor, P. R.; Lee, T. J.; Rice, J. E.; Almlöf, J. *Chem. Phys. Lett.* **1989**, 163, 359; Rice, J. E.; Taylor, P. R.; Lee, T. J.; Almlöf, J. *J. Chem. Phys.* **1991**, 94, 4972.
43. Hirschfelder, J.D.; Brown, W.B.; Epstein, S.T. *Adv. Quantum Chem.* **1964**, 1, 256.
44. Bartlett, R. J.; Silver, D. M. *Int. J. Quantum Chem. Symp.* **1975**, 9, 183.
45. Rice, J. E.; Amos, R. D.; Colwell, S. M.; Handy, N. C.; Sanz, J. *J. Chem. Phys.* **1990**, 93, 8828.
46. Sasagne, K., Aiga, F.; Itoh, R. *J. Chem. Phys.* **1993**, 99, 3738.
47. Bishop, D. M. *Chem. Phys. Lett.* **1978**, 69, 5438; *J. Chem. Phys.* **1989**, 90, 3192; Mizrahi, N.; Shelton, D.P. *Phys. Rev. A* **1985**, 31, 3145.
48. Jørgensen, P. *Am. Rev. Phys. Chem.* **1975**, 26, 239, and reference therein.
49. Karna, S. to be published.
50. Rice, J. E.; Handy, N.C. *Int. J. Quantum Chem.* **1992**, 43, 91.
51. Christiansen, O; Jørgensen, P. *Chem. Phys. Lett.* **1993**, 207, 367.
52. Sim, F.; Chin, S.; Dupuis, M.; Rice, J. *J. Phys. Chem.* **1993**, 97, 1158.
53. Comeau, D.; Bartlett, R.J. *Chem. Phys. Lett.* **1993**, 207, 414; Stanton, J.F,; Bartlett, R.J. *J. Chem. Phys.* **1993**, 98, 7029.
54. Bishop, D. M.; Kirtman, B.; Kurtz, H. A.; Rice, J. E. *J. Chem. Phys.* **1993**, 98, 8024.
55. Kirtman, B. unpublished.
56. Gwaltney, S; Nooijen, M.; Bartlett, R.J. *Chem. Phys. Lett.*, in press.

57. Wu, J.; Zerner, M.C. *J. Chem. Phys.* **1994**, 100, 7487.

58. Willetts, A.; Rice, J.E.; Burland, D.M.; Shelton, D.P. *J. Chem. Phys.* **1992**, 97, 7590.

59. Champagne, B.; Mosley, D.H.; André, J. M. *J Chem. Phys.* **1994**, 100, 2034.

60. Colwell, S. M.; Murray, C.W.; Handy, N.C.; Amos, R.D. *Chem. Phys. Lett.* **1993**, 210, 261.

61. Oliphant, N.; Sekino, H.; Beck, S.; Bartlett, R.J. to be published.

RECEIVED December 29, 1995

Chapter 3

Calculation of Nonlinear Optical Properties of Conjugated Polymers

Bernard Kirtman

Department of Chemistry, University of California, Santa Barbara, CA 93106

We present a comprehensive *ab initio* finite oligomer method for calculating the nonlinear optical properties of conjugated polymers. It is shown that extrapolation to the infinite polymer limit can be accurately carried out with the (large) effect of electron correlation taken into account. Other important aspects treated include frequency-dependence, vibrational distortion and environmental interactions. Some brief speculation about future directions is given.

There has been a recent explosion of interest (*1-4*) in nonlinear optical (NLO) materials because of their potential utilization in optical communication systems. Theoretical calculations that reliably determine the origin and magnitude of dynamic hyperpolarizabilities, which are the properties that govern NLO processes, can play an important role in designing these materials. From a practical standpoint polymers are amongst the most promising candidates for new applications. This paper discusses the progress that we have made in the *ab initio* computation of polymer hyperpolarizabilities, with an emphasis on conjugated systems. Although the focus here will be on *ab initio* treatments, since they are more reliable, we note that semi-empiricism can also be useful especially in circumstances where *ab initio* calculations are otherwise not feasible.

Our general procedure for determining polymer properties is straightforward. Calculations are carried out on finite oligomers of increasing size and, then, extrapolated (*5*) to the infinite polymer limit. This finite oligomer method works very well when there are no fields present. It also works very well for NLO properties, as we will see, but the extrapolation to get rid of end effects must be done with care since the convergence with increasing chain length is slow (*6-10*).

An alternative to the finite oligomer method is a band structure or crystal orbital (CO) approach (*11-16*) which, in principle, would circumvent the extrapolation problem. However, the interaction potential due to the electric field destroys the translational symmetry of a periodic polymer and introduces new complications, in addition to the usual questions concerning convergence of lattice sums and sampling of points in *k*-space. Although this remains an active field, thus far calculations have been limited to linear polarizabilities at the Hartree-Fock level of theory. One should bear in mind the possibility that hybrid formulations could capture the advantageous features of

0097–6156/96/0628–0058$15.25/0

both the crystal orbital and finite oligomer methods. Such a formulation has been developed (*17*) for vibrational properties of polymers and has also been suggested (*18*) for hyperpolarizabilities in conjunction with the local space approximation.

A challenging feature of NLO properties in real materials is the fact that there are so many important aspects. It is well-known, of course, that the frequency-dependence of these properties (as well as the background optical absorption associated with the linear polarizability) is critical for the operation of optical devices. This particular facet will be considered after first discussing extrapolation of the Hartree-Fock results to the infinite chain limit and the effect of electron correlation. Such aspects as electron correlation, vibrational distortion and environmental interactions, as well as impurities and disorder can be of major significance. Electron correlation can change the hyperpolarizability by as much as an order of magnitude, through both indirect (geometry) and direct effects. Although it might seem that the size convergency problem would pose an insurmountable barrier to treating electron correlation in the finite oligomer method that turns out not to be the case (*19*).

The vibrational distortion contribution to the hyperpolarizability arises because the interaction with an electric field depends upon the instantaneous nuclear configuration. As a result there is a field-dependent shift in the equilibrium geometry and in the vibrational force field leading to a so-called vibrational hyperpolarizability. Preliminary investigations on small oligomers have shown (*9*), unexpectedly, that the vibrational term can be larger than its electronic counterpart in some NLO processes. Our first efforts to follow-up on these results, with formal theory and computations, will be described after the section on frequency-dependence. It will be seen that certain collective modes can contribute significantly to the vibrational hyperpolarizability despite being optically inactive in the infinite chain limit.

The influence of environment on NLO properties is an aspect that has barely been touched. In the case of polyacetylene, the geometrical arrangement of the polymer chains and the interchain distances in the solid state (stretched fiber) are known experimentally. Using this information supermolecule calculations on the effect of interchain interactions have been carried out. From the results reported here we will see that the environmental effect on the hyperpolarizability is quite large and contrary to what is predicted by semiempirical studies (*20*).

The last section is devoted to a discussion of future directions including extensions to polymers containing larger monomeric units and to the investigation of other factors, not yet considered, that may strongly influence NLO properties. We also speculate about how insights gained from our calculations, and others of a similar nature, might lead to better NLO materials.

We begin, in the next section, with the first step along the pathway towards a comprehensive treatment for conjugated polymers; namely, calculation of the static longitudinal hyperpolarizability at the coupled perturbed Hartree-Fock (CPHF) level. In this context we deal with the critical problem of extrapolating to the infinite chain limit.

Static Longitudinal Hyperpolarizability at the CPHF Level/Extrapolation to the Infinite Chain Limit

The value of the static hyperpolarizability determined at the CPHF level of approximation is our fundamental reference quantity. Besides being the starting point for further calculations the behavior of this quantity with increasing chain length provides a rough template for the corresponding evolution of changes in the hyperpolarizability due to frequency dispersion, electron correlation, etc. This fact will be used in later sections to facilitate extrapolation of these effects to the infinite chain limit.

Very recently (*10*), CPHF static hyperpolarizabilities were reported for the linear polyenes C_4H_6 through $C_{44}H_{46}$. In addition, it was shown that these results could be extrapolated to yield a value for the infinite polyacetylene (PA) polymer which is comparable in accuracy to that ordinarily achieved for small molecules. Although PA was used in this instance, the method for extrapolating is generally applicable.

From the above treatment, as well as previous calculations (*6-9*), some useful guiding rules and practices have emerged. One guiding rule is that the methods employed need to meet the desired accuracy for the large oligomers but not necessarily the smaller ones. After all, only the asymptotic limit is of interest. In practice this means that a split valence 6-31G basis will usually suffice (*6,9*). It also means that we can employ the geometry of the infinite polymer, obtained either from a CO calculation or extrapolation of small oligomer results, without reoptimizing for the individual oligomers. Finally, the calculations can be restricted to the longitudinal component which will dominate for the infinite chain.

A second guiding rule is that the ratio of similar calculations will usually converge much faster with chain length than either one separately. Thus, the effect of a particular basis set augmentation on the hyperpolarizability can be checked by considering only relatively small oligomers. In the next section it is shown that different correlation treatments can be tested in a like manner. More importantly, we will use this feature as the basis for determining the (static) correlated hyperpolarizability of the infinite chain which, in some instances, can be found even without an extrapolation simply by taking the ratio to the CPHF value. A similar approach also leads to improved convergence for frequency dispersion in the non-resonant regime.

As far as extrapolation is concerned, the guiding rule we have formulated is that the result must be stable with respect to variations in the parameters characterizing the treatment of the data set. Ordinarily, the infinite chain value is found by a least squares fit to a postulated functional form. The shorter oligomers in the data set have only limited bearing on the behavior of this function in the asymptotic region. In fact, including these oligomers can lead to a poorer fit for the longer chains. So the question is - which short chains, if any, should be eliminated? In practice, an initial calculation is done using the minimum number of long chains necessary to define the fitting function. Then the shorter oligomers are added one at a time. Typically, there will be large oscillations, at first, in the extrapolated value because of the few degrees of freedom available. As the number of oligomers included continues to increase the oscillations diminish while the fit remains good. This is the stable region that we seek. Eventually three things happen: (1) the overall quality of the fit deteriorates; (2) there is a steady upward or downward trend in the calculated infinite chain limit; and (3) the deviation for the longest chain, in particular, increases. This last item (i.e. (3)) is the quality of fit criterion that is used to mark the emergence from the stable region and, at that point, we cease consideration of the remaining oligomers.

The behavior just described is illustrated in Table I. Here, for the linear polyenes $C_{2N}H_{2N+2}$, we have fit the static CPHF longitudinal hyperpolarizability per repeat unit to a polynomial of order k in $1/N$. Thus, the constant term in the polynomial

Table I. Static CPHF longitudinal hyperpolarizability per C_2H_2 unit (in 10^4 a.u.) of infinite polyene chain obtained by extrapolation from the finite oligomers $C_{2N}H_{2N+2}$. Extrapolations were done using least squares fit to a polynomial of order k in $1/N$. N_{max} gives the maximum chain length in the data set; smaller oligomers are successively included to give a total of $k+1+p$ points. Reproduced with permission from ref. 10.

A. $N_{max} = 22$

k\p	0	1	2	3	4	5	6	7	8	9	10
1	578.3	577.1	561.3	558.2	552.9	543.7	535.1	524.9	513.0	499.5	484.2
2	602.3	828.6	666.5	651.6	676.8	670.4	669.5	669.1	665.7	659.9	651.8
3	-2106.	2184.	1058.	554.4	685.7	680.2	675.1	686.3	696.7	702.2	705.4
4	-37694.	7437.	5032.	597.9	694.6	705.2	640.4	639.2	664.6	683.2	692.9
5	-313597.	3141.	25853.	3493.	1096.	986.7	710.4	589.2	607.8	649.3	660.1

B. $N_{max} = 21$

k\p	0	1	2	3	4	5	6	7	8	9	10
1	575.9	551.2	551.9	547.4	537.6	528.8	518.2	505.9	492.0	476.1	458.0
2	1030.	608.0	622.1	673.5	666.4	666.9	667.4	663.9	657.5	648.8	636.9
3	5802.	975.9	261.5	630.8	653.5	660.3	681.5	696.6	703.3	706.7	709.6
4	43454.	8766.	-848.5	266.7	545.1	550.0	597.1	650.5	680.3	693.2	710.3

C. $N_{max} = 20$

k\p	0	1	2	3	4	5	6	7	8	9	10
1	528.1	545.0	541.9	531.0	522.0	511.1	498.3	483.8	467.5	448.8	427.6
2	234.7	544.0	670.3	662.0	664.8	666.3	662.4	655.2	645.6	632.7	614.5
3	-3058.	-933.1	466.6	600.4	639.2	677.9	698.8	706.1	709.1	711.5	707.8
4	-18589.	-10859.	-1653.	49.1	363.5	533.7	636.9	681.2	696.5	715.3	729.1

corresponds to the infinite chain limit. The integer *p* is the number of oligomers added beyond the minimum necessary to determine the fitting parameters, while N_{max} gives the chain length of the largest oligomer considered. The case N_{max}=22, k=2 is typical. Note that the extrapolated value oscillates for p=0-4; is stable for p=5-7; and, then, gradually diminishes as the fitting error (not shown) for $C_{44}H_{46}$ increases.

Although it is always desirable to make use of the longest chain for which calculations have been done, we are also interested in the effect of removing this oligomer. The change in the calculated infinite chain limit provides a measure of the uncertainty in its value (see below).

We have yet to discuss the fitting function or the precise definition of the quantity to be fit. Two different prescriptions have been employed to convert calculated hyperpolarizabilities to the hyperpolarizability per repeat unit, which is the quantity that should saturate as the chain length is increased. One of these is simply to divide the total value by the number of repeat units; the other is to take the difference between successively larger oligomers. In general, the latter is preferable because end effects are minimized and, therefore, convergence is faster. Sometimes, however, the differences between successive oligomers may behave erratically for long chains due to the presence of more than one series of oligomers (for example, odd N vs. even N in PA) or the inaccuracy of numerical differentiation in finite field treatments. In that event one must utilize the alternative procedure, or analytical differentiation if numerical inaccuracy is the problem. Finally, some workers (*6*) have found it convenient to replace the hyperpolarizability per repeat unit by its logarithm. This will make little difference if the stability (plus quality of fit) criteria are satisfied and the fitting function has sufficient flexibility as described below.

Several different forms have been proposed to describe the asymptotic behavior in the long chain limit. They include: (1) the simple power series in 1/N previously mentioned, (2) Padé approximants (*21,22*), and (3) the exponential function (*15*) A-Bexp(-cN). None of these are derived from first principles. A perturbation treatment (*23*) of the infinite chain shows that the leading correction to the Hartree-Fock energy per unit cell is proportional to 1/N, which is consistent with (1) and (2). Our choice of (1) is based on the principle of Occam's razor; however, as long as terms can be added in a systematic manner, stability criteria can be applied to any form that satisfies the boundary conditions. For a power series in 1/N stability is required with respect to changing the order of the series.

The methodology described in this section has been applied to the CPHF static hyperpolarizability of polyacetylene using the linear polyene oligomers through $C_{44}H_{46}$. From the subset of extrapolations for $k \leq 5$ and $20 \leq N_{max} \leq 22$, that meet the stability and quality of fit criteria, we extract an infinite chain value which falls in the range $(691 \pm 39) \times 10^4$ a.u. There is some arbitrariness in determining the region of stability, but this result is not particularly sensitive to the precise numerics. The uncertainty of 5.6% reported here is comparable to that normally achieved for small molecules even though the hyperpolarizability per repeat unit for the largest oligomer considered is only 43% of the infinite polymer value.

Effect of Electron Correlation on the Static Hyperpolarizability

As observed earlier, electron correlation can have a profound effect on NLO properties. However, to calculate this effect directly for long chain oligomers, such as $C_{44}H_{46}$, is a daunting task even at the crudest level of approximation. Fortunately, indirect ratio methods can be used to circumvent this problem. One of the first issues that needs to be examined is the level of the correlation treatment. Assuming for the moment that a 6-31G basis is sufficient, we have carried out a set of correlation calculations for small linear polyenes using Moller-Plesset second order (MP2) and fourth order (MP4) perturbation theory. The results obtained at the restricted Hartree-Fock, i.e. RHF/6-31G, geometry are shown in Table II. It appears that the perturbation treatment converges rapidly. The full MP4 values including single (S), double (D), triple (T), and quadruple (Q) excitations are less than MP2, but by a relatively small amount in each case. The breakdown of contributions according to excitation level is arbitrary; using the results reported in the table one can say that the small difference between MP2 and MP4(SDTQ) is due to the effect of single excitations and quadruples (to a lesser extent) being partially canceled by doubles and triples (in the larger polyenes). A coupled cluster doubles calculation, augmented by fourth order singles and triples, was done just for butadiene and it gives a result close to (difference = 1.8 a.u.) MP4(SDTQ). Furthermore, the convergence of the MP4(SDTQ)/MP2 ratio with increasing chain length is also rapid. The infinite polymer limit of this ratio is 0.92, as obtained from an extrapolation based on our usual procedures. We conclude that the correlation effect can be accurately determined from an MP2 calculation multiplied by the scale factor of 0.92. This numerical value is strictly valid only at the RHF/6-31G geometry but it is insensitive to geometry variations. Of course, the above treatment is specific to polyacetylene. It remains to be seen whether or not an analogous procedure will work for other polymers.

A second point that must be considered is the adequacy of the 6-31G basis when correlation is included. For this purpose the effect of several different basis set augmentations on the MP2 longitudinal hyperpolarizability (γ_L) has been examined, again using the linear polyenes as an example. We report the ratio with respect to the 6-31G basis set calculation in Table III (all results were determined at the RHF/6-31G geometry). The 6-31G+PD basis contains diffuse P and D functions on carbon with exponents optimized for the CPHF hyperpolarizability (*6*) in butadiene. This same basis has been employed for correlated hyperpolarizabilities in butadiene (*24*) and for CPHF calculations in longer polyenes (*6*). As expected, the addition of diffuse P and D functions causes γ_L to increase but this effect diminishes rapidly with chain length. For comparison, the analogous ratios for the RHF γ_L are included in parentheses in the table. The RHF values were extrapolated (*6*) using the linear polyenes through $C_{16}H_{18}$ to give an infinite chain ratio of 0.93 (separate extrapolations were done for the two basis sets and, then, the ratio was taken). For the MP2 calculations it appears that the limit will be somewhat closer to unity.

Longitudinal hyperpolarizabilities obtained using the standard 6-31G* and 6-31G** bases are also given in Table III. These bases contain the tight polarization functions on carbon (6-31G*) and on hydrogen as well (6-31G**) that are normally utilized in field-free calculations. Consequently, the ratio is less than unity but the effect again diminishes as the chain is lengthened. The additional polarization functions on hydrogen (cf. 6-31G** with 6-31G*) were found to give only a small increase in the ratio for butadiene and hexatriene so that, for the infinite polymer, there will be close agreement between the 6-31G and 6-31G** basis sets.

Although future studies of more extended bases (6-31G + PD and larger) could lead to some improvement, we conclude from the above that the 6-31G basis will give MP2 results that are accurate to better than 10%. Our MP2/6-31G values of γ_L per repeat unit are reported in Table IV for the linear polyenes at two different geometries.

Table II. Comparison of MP2 and MP4 static γ_L (in units of 10^3 a.u.) for short polyene chains. All calculations were done in the 6-31G basis at the RHF/6-31G geometry. Reproduced with permission from ref. 19.

		MP4				
	MP2	(D)	(DQ)	(SDQ)	(SDTQ)	MP4/MP2
C_4H_6	19.2	29.6	25.2	19.6	18.5	0.964
C_6H_8	119.5	145.1	139.4	114.8	111.8	0.936
C_8H_{10}	434.4	488.3	464.1	394.7	403.7	0.929
$C_{10}H_{12}$	1152.	1196.	1131.	989.3	1065.	0.925

Table III. Variation of MP2 static longitudinal hyperpolarizability , γ_L, with basis set for small linear polyenes. The value given is the ratio with respect to a 6-31G calculation. For the 6-31G+PD basis the corresponding RHF ratios are included in parentheses. Reproduced with permission from ref. 19.

	6-31G+PD[a]	6-31G*[b]	6-31G**[b]
C_4H_6	2.358 (3.915)	0.864	0.874
C_6H_8	1.679 (1.942)	0.895	0.910
C_8H_{10}	1.417 (1.445)	0.921	--
$C_{10}H_{12}$	1.323 (1.327)	0.937	--
$C_{12}H_{14}$	1.227 (1.225)	0.942	--

[a] Diffuse P and D functions on carbon chosen to optimize RHF second-order hyperpolarizability.
[b] See text.

Table IV. MP2/6-31G values of γ_L (in 10^3 a.u.) for the $C_{2N}H_{2N+2}$ oligomers of PA. Also included is the ratio, denoted by MP2/RHF, with respect to an RHF/6-31G calculation. Reproduced with permission from ref. 19.

	RHF geometry[a,b]		MP2 geometry[a,c]	
N	MP2	MP2/RHF	MP2	MP2/RHF
2	9.62	3.05	12.69	4.02
3	39.83	2.20	54.18	2.99
4	108.58	2.00	152.21	2.80
5	230.40	1.92	332.66	2.77
6	414.29	1.89	607.99	2.78
7	685.05	1.87	974.41	2.81
8	965.00	1.88	1441.19	2.82
9	1309.61	1.88	1942.71	2.84
10	1657.00	1.85	2534.02	2.83
11	2090.00	1.90	3135.57	2.85
12	2474.01	1.89	3713.89	2.84

[a]This is the geometry of the MP2 calculation. All RHF calculations are at the RHF/6-31G geometry.

[b] The RHF geometry is that of the individual polyene in the 6-31G basis.

[c] The MP2 geometry is that of the infinite chain as given in Ref. (25) for the 6-31G basis.

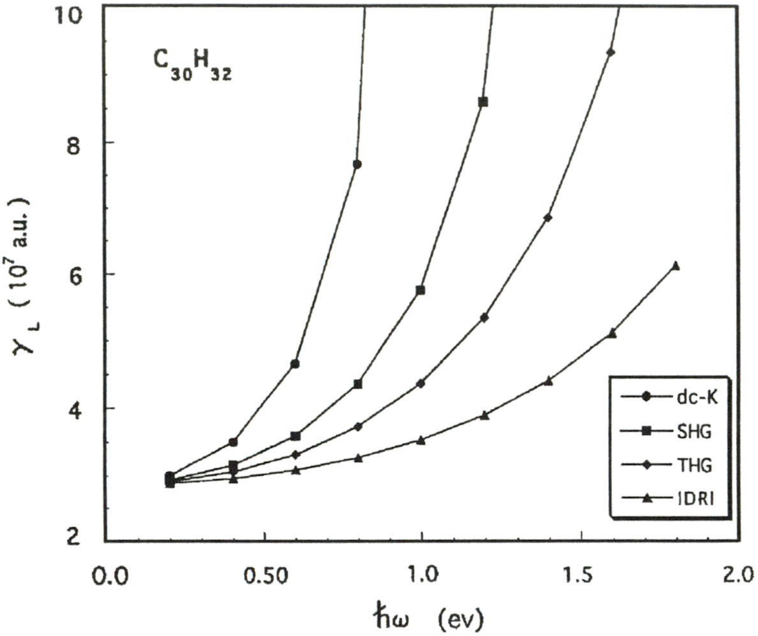

Figure 1 TDHF/6-31G longitudinal hyperpolarizabilities for $C_{30}H_{32}$ in the region of the spectrum up to the first absorption. dc-K is the dc-Kerr effect; SHG is second harmonic generation; THG is third harmonic generation; and IDRI is intensity dependent refractive index.

Of the latter, one is the "frozen" RHF/6-31G geometry optimized for the individual polyene while the other is the MP2/6-31G geometry (*25*) optimized for the infinite polymer (a C-H bond distance of 1.0725 Å and a C-C-H bond angle of 124° was adopted for the terminal hydrogens; other reasonable choices might have a small effect on the convergence with chain length). Included in the table is the ratio taken with respect to an RHF/6-31G calculation (at the RHF/6-31G geometry). This ratio is seen to converge rapidly as the chain is lengthened. In fact, beyond N=4 it is *constant* within the uncertainty of the finite field method of determination.

Electron correlation at the MP2 level clearly causes a substantial increase in γ_L even at the "frozen" RHF geometry. An overall increase by a factor of 2.84 (based on the largest oligomer treated) occurs when the change in geometry, which decreases the bond length alternation, is also taken into account. (A study comparing correlation effects in polyacetylene with polyyne and polypyrrole has now been undertaken by Toto, J.L.; Toto, T.T.; deMelo, C.P., Federal University of Pernambuco, personal communication, 1995.) Using the scaling constant of 0.92 to convert to the fully correlated result, along with the RHF static γ_L given previously, we obtain 181 x 10^5 a.u. as our best estimate of the static γ_L per repeat unit for the infinite polymer.

Frequency-dependence

So far, only *static* hyperpolarizabilities have been considered. For practical applications it is the nonlinear response to temporally oscillating laser fields that is of interest. In order to calculate this response, which is governed by the *dynamic* hyperpolarizabilities, a suitable starting point is the time-dependent analogue of the static CPHF treatment, often referred to simply as the time-dependent Hartree-Fock (TDHF) approximation (*26*).

Building upon the work of Karna, *et al.* (*27*) for small oligomers we have undertaken TDHF/6-31G calculations on the linear polyenes thru $C_{30}H_{32}$. The results for $C_{30}H_{32}$, in the region of the spectrum up to the first absorption, are shown in Figure 1. Characteristically, the dispersion is greatest for third harmonic generation (THG) and successively decreases in the order:

$$\text{THG} > \text{dc-SHG} > \text{IDRI (or DFWM)} > \text{dc-Kerr} \qquad (1)$$

Here dc-SHG is field induced second harmonic generation; IDRI is the intensity-dependent refractive index, also known as degenerate four wave mixing (DFWM); and dc-Kerr is the field-induced Kerr effect. In the usual notation for the dynamic hyperpolarizabilities, i.e. $\gamma_L(-\omega_\sigma;\omega_1,\omega_2,\omega_3)$ with $\omega_\sigma = \omega_1 + \omega_2 + \omega_3$, these four processes correspond to the cases: $\omega_1 = \omega_2 = \omega_3 = \omega$ (THG); $\omega_1 = \omega_2 = \omega$, $\omega_3 = 0$ (dc-SHG); $\omega_1 = \omega_2 = \omega$, $\omega_3 = -\omega$ (IDRI); and $\omega_1 = \omega$, $\omega_2 = \omega_3 = 0$ (dc-Kerr). It has been shown (*28*) that, in the low frequency limit, the ratio of the dynamic to the static hyperpolarizability may be written:

$$\gamma_L(-\omega_\sigma;\omega_1,\omega_2,\omega_3)/\gamma_L(0;0,0,0) = 1 + A\omega_D^2 \qquad (2)$$

where $\omega_D^2 = \omega_\sigma^2 + \omega_1^2 + \omega_2^2 + \omega_3^2$ and A is a constant independent of the process. This relationship determines the order given in equation 1. We find that it is satisfied

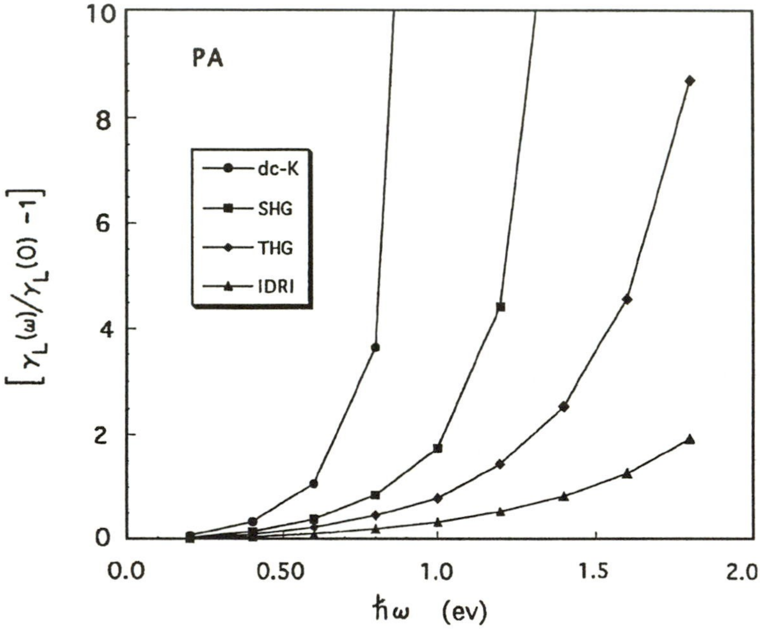

Figure 2 TDHF/6-31G dispersion curves for PA in the low frequency region obtained by extrapolation of calculations on the finite oligomers through $C_{30}H_{32}$. dc-K is the dc-Kerr effect; SHG is second harmonic generation; THG is third harmonic generation; and IDRI is intensity dependent refractive index.

(all processes) to within an accuracy of $2x10^{-3}$ in $C_{30}H_{32}$ for frequencies up to $\hbar\omega$ ~ 0.3 eV. For higher frequencies the dc-Kerr and dc-SHG processes should satisfy equation 2 after adding a $B\omega_D{}^4$ term (*28*) to the *rhs*, where B is a constant. With the additional term these two processes are obtained to an accuracy of $2x10^{-3}$ in $C_{30}H_{32}$ for all frequencies less than $\hbar\omega$ ~0.8 eV.

Dynamic RHF hyperpolarizabilities converge slowly with increasing chain length like their static counterpart. Can this situation be improved by taking the ratio of the two in analogy with our successful treatment of electron correlation? Some typical results are displayed in Table V for dc-SHG at two different frequencies corresponding to a dispersion of 10% ($\hbar\omega$ = 0.4 eV) and 52% ($\hbar\omega$ = 0.8 eV) in $C_{30}H_{32}$. In both cases the ratio approaches the asymptotic limit much more rapidly than the hyperpolarizability itself. Using the methods described earlier to extrapolate this ratio we have constructed the dispersion curves for the infinite polymer shown in Figure 2.

Of course, we are interested in the entire spectrum rather than just the low frequency region. However, the self-consistent-field iterations in the TDHF procedure do not converge at frequencies near, and beyond, the first absorption. This difficulty can be overcome by the introduction of radiative damping, as in the uncoupled Hartree-Fock sum over states treatment of Shuai and Brèdas (*29*). The computer program modifications to do so are currently being implemented (in collaboration with S. P. Karna). In the simplest version the linewidth is taken to be proportional to the orbital excitation energy and the proportionality constant is varied to yield stable results. Alternatively, one can use the random phase approximation, although the computations are considerably more extensive.

Electron correlation will critically affect dynamic hyperpolarizabilities just as it does the corresponding static property. Ordinary finite field or analytical differentiation techniques cannot be utilized to compute this effect. Several appropriate methods have been developed (*30-33*), although they are often tedious to apply. As a shortcut, a hybrid approach (*34*) has been proposed in which it is assumed that the percentage correction due to dispersion at the RHF level is also valid for a correlated treatment. If that is the case, then it is sufficient to know the TDHF dispersion plus the *static* value of the correlated hyperpolarizability. This approximation has been tested recently (*35*) for the optical Kerr effect in *trans*-butadiene using an equations-of-motion coupled-cluster singles and doubles (EOM-CCSD) frequency-dependent correlation treatment (*33*). It was found that the TDHF calculation overestimates dispersion for γ_L by as much as 32% at 694.3 nm. Computations are underway to determine whether this discrepancy grows with increasing chain length as speculated (*35*) and, if so, how rapidly. We also plan to examine other NLO processes, particularly dc-SHG.

Vibrational Hyperpolarizabilities

We have noted previously that vibrational hyperpolarizabilities can be comparable to, or larger than, their electronic counterparts depending upon the system and the NLO property. Two relevant examples are the longitudinal dc-Kerr effect and IDRI in trisilane (*9*). At the double harmonic level of approximation (see below) the ratio of the vibrational to the electronic term has been estimated to be 0.50 and 3.12 respectively. The corresponding ratios in *trans*-butadiene, based on a more accurate treatment, are 0.192 and 0.433.

Table V. TDHF/6-31G longitudinal hyperpolarizability (in 10^4 a.u.) for dc-SHG in $C_{2N}H_{2N+2}$ oligomers of PA at $\hbar\omega$=0.40 and 0.80 eV. Also included is the ratio with respect to the static longitudinal hyperpolarizability $\gamma_L(0)$.

	$\hbar\omega = 0.40$ eV		$\hbar\omega = 0.80$ eV	
N	γ_L(dc-SHG)[a]/*N*	γ_L(dc-SHG)[a]/ $\gamma_L(0)$	γ_L(dc-SHG)[a]/*N*	γ_L(dc-SHG)[a]/ $\gamma_L(0)$
2	0.32	1.0272	0.35	1.1154
3	1.88	1.0358	2.09	1.1547
4	5.67	1.0446	6.50	1.1970
5	12.64	1.0532	14.87	1.2394
6	23.26	1.0611	28.06	1.2801
7	37.59	1.0683	46.41	1.3187
8	55.17	1.0747	69.51	1.3541
9	75.26	1.0804	96.58	1.3864
10	97.07	1.0854	126.58	1.4153
11	119.89	1.0898	158.58	1.4414
12	143.04	1.0936	191.58	1.4647
13	166.07	1.0970	224.89	1.4855
14	188.52	1.0999	257.75	1.5038
15	210.20	1.1025	289.85	1.5203

[a] γ_L(dc-SHG) = $\gamma_L(-2\omega; \omega, \omega, 0)$

A perturbation method for calculating the dynamic (or static) vibrational hyperpolarizabilities of a general polyatomic molecule has recently been formulated by Bishop and Kirtman (*36-38*). Their procedure is based on the double harmonic initial approximation and accounts for terms of the following order: (0,0), (1,0), (2,0), (0,1) and (1,1), where the first index refers to electrical anharmonicity, and the second index to mechanical anharmonicity. Contributions due to third derivatives of the electrical properties with respect to normal coordinates are omitted from the (2,0) term. In principle, these could be included along with the (0,2) term, but the computation of the parameters involved represents a formidable obstacle.

A simple procedure (*39*) for approximating the perturbation expressions has also been developed. It is based on the evaluation of electrical properties in the presence of a *static* field with and without reoptimizing the geometry. For example, by fitting the change in the linear polarizability ($\alpha_{\alpha\beta}$) due to the geometry relaxation one obtains the dc-Kerr hyperpolarizability $\gamma^{v(r)}_{\alpha\beta\gamma\delta}(-\omega;\omega,0,0)_{\omega\to\infty}$ from the coefficient of the term quadratic in the field (i.e., the term proportioned to $F_\gamma F_\delta$). Here the superscript v(r) denotes the relaxation part of the total vibrational contribution which, it turns out (see below), is due to the lowest order perturbation corrections of each type that appear in the complete treatment (Bishop, D.M.; Hasan, M.; Kirtman, B., *J. Chem. Phys.*, in press). We use $\omega \to \infty$ to indicate that the optical frequency is allowed to approach infinity. The terms kept in this "infinite (optical) frequency" limit are similar to those designated as "enhanced" by Elliott and Ward (*40*) and are expected to be dominant (a set of confirmatory tests has now been carried out for several small molecules by Bishop, D.M.; Dalskov, E.K., Ottawa University, personal communication, 1995). For the above example our treatment yields the (1,0) + (0,1) perturbation contributions to the vibrational dc-Kerr effect in the infinite frequency approximation.

The linear term in the static field expansion of the first hyperpolarizability ($\beta_{\alpha\beta\gamma}$) gives $\gamma^{v(r)}_{\alpha\beta\gamma\delta}(-2\omega;\omega,\omega,0)_{\omega\to\infty}$, which yields the vibrational dc-SHG to the same order as above (although, in this case the (1,0) and (0,1) contributions vanish). Finally - again to the order (1,0) + (0,1) - the vibrational IDRI is a linear combination of the dc-Kerr and dc-SHG vibrational hyperpolarizabilities plus the relaxation component of the static vibrational hyperpolarizability. The latter is evaluated from the cubic term in the static field expansion of the dipole moment induced by vibrational relaxation.

The anharmonic frequency-dependent perturbation theory expressions for the vibrational hyperpolarizability have, thus far, been evaluated only for small molecules (*37, 41*). However, there are double harmonic calculations suggesting that vibrational hyperpolarizabilities may be quite significant in certain polymers. For instance, in *trans*-polysilane (*41*) the static linear vibrational polarizability, obtained by extrapolating double harmonic results for the oligomers $Si_{2N}H_{4N+2}$ with N=1,2,...8, is 50% of its electronic counterpart. In small polysilane oligomers (*9*) the corresponding hyperpolarizability ratio for IDRI (532nm) is much larger (by a factor of 4-6), leading one to speculate that the same may be true for the infinite polymer.

A second example is provided by the double harmonic calculations of Zerbi and co-workers (*43*) on oligomers of polyacetylene through $C_{12}H_{14}$. Although these chain lengths are not sufficient to yield a satisfactory estimate for the infinite polymer, the particular contribution to the static vibrational hyperpolarizability that they evaluate is comparable to the total electronic term. Furthermore, it can be shown (*44*) that in the "infinite (optical) frequency" approximation this same contribution, after multiplication by 2/3, yields the vibrational IDRI. It also enters into the vibrational dc-Kerr effect with a multiplicative factor of 1/3 (although, in this case, there is another term).

A detailed analysis (*44,45*) of the static linear vibrational polarizability, $\alpha^v(0)$, of polyacetylene indicates that low frequency, collective acoustic modes may play an important role in the vibrational *hyper*polarizability. For polyacetylene, it turns out that $\alpha^v(0)$ is dominated in zeroth-order by a term that arises from a transverse acoustic mode known as the TAM. As the finite oligomer chain is lengthened, both the infrared intensity and vibrational frequency associated with this mode approach zero but the ratio, which determines the contribution to $\alpha^v(0)$, remains finite. The symmetry of the TAM is such that it will also contribute to the zeroth-order vibrational dc-SHG and dc-Kerr hyperpolarizabilities. In fact, at this level of approximation, no other symmetries are involved in the vibrational dc-SHG.

There is a second collective acoustic mode that cannot contribute to $\alpha^v(0)$, since it is totally symmetric, yet can give rise to vibrational IDRI and dc-Kerr hyperpolarizabilities (no other symmetries can contribute to the former in the double harmonic approximation). This mode, known as the longitudinal acoustic/accordion mode (LAM), creates a Raman intensity parallel to the chain. The intensity and frequency of the LAM are similar to the TAM in their behavior with increasing chain length. Again, both properties approach zero but the ratio remains finite and, in general, will create a vibrational hyperpolarizability.

The importance of these acoustic modes will depend to a large extent on the polymer architecture because of the symmetry requirements. Crude vibrational hyperpolarizability calculations for polyyne indicate that the LAM is significant, but not dominant, whereas the TAM cannot contribute because the chain is rigorously linear. The quantitative behavior in other instances remains to be established; our preliminary results suggest that the LAM is important in fully saturated polymers (e.g. polysilane), but not in π-conjugated cases. If the TAM is important then, of course, the influence of interchain interactions (see next section) will have to be carefully evaluated.

Interchain Interactions in Polyacetylene

The effect of the medium on NLO materials is a subject that has recently begun to attract attention. It is well-known (*46*) that the first hyperpolarizability of donor-acceptor π-conjugated chromophores is strongly affected by solvation. However, the influence of the solid state environment on the second hyperpolarizability of conjugated polymers is a matter of debate which, thus far, has focused specifically on polyacetylene.

Stretched fibers of PA have a definite geometric structure (*47*) wherein each polymer chain is surrounded by a regular hexagon of neighboring chains (see Fig.3). Using an idealized geometry for the polymer, McWilliams and Soos (*20*) have calculated the effect of interchain interactions on the THG spectrum by means of a semi-empirical valence bond configuration interaction method. They conclude that the interactions play a crucial role in determining the spectrum. However, this finding has been challenged (*48*) on the grounds that the short chains employed in the calculations are not representative of the infinite polymer. It has also been suggested elsewhere (*49*), on the basis of a density functional treatment, that chain interactions could significantly alter the degree of bond alternancy and, thereby, affect the NLO properties.

We have now obtained the first *ab initio* results (see Table VI) for the solid state medium effect on the static hyperpolarizability of PA in stretched fibers. These calculations employed the supermolecule approach on up to three interacting butadiene or hexatriene molecules and were carried out in a 6-31G basis. Our conclusions are preliminary in nature because only small oligomers were treated, yet they appear to be robust as seen below.

Figure 3 shows that nearest neighbor chains may be oriented either parallel (e.g. *a* and *b*) or perpendicular (e.g. *a* and *c*) to one another. In either case the interaction leads to a 30 ± 5 % decrease (see Table VI) in the static γ_L per chain regardless of the chain length (C_4H_6,C_6H_8) or the level of approximation (RHF, MP2, MP4(SDQ), MP4(SDTQ), CCSD). This is a pure electronic effect since the calculated change in geometry is negligible.

The set of three nearest neighbor chains corresponds to a configuration (e.g. *a,b,c*) with one parallel, and two perpendicular, interactions. There is a further substantial reduction in the static γ_L per chain resulting in an overall decrease of 47 ± 3%. This contrasts markedly with the findings of McWilliams and Soos (*20*) who obtain a small decrease in the same quantity due to interaction with the first nearest neighbor and, then, a sharp increase when the remaining neighbors are included. Further *ab initio* calculations building up to the full complement of nearest neighbors are in progress.

Since the basis set we have employed is small there could be a large superposition error. In order to test this possibility the calculations for a single C_6H_8 chain were repeated in the basis set of two or three chains with the following results: double(||) = 55.6 a.u.; double ($\perp$) = 54.6 a.u.; and triple (all ||) = 54.8 a.u. From the small variation with respect to the single chain basis set value of 54.7 a.u. we conclude that the superposition error is negligible.

Finally, we examine the question of pair additivity using the butadienes as an example. From the hyperpolarizabilities determined for a pair of chains the pair interaction term, Δ, can be determined as shown in Table VII. Then, one can predict the value for three interacting chains, assuming perfect additivity, and compare with that found by direct computation. For three parallel chains (two pairs of nearest neighbors; one pair of second nearest neighbors), neglecting the second nearest neighbor interaction leads to exact additivity (as might have been expected). However, in the nearest neighbor arrangement (two $\perp$ pairs; one || pair) the simple pairwise additivity model breaks down.

Discussion

The *ab initio* finite oligomer method for determining NLO properties of conjugated polymers is at an early stage of elaboration. Our initial calculations, done primarily on PA, have established the feasibility of this approach. Although the PA repeat unit is small, the current pace of hardware and software advances suggests that a repeat unit orders of magnitude larger will be computationally accessible in the near future. Some of the more promising areas of software development for spatially extended systems include: (1) various techniques for rapidly approximating most two electron integrals

Table VI. Static γ_L per chain (in 10^3 a.u.) for interacting nearest neighbor butadienes (or hexatrienes) in solid state PA geometry. All calculations were done in the 6-31G basis.

	RHF	MP2	MP4 (SDQ)	MP4 (SDTQ)	CCSD	CCSD(T)
single chain						
RHF geom	6.3(54.7)[a]	19.2(121.)[a]	20.0	18.7	16.8	14.7
MP2	–	23.6	24.4	22.0	19.9	16.8
double chain						
RHF geom (‖)	4.1(34.8)[a]	14.2(84.0)[a]	14.7	13.6	–	–
MP2 geom (‖)	–	16.7	17.2	15.5	14.9	–
RHF geom (⊥)	4.6(36.9)[a]	14.1	14.7	13.6	–	–
triple chain[b]						
RHF geom	3.2	10.7				

[a] values for C_6H_8 in parentheses
[b] see text

Table VII. RHF/6-31G static longitudinal hyperpolarizabilities (in 10^3 a.u.), $\Delta\gamma_L$, due to interaction between butadiene molecules arranged in the solid state PA geometry

		$\Delta\gamma_L$ (no interaction)	γ_L(actual)	$\Delta\gamma_L$	perfect pairwise additivity
($2 \times C_4H_6$)	∥	12.6	8.2	4.4	–
	⊥	12.6	9.2	3.4	–
($3 \times C_4H_6$)	∥	18.9	10.2	8.7	8.8 (2 x 4.4)
	⊥	18.9	9.6	9.3	11.2 (2 x 3.4 + 4.4)

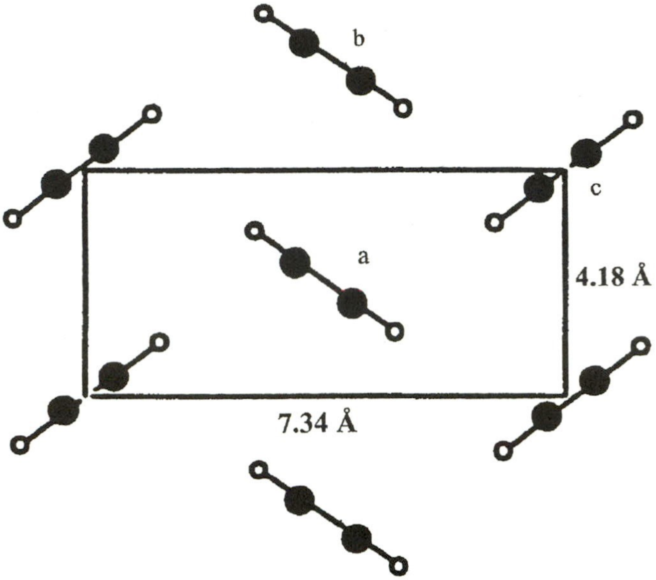

Figure 3 Three dimensional geometry of stretched fibers of PA (see Ref. 19). Chains *a* and *b* are in the parallel configuration; chains *a* and *c* are in the perpendicular configuration.

and, at the Hartree-Fock level, (2) improved numerical methods (*50*; also Yang, W., Duke University, personal communication 1995) for solving the SCF equations and (3) special procedures for combining fragments (*51*) or lengthening chains (*52*). At the correlated level, various density functional approaches (see other contributions in this volume) and other "local" approximations (*53*) may ultimately prove to be competitive with the scaling employed here. Finally, for certain aspects (such as vibrational distortion), there might be useful ways (work in progress with Champagne, B. and Andre, J. M.) to hybridize the finite oligomer method with band structure techniques.

Our goal is a comprehensive treatment of NLO activity in polymers. With this in mind we have considered the roles of vibrational distortion and the solid state medium, as well as the effects of electron correlation and frequency dispersion. There are still other facets that remain to be examined, especially structural defects and impurities. Again, the finite oligomer method seems well-suited for this purpose.

As we have seen, both vibrational distortion and interactions with the environment can have a major influence on the hyperpolarizability. This implies some novel possibilities for designing π-conjugated polymers with a desired NLO response. For example, the contribution from individual vibrational modes depends upon a number of features that are potentially tunable by chemical modification. These include infrared activity due to the dipole generated in the longitudinal direction (*42,44*); Raman activity arising from the induced longitudinal polarizability (*44*); and harmonic (as well as anharmonic) force constants (*37*). The environmental contribution could be modulated by specific, as well as non-specific, side-chain interactions. Finally, the possibilities for tuning the effect of electron correlation await further understanding of this phenomenon.

Acknowledgments

The author wishes to warmly thank those who have collaborated with him on various aspects of the research reported here; namely (in alphabetical order) Jean-Marie Andrè, David M. Bishop, Benoît Champagne, Muhammad Hasan, Seung-Joon Kim, Celso P. deMelo, Kathleen A. Robins, Joseph L. Toto and Teressa T. Toto. Acknowledgment is made to the donors of the Petroleum Research Fund, administered by the American Chemical Society, for partial support of this research. Partial support was also provided by the National Science Foundation US-Brazil Cooperative Science Program under award INT-9217464.

Literature Cited

1. R.F. Service, Science **1995** *267*, 1918.
2. Optical Nonlinearities in Chemistry; Burland, D. M., Ed.; *Chemical Reviews*, **1994**; Vol. 94(1).
3. Modern Nonlinear Optics; Evans, M.; Kielich, S., Eds.; *Advances in Chemical Physics*; **1993**; Vol. 85.
4. Molecular Nonlinear Optics; Ratner, M. A., Ed.; *International Journal of Quantum Chemistry*; **1992**; Vol. 43(1).
5. Kirtman, B.; Nilsson, W. B.; Palke, W. E. *Solid State Commun.* **1983**, *46*, 791.
6. Hurst, G. J. B.; Dupuis, M.; Clementi, E. *J. Chem. Phys.* **1988**, *89*, 385.
7. Kirtman, B.; Hasan, M. *Chem. Phys. Lett.* **1989**, *157*, 123. See also ref. 16.
8. Archibong, E. F.; Thakkar, A. J. *J. Chem. Phys.* **1993**, *98*, 8324; Chopra, P.; Carlacci, L.; King, H. F.; Prasad, P. N. *J. Phys. Chem.* **1989**, *93*, 7120; Jaszunski, M.; Jorgensen, P.; Koch, H.; Aagren, H. *J. Chem. Phys.* **1993**, *98*, 7229.
9. Kirtman, B.; Hasan, M. *J. Chem. Phys.* **1992**, *96*, 470.
10. Kirtman, B.; Toto, J. L.; Robins, K. A.; Hasan, M. *J. Chem. Phys.* **1995**, *102*, 5350.
11. Genkin, V. M.; Mednis, P. M. *Zh. Eksp. Teor. Fiz.* **1968**, *54*, 1137; *Sov. Phys. JETP* **1968**, *27*, 609.
12. Barbier, C. *Chem. Phys. Lett.* **1987**, *142*, 349.
13. Ladik, J. *J. Molec. Struct. (Theochem)* **1989**, *199*, 55; Ladik, J.; Dalton, L. *J. Molec. Struct. (Theochem)* **1991**, *231*, 77. See also Ladik, J. in this volume.
14. Otto, P. *Phys. Rev.* **1992**, *B46*, 10876.
15. Champagne, B.; Moseley, D. H.; Andrè, J. M. *Int. J. Quantum Chem.* **1993**, *S27*, 667.
16. Champagne, B.; Ohrn, Y. *Chem. Phys. Lett.* **1994**, *217*, 551.
17. Cui, C. X.; Kertesz, M. *J. Chem. Phys.* **1990**, *93*, 5257; Cui, C. X.; Kertesz, M.; Dupuis, M. *J. Chem. Phys.* **1990**, *93*, 5890.
18. Kirtman, B. *Int. J. Quantum Chem.* **1992**, *43*, 147.
19. Toto, T. T.; Toto, J. L.; deMelo, C. P.; Hasan, M.; Kirtman, B. *Chem. Phys. Lett.* **1995**, in press.
20. Mc Williams, P. C. M.; Soos, Z. G. *J. Chem. Phys.* **1991**, *95*, 2127.
21. Weniger, E. J.; Liegener, C. M. *Int. J. Quantum Chem.* **1990**, *38*, 55.
22. Cioslowski, J. *Chem. Phys. Lett.* **1988**, *153*, 446.
23. Cioslowski, J.; Lepetit, M. *J. Chem. Phys.* **1991**, *95*, 3536.
24. Sekino, H.; Bartlett, R. J. *Int. J. Quantum Chem.* **1992**, *43*, 119. See also refs. 34 and 35.

25. Suhai, S. *Chem. Phys. Lett.* **1983**, *96*, 619.
26. Karna, S. P.; Dupuis, M. *J. Comput. Chem.* **1991**, *12*, 487 give the most recent formulation. See also earlier references cited therein.
27. Karna, S. P.; Talapatra, G. B.; Wijekoon, W. M. K. P.; Prasad, P. N. *Phys. Rev.* **1992**, *A45*, 2763.
28. Bishop, D. M. *J. Chem. Phys.* **1989**, *90*, 3192.
29. Shuai, Z.; Bredas, J. L. *Phys. Rev.* **1991**, *B44*, 5962; **1992**, *B46*, 4395.
30. Rice, J. E.; Handy, N. C. *J. Chem. Phys.* **1991**, *94*, 4959.
31. Rice, J. E.; Handy, N. C. *Int. J. Quantum Chem.* **1992**, *43*, 91.
32. Christiansen, O.; Jorgensen, P. *Chem. Phys. Lett.* **1993**, *207*, 367 and references cited therein.
33. Stanton, J. F.; Bartlett, R. J. *J. Chem. Phys.* **1993**, *99*, 5178.
34. Sekino, H.; Bartlett, R. J. *J. Chem. Phys.* **1991**, *94*, 3665.
35. Sekino, H.; Bartlett, R. *J. Chem. Phys. Lett.* **1995**, *234*, 87.
36. Kirtman, B.; Bishop, D. M. *Chem. Phys. Lett.* **1990**, *175*, 601.
37. Bishop, D. M.; Kirtman, B. *J. Chem. Phys.* **1991**, *95*, 2646.
38. Bishop, D. M.; Kirtman, B. *J. Chem. Phys.* **1992**, *97*, 5255.
39. Bishop, D. M.; Hasan, M.; Kirtman, B. *J. Chem. Phys.*, in press.
40. Elliott, D. S.; Ward, J. F. *Mol. Phys.* **1984**, *51*, 45.
41. Bishop, D. M.; Kirtman, B. *J. Chem. Phys.* **1991**, *95*, 2646; Bishop, D. M.; Kirtman, B.; Kurtz, H. A.; Rice, J.E. *ibid.* **1993**, *98*, 8024; Cohen, M. J.; Willetts, A. J.; Amos, R. D.; Handy, N. C. *ibid.* **1994**, *100*, 4467; Bishop, D. M.; Pipin, J.; Kirtman, B. *ibid.* **1995**, 102, 6778.
42. Champagne, B.; Perpete, E.; Andrè, J. M.; Kirtman, B. *J. Chem. Soc. Faraday Trans.* **1995**, *91*, 1641.
43. Castiglioni, C.; Gussoni, M.; Del Zoppo, M.; Zerbi, G. *Solid State Commun.* **1992**, *82*, 13; Del Zoppo, M.; Castiglioni, C.; Veronelli, M.; Zerbi, G. *Synth. Met.* **1993**, *57*, 3919.
44. Kirtman, B.; Champagne, B.; Andrè, J. M. *J. Chem. Phys.*, submitted.
45. Champagne, B.; Perpète, E. A.; Andrè, J. M. *J. Chem. Phys.* **1994**, *101*, 10796.
46. Bourhill, G.; Bredas, J. L.; Cheng, L. T.; Marder, S. R.; Meyers, F.; Perry, J. W.; Tiemann, B. G. *J. Am. Chem. Soc.* **1994**, *116*, 2619; Stahelin, M.; Moylan, C. R.; Burland, D. M.; Willetts, A.; Rice, J. E.; Shelton, D. P.; Donley, E. A. *J. Chem. Phys.* **1993**, *98*, 5595; Robinson, D. W.; Long, C. A. *J. Phys. Chem.* **1993**, *97*, 7540; Stahelin, M.; Burland, D. M.; Rice, J. E. *Chem. Phys. Lett.* **1992**, *191*, 245; etc.
47. Kahlert, H.; Leitner, O.; Leising, G. *Synth. Met.* **1987**, *17*, 467.
48. Guo, D.; Mazumdar, S. *J. Chem. Phys.*, *97*, 2170. See also McWilliams, P. C. M.; Soos, Z. G.; Hayden, G. W. *ibid.* **1992**, *97*, 2172.
49. Springborg, M. *Phys. Rev.* **1989**, *B40*, 5774.
50. Yang, W. 1995 Sanibel Symposium, February 25 - March 4; Daw, M. S. *Phys. Rev.* **1993**, *B47*, 10895; Friesner, R. A. *Chem. Phys. Lett.* **1985**, *116*, 39; *J. Chem. Phys.* **1986**, *85*, 1462; *ibid.* **1987**, *86*, 3522.
51. Robins, K. A.; Kirtman, B. *J. Chem. Phys.* **1993**, *99*, 6777.
52. Maekawa, K.; Imamura, A. *ibid.* **1993**, *98*, 534.
53. Saebo, S.; Pulay, P. *Annu. Rev. Phys. Chem.* **1993**, *44*, 213 and references cited therein; Grafenstein, J.; Stoll, H.; Fulde, P. *Chem. Phys. Lett.* **1993**, *215*, 611; Kirtman, B. *Int. J. Quantum Chem.* **1995**, *55*, 103.

RECEIVED December 29, 1995

Chapter 4

Sum-Over-State Representation of Nonlinear Response Properties in Time-Dependent Hartree–Fock Theory

The Role of State Truncation

Hideo Sekino and Rodney J. Bartlett

Quantum Theory Project, Departments of Chemistry and Physics, University of Florida, Gainesville, FL 32611–8435

Non-linear response properties evaluated by Time Dependent Hartree Fock (TDHF) theory is represented in a Sum-Over-State (SOS) formulation. The formalism is completely equivalent to that obtained by the original formulation of TDHF using perturbed density matrices if all intermediate, excited states are taken into consideration. Severely truncated SOS expressions are used in "few" state models for the design of NLO materials. Although a minor truncation in the intermediate states has little effect on the calculated hyperpolarizabilities over the range of calculated frequencies, a more drastic truncation to about 50% of the states results in a breakdown in the quantitative use of the TDHF method, much less a truncation to only a handful of states. Furthermore, the individual components can be drastically in error, even with modest truncations. However, using TDHF to estimate dispersion effects for higher, correlated results by taking the percentage (i.e., TDHF(ω)/TDHF(0) $\times$ the correlation correction) still applies adequately, despite severely truncating the sum-over-states.

The theoretical investigation of non-linear optical (NLO) properties of molecules is one of the most challenging subjects in quantum chemistry. However, because NLO properties, which depend upon frequency dependent hyperpolarizabilities, are highly sensitive to the frequency of the applied optical field, static theories are generally not thought to be able to provide a meaningful prediction of the quantity. Hence, we previously formulated and implemented high-order Time Dependent Hartree Fock (TDHF) theory (also known as the Random Phase Approximation (RPA)) to provide non-linear dynamic response properties such as frequency dependent hyperpolarizabilities [1]. This permits any-order frequency dependent hyperpolarizabilities to be

0097–6156/96/0628–0079$15.75/0

efficiently evaluated analytically. Since our original development, this method has been widely used for theoretical predictions of NLO properties [2,3,4,5].

The analytic TDHF theory was originally formulated using an atomic orbital (AO) based algorithm [1,5] and has been recently implemented onto parallel machines using direct AO algorithms [6,7]. The method has also been reformulated using a molecular orbital (MO) based algorithm [8,9], which is highly efficient for systems of intermediate size. The MO based method further enables the use of diagrammatic techniques to analyze each term emerging in high order TDHF theory [9].

Equivalent equations can also be derived by diagonalization of the closed form resolvent operator discussed below, or, equivalently, by propagator techniques [10,11,12,13,14] where the non-linear optical properties are represented as high-order response functions by sum-over-state type formulas. We have shown that by transforming the excitation and deexcitation manifold, high-order response functions can be obtained using TDHF amplitudes and an operator representing the external perturbation [9].

The advantage of the SOS formulation is that it matches the traditional interpretation of NLO phenomena using intermediate, excited states, especially when the number of the excited states is limited. The SOS formulation is also computationally convenient, as every component of any property at arbitrary frequency can be immediately evaluated once the response function is obtained. The problem is that the number of terms in the SOS response function becomes huge for even small systems. Since the transformation to SOS form requires a full diagonalization of a matrix with dimension 2(M-n)n where M is the dimension of the basis set and n is the number of occupied orbitals, the formulation rapidly ceases to be practical. However, if only a few of the intermediate, excited states are physically important and are expected to contribute significantly, the SOS formulation with truncated intermediate states might be a useful approximation to the full response function. This is the basis for "few" state models of NLO behavior [15,16]. However, it is not yet clear how reliable such truncated SOS expressions can be. A previous semi-empirical study considered truncations of the *determinantal* space, and its deleterious effect on predicted hyperpolarizabilities, but *not* truncations among the actual excited, intermediate states [17].

In this paper, we investigate this question numerically by investigating various SOS truncations to the exact TDHF solution represented on TDHF=RPA intermediate states. We recognize that electron correlation is essential in providing predictive values for molecular hyperpolarizabilities [18], but the TDHF=RPA provides a consistent model that can be used to unambiguously assess the effect of the truncation.

Theory

The external perturbation of an oscillating electric field is $\sum_j \vec{E}_j \cdot \vec{O}_j e^{\pm i\omega t}$, where $|\vec{E}_j|$ is the field amplitude and $\vec{O}_j$ is the property under consideration. The TDHF amplitude vector, **U**, is expanded by order and the process involved,

$$\begin{aligned}\mathbf{U}(E) = \mathbf{U}^{(0)} + E\mathbf{U}(\pm\omega_1)e^{\pm i\omega_1 t} + \frac{E^2}{2}\mathbf{U}(\pm\omega_1, \pm\omega_2)e^{\pm i(\omega_1 \pm \omega_2)t} \\ + \frac{E^3}{6}\mathbf{U}(\pm\omega_1, \pm\omega_2, \pm\omega_3)e^{\pm i(\omega_1 \pm\omega_2 \pm\omega_3)t} \\ + \dots\end{aligned} \tag{1}$$

The density matrices, obtained through **d**(*E*)=**U**(*E*)**U**†(*E*), are expanded in the same manner,

$$\begin{aligned}\mathbf{d}(E) = \mathbf{d}^{(0)} + E\mathbf{d}(\pm\omega_1)e^{\pm i\omega_1 t} + \frac{E^2}{2}\mathbf{d}(\pm\omega_1, \pm\omega_2)e^{\pm i(\omega_1 \pm \omega_2)t} \\ + \frac{E^3}{6}\mathbf{d}(\pm\omega_1, \pm\omega_2, \pm\omega_3)e^{\pm i(\omega_1 \pm\omega_2 \pm\omega_3)t} + \dots\end{aligned} \tag{2}$$

The dynamic non-linear response properties are given as a contraction of the perturbation property **O** with the density matrix which comprises the non-linear perturbation process under consideration. For the Hartree-Fock case, it is given as a contraction between the property corresponding to the perturbation and the one-body TDHF density matrix corresponding to the process such as $\mathbf{d}(\pm\omega,\pm\omega)$, $\mathbf{d}(0,\pm\omega)$,$\mathbf{d}(\pm\omega,\pm\omega,\pm\omega)$... for Second Harmonic Generation (SHG), Electro Optic Pockels Effect (EOPE), Third Harmonic Generation (THG)..., etc. We then have

$$\begin{aligned}\alpha(-\omega_\sigma;\omega_1) &= -Tr\{\mathbf{Od}(\pm\omega_1)\} \\ \beta(-\omega_\sigma;\omega_1,\omega_2) &= -Tr\{\mathbf{Od}(\pm\omega_1,\pm\omega_2)\} \\ \gamma(-\omega_\sigma;\omega_1,\omega_2,\omega_3) &= -\ Tr\{\mathbf{Od}(\pm\omega_1,\pm\omega_2,\pm\omega_3)\} \\ &\dots\dots\end{aligned} \tag{3}$$

As we only require the virtual occupied block of **U**, namely $\mathbf{U}^{vo}$, the n-th order TDHF equations are

$$\begin{bmatrix}\mathbf{A} & \mathbf{B}\\ \mathbf{B} & \mathbf{A}\end{bmatrix}\begin{bmatrix}\mathbf{U}^{vo}(+\omega_1,+\omega_2,\dots)\\ \mathbf{U}^{vo}(-\omega_1,-\omega_2,\dots)\end{bmatrix} \pm \omega_\sigma \begin{bmatrix}\mathbf{U}^{vo}(+\omega_1,+\omega_2,\dots)\\ \mathbf{U}^{vo}(-\omega_1,-\omega_2,\dots)\end{bmatrix} + \begin{bmatrix}h^{vo}(+\omega_1,+\omega_2,\dots)\\ h^{vo}(-\omega_1,-\omega_2,\dots)\end{bmatrix} = \begin{bmatrix}\mathbf{0}\\ \mathbf{0}\end{bmatrix} \tag{4}$$

Here, **A** and ***B*** are ordinary Random Phase Approximation (RPA) matrices and $\omega_\sigma = +\omega_1 + \omega_2 + \dots$ The TDHF amplitude $\mathbf{U}^{vo}$ and $\mathbf{h}^{vo}$ are vectors of dimension *nN* when *n* is the number of occupied orbitals and *N* the number of virtuals. The constant vector, $\mathbf{h}^{vo}$, is determined solely from the lower-order solutions.

We can formally solve the equations,

$$\begin{bmatrix}\mathbf{U}^{vo}(+\omega_1,+\omega_2,\dots)\\ \mathbf{U}^{vo}(-\omega_1,-\omega_2,\dots)\end{bmatrix} = -\mathbf{P}^{-1}(\omega_\sigma)\begin{bmatrix}\mathbf{h}^{vo}(+\omega_1,+\omega_2,\dots)\\ \mathbf{h}^{vo}(-\omega_1,-\omega_2,\dots)\end{bmatrix} \tag{5}$$

where

$$\mathbf{P}(\omega) = \begin{bmatrix}\omega\mathbf{I}+\mathbf{A} & \mathbf{B}\\ \mathbf{B} & -\omega\mathbf{I}+\mathbf{A}\end{bmatrix} \tag{6}$$

and $\mathbf{U}^{vo}(+\omega_1,+\omega_2,\dots)$ and $\mathbf{U}^{vo}(-\omega_1,-\omega_2,\dots)$ are a pair of TDHF amplitude vectors.

We now introduce a set of excitation and deexcitation operators $\{Q_J^\dagger, Q_J\}$ defined by

$$Q_J^\dagger = \sum_{a,i} a^\dagger i Z_{ai}^J + i^\dagger a Y_{ai}^J \qquad J = 1, 2, ..., nN \tag{7}$$

$\mathbf{U}_{RPA} = \begin{bmatrix} \mathbf{Z}^J & \mathbf{Y}^J \\ \mathbf{Y}^J & \mathbf{Z}^J \end{bmatrix}$ is a pair of the RPA solutions

$$\begin{bmatrix} \mathbf{A} & \mathbf{B} \\ \mathbf{B} & \mathbf{A} \end{bmatrix} \mathbf{U}_{RPA} = \pm\omega_\mathbf{J} \mathbf{U}_{RPA} \tag{8}$$

The amplitudes and the constant terms may be transformed onto the new basis $\{Q_J^\dagger, Q_J\}$,

$$\begin{bmatrix} \tilde{\mathbf{U}}^{vo}(+\omega_1, +\omega_2, ...) \\ \tilde{\mathbf{U}}^{vo}(-\omega_1, -\omega_2, ...) \end{bmatrix} = \begin{bmatrix} \mathbf{Z}^\dagger & \mathbf{Y}^\dagger \\ \mathbf{Y}^\dagger & \mathbf{Z}^\dagger \end{bmatrix} \begin{bmatrix} \mathbf{U}^{vo}(+\omega_1, +\omega_2, ...) \\ \mathbf{U}^{vo}(-\omega_1, -\omega_2, ...) \end{bmatrix} \tag{9a}$$

and

$$\begin{bmatrix} \tilde{\mathbf{h}}^{vo}(+\omega_1, +\omega_2, ...) \\ \tilde{\mathbf{h}}^{vo}(-\omega_1, -\omega_2, ...) \end{bmatrix} = \begin{bmatrix} \mathbf{Z}^\dagger & \mathbf{Y}^\dagger \\ \mathbf{Y}^\dagger & \mathbf{Z}^\dagger \end{bmatrix} \begin{bmatrix} \mathbf{h}^{vo}(+\omega_1, +\omega_2, ...) \\ \mathbf{h}^{vo}(-\omega_1, -\omega_2, ...) \end{bmatrix} \tag{9b}$$

The equations for the TDHF amplitudes then become

$$(\mathbf{D} \pm \omega_\sigma \mathbf{I}) \begin{bmatrix} \tilde{\mathbf{U}}^{vo}(+\omega_1, +\omega_2, ...) \\ \tilde{\mathbf{U}}^{vo}(-\omega_1, -\omega_2, ...) \end{bmatrix} + \begin{bmatrix} \tilde{\mathbf{h}}^{vo}(+\omega_1, +\omega_2, ...) \\ \tilde{\mathbf{h}}^{vo}(-\omega_1, -\omega_2, ...) \end{bmatrix} = \begin{bmatrix} \mathbf{0} \\ \mathbf{0} \end{bmatrix} \tag{10}$$

where $\mathbf{D}$ is a diagonal matrix which contains the TDHF=RPA energies $\{\omega_J\}$ as its diagonal elements.

Thus, the TDHF amplitudes are easily solved to be

$$\tilde{U}^J(\pm\omega_1, \pm\omega_2, ...) = -\frac{\tilde{h}^J(\pm\omega_1, \pm\omega_2, ...)}{sgn(J)\omega_\sigma + \omega_J} \tag{11}$$

In other words, the inverse matrix in Eqn. 5 of the 2*nN* dimension can be represented, after the transformation, as

$$\mathbf{P}(\omega)^{-1} = \begin{bmatrix} \mathbf{Z} & \mathbf{Y} \\ \mathbf{Y} & \mathbf{Z} \end{bmatrix} \begin{bmatrix} (\omega\mathbf{I} + \mathbf{D})^{-1} & 0 \\ 0 & (\omega\mathbf{I} - \mathbf{D})^{-1} \end{bmatrix} \begin{bmatrix} \mathbf{Z}^\dagger & \mathbf{Y}^\dagger \\ \mathbf{Y}^\dagger & \mathbf{Z}^\dagger \end{bmatrix} \tag{12}$$

This gives the RPA spectral expansion of the general form

$$\mathbf{P}(\omega)^{-1} = \sum_{k\neq 0} \Big\{ |Z^k\rangle(\omega + \omega_k)^{-1}\langle Z^k| + |Y^k\rangle(\omega + \omega_k)^{-1}\langle Z^k| + |Y^k\rangle(\omega - \omega_k)^{-1}\langle Y^k| + |Z^k\rangle(\omega - \omega_k)^{-1}\langle Y^k| \Big\} \tag{13}$$

in terms of the eigenvectors for the RPA excited states.

In this spectral representation of the RPA propagator, the poles are equivalent to the square root of the eigenvalues for the matrix,

$$(\mathbf{A}+\mathbf{B})^{\frac{1}{2}}(\mathbf{A}-\mathbf{B})(\mathbf{A}+\mathbf{B})^{\frac{1}{2}} \tag{14}$$

and are, therefore, positive definite for real solutions. We can obtain *any order* of TDHF amplitudes through Eqn. (13) with this matrix $P(\omega)^{-1}$, and, therefore *any* high-order dynamic property from Eqn. (10) once the RPA equations are solved.

However, we need to solve the full RPA equations to obtain the exact inverse matrix $P(\omega)^{-1}$. One way is full diagonalization of the two *nN* matrices. Another would be the direct solution of the corresponding linear equation $\mathbf{P}(\omega)\tilde{\mathbf{U}}^{vo} = \tilde{\mathbf{h}}^{vo}$ for the $\tilde{\mathbf{U}}^{vo}$. Both are computationally demanding, but the latter is used to provide the untruncated RPA solution, which is our reference. On the other hand, we can approximate the inverse matrix using a limited number of the RPA solutions. In this way, we could potentially calculate approximate hyperpolarizabilities with less computational effort; but, more importantly, we can assess convergence of the computed hyperpolarizability restricted to a limited number of excited states; the main focus of this study. The formalism also allows us to make a physical interpretation of the TDHF hyperpolarizabilities using RPA intermediate states. RPA solutions with eigenvalues close to the value ω_σ are expected to be important. For example, in the Second Harmonic Generation (SHG) case, the RPA solutions with eigenvalues around $\omega_\sigma = 2\omega$ should have a significant contribution.

Eqn. 4 could be solved differently by eliminating either $\mathbf{U}^{vo}(+\omega_1,+\omega_2,\ldots)$ or $\mathbf{U}^{vo}(-\omega_1,-\omega_2,\ldots)$ from the equation. The resulting equation contains not only terms proportional to $\omega_\sigma{}^2$ but also terms proportional to ω_σ. However, such a non-symmetrical transformation shows numerical instability for near zero frequency cases. We find that the symmetrical transformation of the RPA matrix described above to be the most stable, numerically.

Calculation

To investigate the effect of truncating intermediate states, we choose two examples. Trans-butadiene is a prototype example for long polyenes where enhanced second hyperpolarizabilities, γ, are expected through the longitudinal component γ_{xxxx}. We use a [3s3p1d/2s] 6–31G basis which includes polarization and diffuse functions [19]. Correlated results are presented elsewhere [20]. The intermediate space is truncated for symmetries (A_g, A_u, B_u, B_g), indicating the number of roots in each block ordered by energy as (4,3,3,1), e.g. Second, we calculate hyperpolarizabilities of the water molecule which has a first hyperpolarizability, β, as well as second hyperpolarizability, γ. This allows us to look at convergence for β. A polarizability consistent basis set [21] augmented by a set of d-type Cartesian Gaussian functions (ζ_d=0.1) is used. The intermediate space is truncated for symmetries (A_1, B_1, B_2, A_2). Correlated results are presented in reference [18].

Results and Discussion

Orientationally averaged polarizabilities and second hyperpolarizabilities of trans-butadiene are summarized in Table 1 to Table 7. We observe that truncation of the RPA intermediate states does not have too much influence until the reduction goes beyond half of the full RPA space. When a severe truncation is imposed, the resulting hyperpolarizabilities deviate considerably from that obtained by the full space calculation. The deviation γ from the full space calculation is not monotonic though α obviously is, as all contributions to the ground state polarizability have the same sign. For example, the truncation [35,17,35,17] provides the static polarizability 34.97 and second hyperpolarizability 19304, while for the more severe truncation [3,1,3,1], the values are 18.68 and 6647, respectively, in comparison with the full space calculations 53.24 and 14812. The [11,5,11,5] truncation provides a γ value of 14613, which is, fortuitously, close to the full space value, as both larger and smaller truncations deviate markedly from the correct value. For example, the α value in the same truncation is 23.92, which is far from the one in the full space calculation. It seems that there is no justification whatsoever for truncating the intermediate states beyond half of the space [175,84,175,84]. However, the figures (Fig. 1 to Fig. 3) indicate that the overall behavior of the hyperpolarizabilities calculated in the reduced space is quite similar over the frequency range considered. This means that the *percentage correction* by the TDHF method is still useful for the prediction of the hyperpolarizabilities by high-level correlated calculations of different non-linear optical processes, even in a drastically truncated scheme.

To consider a molecule with a β value, the polarizabilities and first and second hyperpolarizabilities of the water molecule are summarized in Table 8 to Table 12 as well as Fig. 4 to Fig. 7. The same trends are observed. That is, a modest truncation does not affect the calculated polarizabilities or hyperpolarizabilities, but the calculations with truncation beyond half of the full space provide values that deviate considerably from the full space calculation. Again, despite the large deviation of the absolute values calculated in truncated spaces compared to full space values, the overall curves in the frequency range of hyperpolarizabilities are quite similar to that of the full space calculation.

In Tables 13 to 16, we summarize all components of the THG γ for trans-butadiene for both full space and drastically truncated space calculations. While the longitudinal component, γ_{xxxx}, is not affected drastically by the truncation (3,1,3,1), the out-of-plane component, γ_{zzzz}, in the truncated calculation is totally off from the full space calculation. For non-zero frequencies, Kleinman symmetry no longer holds. This is demonstrated by those components that would be identical assuming Kleinman symmetry, which take entirely different values near resonance (ω=0.075). The truncated space calculations seem to fail to describe even the trend in the deviation of each component. The longitudinal component, γ_{xxxx}, of the Third Harmonic Generation (THG) tensor blows up at one-third of the frequency corresponding to the lowest RPA state as expected (ω=0.75). We note that the dominant component,

Table 1 Polarizability and Second Hyperpolarizability of Trans-Butadiene : Full Space (349,168,350,168)

Top row corresponds to the parallel component,[a] and second row corresponds to the perpendicular component.[b]

	0.	0.04	0.043	0.05	0.06	0.0656	0.075
α	53.24	54.01	54.14	54.46	55.03	55.41	56.14
THG[c]	14812	21453	22956	27917	42604	62581	494427
		7151	7652	9306	14201	20860	164809
DCSHG[c]		17607	18121	19592	22585	24932	30656
		5869	6041	6535	7546	8345	10316
IDRI[c]		16580	16886	17730	19324	20470	22953
		5532	5633	5913	6434	6803	7580
EOKE[c]		15657	15795	16167	16830	17277	18173
		5218	5266	5388	5610	5761	6064

a. $(\gamma_{iijj}+\gamma_{ijij}+\gamma_{ijji})/15$

b. $\{2\gamma_{ijji}-(\gamma_{iijj}+\gamma_{ijij})\}/15$

c. THG = third harmonic generation; DCSHG = dc-induced second harmonic generation; IDRI = intensity dependent refractive index; EOKE = Electro Optical Kerr Effect.

Table 2 Polarizability and Second Hyperpolarizability of Trans-Butadiene : Truncated Space(262,126,262,126)

Top row corresponds to the parallel component, and second row corresponds to the perpendicular component.[a]

	0.	0.04	0.043	0.05	0.06	0.0656	0.075
α	53.22	54.00	54.11	54.44	55.01	55.39	56.12
THG	14814	21456	22959	27920	42608	62586	494434
		7152	7653	9307	14203	20862	164811
DCSHG		17610	18124	19595	22588	24935	30660
		5870	6042	6536	7547	8346	10317
IDRI		16582	16888	17733	19326	20472	22956
		5532	5634	5914	6435	6804	7581
EOKE		15659	15797	16169	16832	17280	18176
		5218	5265	5389	5611	5761	6065

a. See footnotes a,b, and c in Table 1.

Table 3 Polarizability and Second Hyperpolarizability of Trans-Butadiene : Truncated Space(175,84,175,84)

Top row corresponds to the parallel component, and second row corresponds to the perpendicular component.[a]

	0.	0.04	0.043	0.05	0.06	0.0656	0.075
α	52.14	52.92	53.04	53.37	53.94	54.31	55.05
THG	14662	21255	22746	27668	42238	62050	489909
		7085	7582	9223	14079	20683	163303
DCSHG		17439	17949	19409	22379	24708	30392
		5813	5984	6474	7478	8271	10229
IDRI		16416	16720	17558	19138	20276	22741
		5477	5578	5856	6373	6738	7510
EOKE		15502	15639	16009	16666	17110	17999
		5166	5212	5335	5555	5705	6006

a. See footnotes a, b, and c in Table 1.

Table 4 Polarizability and Second Hyperpolarizability of Trans-Butadiene : Truncated Space (87,21,87,21)

Top row corresponds to the parallel component, and second row corresponds to the perpendicular component.[a]

	0.	0.04	0.043	0.05	0.06	0.0656	0.075
α	44.17	44.92	45.04	45.36	45.91	46.27	46.98
THG	16267	23510	25137	30496	46293	67696	528147
		7837	8379	10165	15431	22565	176049
DCSHG		19331	19890	21487	24728	27263	33433
		6437	6623	7156	8243	9100	11209
IDRI		18196	18527	19442	21164	22400	25076
		6088	6202	6513	7091	7501	8366
EOKE		17200	17351	17757	18478	18964	19937
		5727	5776	5910	6148	6310	6635

a. See footnotes a,b, and c in Table 1.

Table 5 Polarizability and Second Hyperpolarizability of Trans-Butadiene : Truncated Space (35,17,35,17)

Top row corresponds to the parallel component, and second row corresponds to the perpendicular component.[a]

	0.	0.04	0.043	0.05	0.06	0.0656	0.075
α	34.97	35.69	35.81	36.11	36.64	36.99	37.68
THG	19304	27406	29218	35179	52725	76499	590304
		9135	9739	11726	17575	25500	16519
DCSHG		22740	23364	25145	28750	31565	38399
		7569	7776	8368	9574	10522	12848
IDRI		21463	21833	22851	24763	26132	29085
		7190	7318	7669	8321	8782	9754
EOKE		20355	20524	20978	21785	22328	23413
		6775	6830	6979	7270	7434	7784

a. See footnotes a, b, and c in Table 1.

Table 6 Polarizability and Second Hyperpolarizability of Trans-Butadiene : Truncated Space (11,5,11,5)

Top row corresponds to the parallel component, and second row corresponds to the perpendicular component.[a]

	0.	0.04	0.043	0.05	0.06	0.0656	0.075
α	23.92	24.56	24.66	24.93	25.40	25.71	26.32
THG	14613	21973	23654	29244	46053	69231	577427
		7324	7885	9748	15351	23077	192476
DCSHG		17701	18268	19895	23227	25858	32333
		5930	6126	6689	7851	8776	11076
IDRI		16547	16880	17804	19554	20819	23579
		5445	5542	5808	6300	6647	7370
EOKE		15551	15702	16110	16836	17327	18311
		5199	5252	5396	5653	5828	6180

a. See footnotes a, b, and c in Table 1.

Table 7 Polarizability and Second Hyperpolarizability of Trans-Butadiene : Truncated Space (3,1,3,1)

Top row corresponds to the parallel component, and second row corresponds to the perpendicular component.[a]

	0.	0.04	0.043	0.05	0.06	0.0656	0.0756
α	18.68	19.24	19.33	19.57	20.00	20.28	20.83
THG	6647	10828	11806	15117	25504	40594	411916
		3609	3935	5039	8501	13531	137305
DCSHG		8382	8703	9631	11558	13101	16965
		2782	2888	3193	3832	4347	5652
IDRI		7727	7914	8435	9432	10159	11759
		2610	2678	2865	3220	3474	4024
EOKE		7173	7257	7485	7892	8168	8724
		2383	2410	2483	2614	2704	2886

a. See footnotes a, b, and c in Table 1.

Table 8 Polarizability and Hyperpolarizabilities of Water : Full Space (89,52,73,41)

	0.	0.04	0.043	0.06	0.0656	0.08	0.1
$\alpha_{\parallel}$ [a]	8.538	8.576	8.583	8.625	8.642	8.694	8.787
$\alpha_{\perp}$ [a]	1.093	1.084	1.083	1.072	1.068	1.055	1.029
SHG[b]	10.22	10.85	10.96	11.74	12.08	13.21	15.66
EOPE[b]		10.42	10.46	10.69	10.78	11.08	11.61
THG[b]	952	1091	1116	1322	1426	1862	4910
DCSHG[c]		1017	1028	1111	1147	1268	1540
IDRI[c]		995	1002	1053	1075	1145	1285
EOKE[c]		972	976	1000	1010	1041	1097

a. $\alpha_{\parallel} = \frac{1}{3}(\alpha_{xx} + \alpha_{yy} + \alpha_{zz})$; $\alpha_{\perp} = \frac{1}{\sqrt{2}}[(\alpha_{xx} - \alpha_{yy})^2 + (\alpha_{xx} - \alpha_{zz})^2 + (\alpha_{yy} - \alpha_{zz})^2]^{\frac{1}{2}}$.

b. SHG = second harmonic generation; EOPE = Electro Optical Pockel Effect; THG = third harmonic generation.

c. See footnote c in Table 1.

Table 9 Polarizability and Hyperpolarizabilities of Water : Truncated Space (67,39,55,31)[a]

	0.	0.04	0.043	0.06	0.0656	0.08	0.1
$\alpha_{\parallel}$	8.499	8.538	8.544	8.586	8.603	8.655	8.724
$\alpha_{\perp}$	1.192	1.183	1.182	1.171	1.671	1.153	1.126
SHG	10.27	10.90	11.01	11.79	12.14	13.27	15.73
EOPE		10.47	10.50	10.74	10.83	11.13	11.67
THG	953	1092	1117	1324	1428	1865	4922
DCSHG		1019	1030	1112	1148	1270	1543
IDRI		996	1003	1054	1076	1146	1287
EOKE		974	977	1002	1012	1042	1099

a. See footnotes a, b, and c in Table 8.

Table 10 Polarizability and Hyperpolarizabilities of Water : Truncated Space (45,26,37,21)[a]

	0.	0.04	0.043	0.06	0.0656	0.08	0.1
$\alpha_{\parallel}$	8.141	8.179	8.184	8.227	8.244	8.296	8.388
$\alpha_{\perp}$	1.531	1.523	1.521	1.511	1.507	1.494	1.468
SHG	8.99	9.62	9.72	10.50	10.84	11.96	14.40
EOPE		9.19	9.22	9.45	9.55	9.84	10.37
THG	974	1116	1141	1351	1457	1900	4994
DCSHG		1041	1052	1136	1173	1296	1573
IDRI		1018	1025	1077	1099	1171	1314
EOKE		995	999	1023	1033	1065	1122

a. See footnotes a, b, and c in Table 8.

Table 11 Polarizability and Hyperpolarizabilities of Water : Truncated Space (22,13,18,10)[a]

	0.	0.04	0.043	0.06	0.0656	0.08	0.1
$\alpha_{\|\|}$	6.364	6.400	6.405	6.445	6.461	6.510	6.597
$\alpha_{\perp}$	2.175	2.168	2.167	2.160	2.157	2.146	2.126
SHG	19.90	20.74	20.88	21.90	22.34	23.79	26.86
EOPE		20.17	20.22	20.52	20.65	21.03	21.73
THG	1167	1327	1356	1591	1710	2202	5615
DCSHG		1243	1255	1350	1391	1530	1840
IDRI		1217	1225	1284	1309	1390	1551
EOKE		1192	1196	1223	1235	1270	1334

a. See footnotes a,b, and c in Table 8.

Table 12 Polarizability and Hyperpolarizabilities of Water : Truncated Space (4,2,3,1)[a]

	0.	0.04	0.043	0.06	0.0656	0.08	0.1
$\alpha_{\|\|}$	1.910	1.929	1.932	1.953	1.961	1.988	2.035
$\alpha_{\perp}$	0.365	0.355	0.353	0.341	0.336	0.321	0.292
SHG	14.48	15.29	15.43	16.44	16.88	18.34	21.50
EOPE		14.74	14.78	15.08	15.21	15.58	16.28
THG	518	613	631	780	859	1205	3996
DCSHG		563	570	627	653	740	946
IDRI		547	552	587	602	651	753
EOKE		532	535	551	558	579	618

a. See footnotes a, b, and c in Table 8.

Table 13 Third Harmonic Generation $\gamma(-3\omega_l;\omega_j,\omega_k,\omega_l)$; Full v.s. Truncated(A_g,A_u,B_u,B_g)

ω = 0. (in a.u.)

Components (i;j,k,l)	Full	(3,1,3,1)
(x;x,x,x)	23,514	26,916
(y;y,y,y)	6,860	1,436
(z;z,z,z)	13,291	95
(x;x,y,y), (x;y,y,x), (x;y,x,y) (y;y,x,x), (y;x,x,y), (y;x,y,x)	2,965	1,607
(y;y,z,z), (y;z,z,y), (y;z,y,z) (z;z,y,y), (z;y,y,z), (z;y,z,y)	3,396	131
(x;,x,z,z), (x;z,z,x), (x;z,x,z) (z;z,x,x), (z;,x,x,z), (z;,x,z,x)	8,837	656
Classical Orientational Average	14,812	6,647

Table 14 Third Harmonic Generation $\gamma(-3\omega_l;\omega_j,\omega_k,\omega_l)$; Full v.s. Truncated(A_g,A_u,B_u,B_g)

ω = 0.043 (in a.u.)

Components (i;j,k,l)	Full	(3,1,3,1)
(x;x,x,x)	41,533	49,054
(y;y,y,y)	8,416	1,838
(z;z,z,z)	17,900	146
(x;x,y,y), (x;y,y,x), (x;y,x,y)	4,612	2,592
(y;y,x,x), (y;x,x,y), (y;x,y,x)	3,667	2,351
(y;y,z,z), (y;z,z,y), (y;z,y,z)	4,385	125
(z;z,y,y), (z;y,y,z), (z;y,z,y)	4,683	224
(x;,x,z,z), (x;z,z,x), (x;z,x,z)	15,985	1,956
(z;z,x,x), (z;,x,x,z), (z;,x,z,x)	13,598	744
Classical Orientational Average	22,956	11,806

Table 15 Third Harmonic Generation $\gamma(-3\omega_l;\omega_j,\omega_k,\omega_l)$; Full v.s. Truncated(A_g,A_u,B_u,B_g)

ω = 0.0656 (in a.u.)

Components (i;j,k,l)	Full	(3,1,3,1)
(x;x,x,x)	146,475	174,747
(y;y,y,y)	11,652	2,474
(z;z,z,z)	30,424	331
(x;x,y,y), (x;y,y,x), (x;y,x,y)	13,841	8,784
(y;y,x,x), (y;x,x,y), (y;x,y,x)	4,463	3,680
(y;y,z,z), (y;z,z,y), (y;z,y,z)	6,839	-302
(z;z,y,y), (z;y,y,z), (z;y,z,y)	8,784	587
(x;,x,z,z), (x;z,z,x), (x;z,x,z)	59,849	11,861
(z;z,x,x), (z;,x,x,z), (z;,x,z,x)	30,579	809
Classical Orientational Average	62,581	40,594

Table 16 Third Harmonic Generation $\gamma(-3\omega_l;\omega_j,\omega_k,\omega_l)$; Full v.s. Truncated(A_g,A_u,B_u,B_g)

ω = 0.075 (in a.u.)

Components (i;j,k,l)	Full	(3,1,3,1)
(x;x,x,x)	1,522,212	1,805,864
(y;y,y,y)	19,002	-983
(z;z,z,z)	47,867	699
(x;x,y,y), (x;y,y,x), (x;y,x,y)	135,603	96,956
(y;y,x,x), (y;x,x,y), (y;x,y,x)	-6,516	726
(y;y,z,z), (y;z,z,y), (y;z,y,z)	16,785	-9,799
(z;z,y,y), (z;y,y,z), (z;y,z,y)	15,668	1,346
(x;,x,z,z), (x;z,z,x), (x;z,x,z)	659,666	164,182
(z;z,x,x), (z;,x,x,z), (z;,x,z,x)	61,846	588
Classical Orientational Average	494,427	411,916

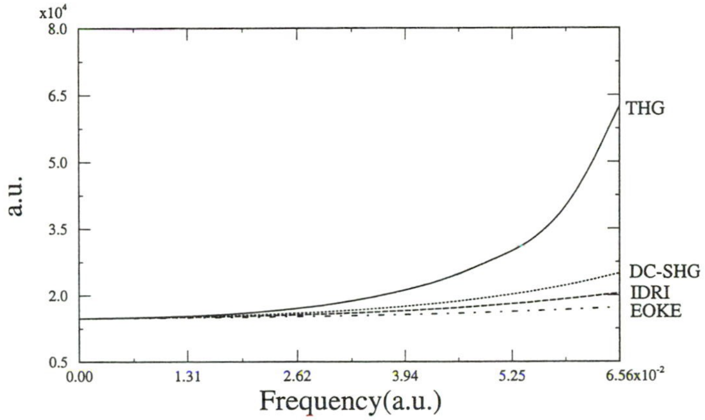

Figure 1. Trans Butadiene γ Full Space [349, 168, 350, 168]

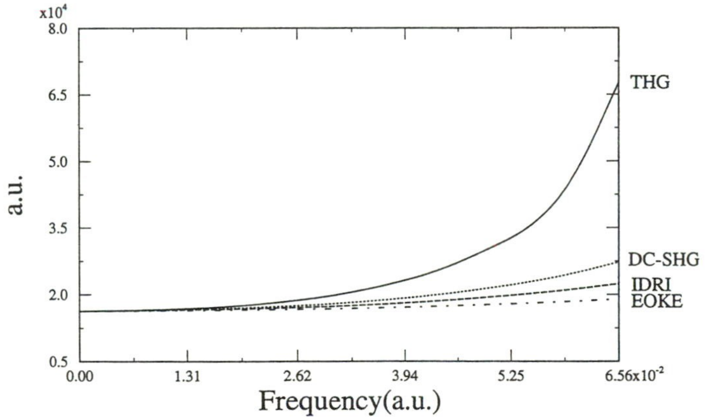

Figure 2. Trans Butadiene γ Reduced Space [87, 21, 87, 21]

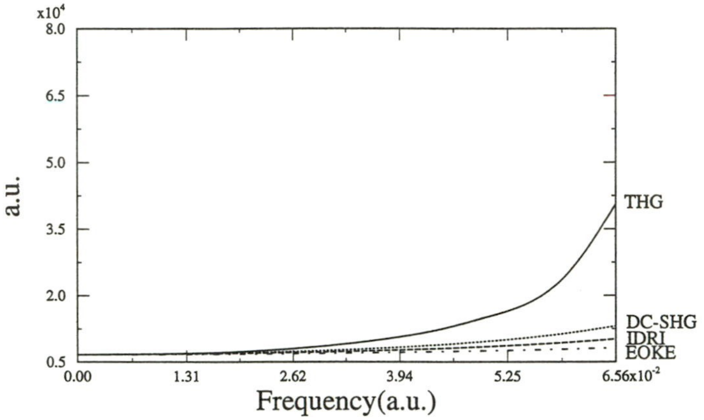

Figure 3. Trans Butadiene γ Reduced Space [3, 1, 3, 1]

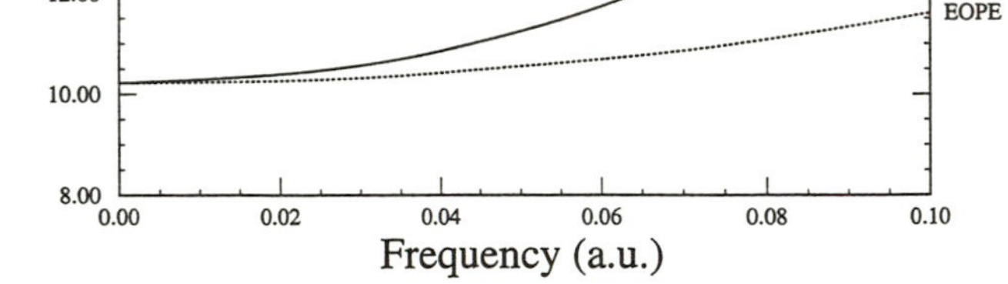

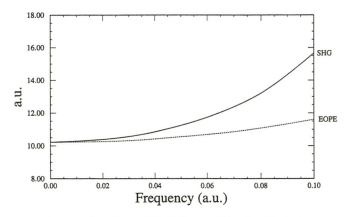

Figure 4. Water-β Full Space [89, 52, 73, 41]

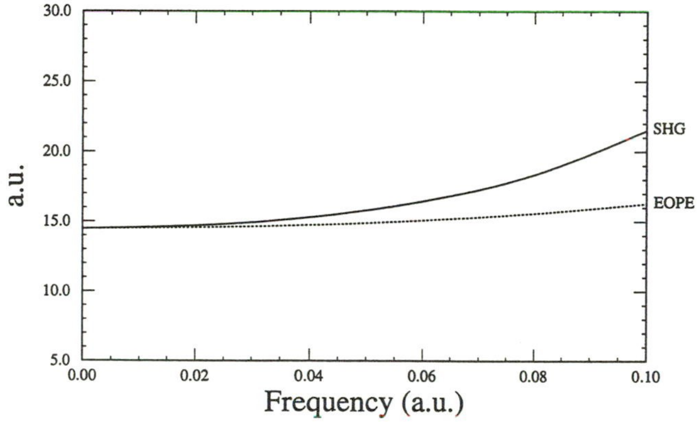

Figure 5. Water-β Reduced Space [4, 2, 3, 1]

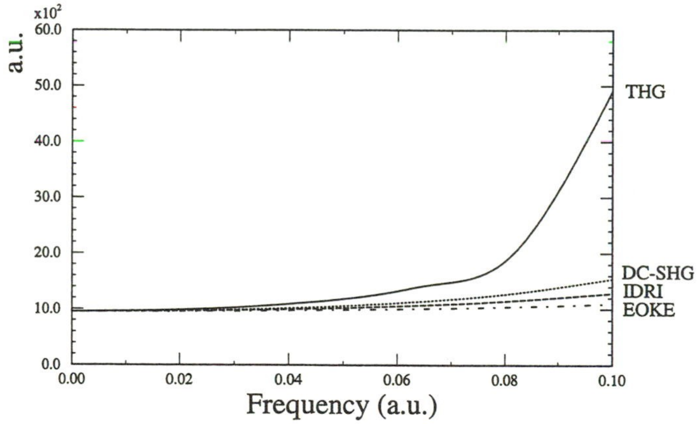

Figure 6. Water γ Full Space [89, 52, 73, 41]

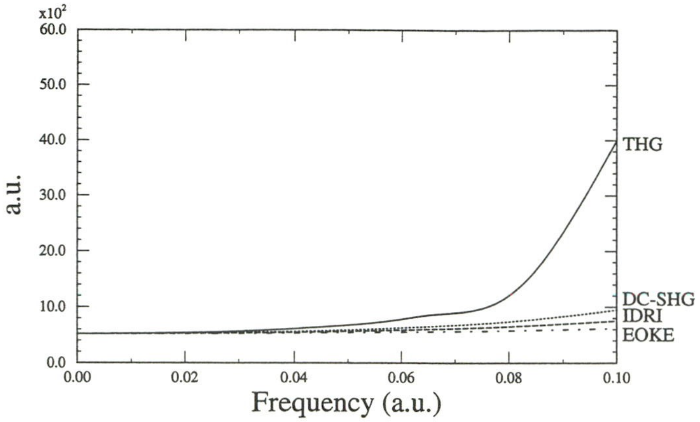

Figure 7. Water γ Reduced Space [4, 2, 3, 1]

γ_{xxxx}, is relatively well described even in the drastically truncated space calculations although the second- and third-largest components are not. The orientationally averaged hyperpolarizabilities are relatively well described in the truncated scheme when the applied frequency approaches the resonance. This is because the γ_{xxxx} term becomes dominant in the averaging sum for the near resonance situation.

In conclusion, while the SOS formalism is a useful tool to qualitatively analyze non-linear optical phenomena through percentage dispersion effects, it is severely limited for the quantitative evaluation of absolute hyperpolarizabilities. In particular, for the lesser components, the truncated calculations are quite unreliable. The argument for using the SOS formula for the longitudinal component may be partially rationalized for near resonance situations, since the longitudinal component becomes overwhelmingly dominant in that case.

These results lend little support for the use of "few state" models for design purposes, insofar as such analyses are contingent upon a reasonably accurate magnitude of the hyperpolarizabilities. Some situations, such as in β materials, where potentially dominant charge transfer excited states caused by adding electron pushing and withdrawing groups are not considered in the two examples discussed, but we would expect similar limitations for that case as well, since there are *many* electronic states whose individual contribution might be small, but whose total is a significant part of the value of the hyperpolarizability.

Acknowledgments

This work is supported by the NLO program under Grant No. AFOSR-F49620–93–I–0118. Support from this program has been responsible for the work in this area of one of us (RJB) dating back to 1978, and for both of us since 1986. We greatly appreciate AFOSR's generous support during that time.

Literature Cited

1. Sekino, H.; Bartlett, R.J. *J. Chem. Phys.* **1986**, 85, 976.
2. Karna, S. P.; Prasad, P.N.; Dupuis, M. *J. Chem. Phys.* **1991**, 94, 1171; Keshari, V.; Karna, S. P.; Prasad, P.N. *J. Phys. Chem. 1993*, 97 3525.
3. Rice, J. E.; Amos, R. D.; Colwell, S. M.; Handy, N. C.; *Sanz, J. J. Chem. Phys.* **1990**, 93, 8828.
4. Aiga, F.; Sasagane, K.; Itoh, R. *J. Chem. Phys.* **1992**, 167, 277.
5. Karna, S.P.; Dupuis, M. *J. Compt. Chem.* **1991**, 12, 487.
6. Ågren, H.; Vahtras, O.; Koch, H.; Jørgensen, P.; Helgaker, T. *J. Chem. Phys.* **1993**, 98, 6417.
7. Karna, S. P. *Chem. Phys. Lett.* **1993**, 214, (1993) 186.
8. Sekino, H.; Bartlett, R. J. *Int. J. Quant. Chem.*, **1992**, 43, 119.
9. Sekino, H. *Can. J. Chem.* **1992**, 70, 677.
10. Dalgaard, E. *Phys. Rev.* **1982**, A26, 3235.
11. Olsen, J.; Jorgensen, P. *J. Chem. Phys.* **1985**, 82, 3235.
12. Sengelov, P.W.; Oddershede, J. *Chem. Phys.* **1988**, 124, 371.
13. Parkinson, W.A.;Oddershede, J. *J. Chem. Phys.* **1991**, 94, 7251.
14. Sasagane, K.; Aiga, F.; Itoh, R. *J. Chem. Phys.* **1993**, 99, 3738.
15. Oudar, J.L.; Chemla, D.S. *J. Chem. Phys.* **1977**, 66, 2664.
16. Lalama, S. J.; Garito, A.F. *Phys. Rev.* **1979**, A20, 1179.
17. Prasad, P. N.; Karna, S.P. *Int. J. Quantum Chem.* 1994, S28, 395.
18. Sekino, H.; Bartlett, R.J. *J. Chem. Phys.* 1993, 98, 3022.
19. Hurst, G.J.; Dupuis, M.; Clementi, E. *J. Chem. Phys.* 1988, 89, 385.
20. Sekino, H. Bartlett, R.J. *J. Chem. Phys.* **1991**, 94, 3665.
21. Sadlej, A. J. *Coll. Czech. Chem. Commun.* **1988**, 53, 1995; *Theoret. Chim. Acta.* **1991**, 79, 123.

RECEIVED December 29, 1995

Chapter 5

Early Theoretical Studies of Third-Order Nonlinear Electric Susceptibilities

Hendrick F. Hameka

Department of Chemistry, University of Pennsylvania, Philadelphia, PA 19104–6323

We review the theoretical work on the third order-nonlinear electric susceptibility γ. We discuss various general theoretical approaches and applications to polyenes and benzene where theoretical results may be compared with experimental data. Various applications to other aromatic molecules are also mentioned. It is concluded that theoretical procedures based on the semiempirical Pariser-Parr-Pople method give satisfactory results for conjugated and aromatic molecules.

I. Introduction

It is not particularly difficult to extend the theoretical description of electric susceptibilities to higher order terms but few efforts in this direction were reported before the experimental discovery of various nonlinear optical effects around 1960. As early as 1962 Franken and Ward (1) presented a review of those early experimental discoveries together with a theoretical analysis. During the next few years there was a great deal of experimental activity in nonlinear optics and in 1965 Ward (2) presented an extensive theoretical analysis of the various nonlinear optical effects and their relation to corresponding higher-order electric susceptibility terms. Bloembergen (3), who is one of the pioneers in nonlinear optics, prefers also to give a unified description of various nonlinear optical effects based on the generalized polarization **P**(r; t) of the medium. This polarization **P** is then represented as a power series expansion in terms of the electric field strength E_i (i = x, y, z):

$$P_i = \Sigma_j \alpha_{ij} E_J + \Sigma_j \Sigma_k \beta_{ijk} E_J E_k + \sum_j \sum_k \sum_l \gamma_{ijkl} E_J E_k E_l + \qquad (1)$$

Here the symbol α denotes the linear polarizability, β denotes the second-order and γ the third-order nonlinear susceptibility, etc.

During the seventies we became interested in one particular nonlinear optical effect namely the electric-field induced second harmonic generation. We introduced the acronym EFISH (4) for this effect and this name has been generally adopted since then. We felt that the EFISH effect might become very important because of its potential application for modulator devices in light wave communication systems. Our

0097–6156/96/0628–0102$15.00/0

expectations were not realized but the theory of the EFISH effect became an interesting problem in molecular quantum theory.

The EFISH effect was first discovered by Terhune et.al. (5, 6) in crystals and later by Mayer (7) in gases and liquids. A theoretical description of the EFISH effect based on the classical Maxwell equations was presented by Kielich (8). We studied the effect from the point of view of photon scattering (9) in order to express it in terms of molecular parameters. Our conclusions agreed with Kielich's work (8). We should differentiate between a temperature dependent EFISH effect for polar molecules depending on the electric dipole orientation of the molecules in the DC electric field and a temperature independent EFISH effect for nonpolar molecules related to the third-order electric susceptibilities γ.

The temperature independent effect is interesting from a theoretical point of view because its magnitude is related to a fourth-order energy perturbation. It seems that this was the first time in the history of quantum chemistry that a fourth-order perturbation term is related to practical applications.

During the early seventies attention was focused on compounds occurring in the human retina. Hermann, Richard and Ducuing (10) reported measurement on β-carotene and Hermann and Ducuing (11) measured the third-order susceptibilities of some other long-chain conjugated hydrocarbons.

Levine and Bethea (12) presented an extensive and careful experimental effort based on static field induced second harmonic generation in liquids to determine both second and third order hyperpolarizabilities β and γ for a wide range of conjugated and nonconjugated organic molecules. Ward and Elliott (13) used the same experimental technique in the gas phase to determine the third-order hyperpolarizability γ for four organic molecules: ethylene, benzene, 1,3-butadiene and 1,3,5-hexatriene. The experimental results of Levine and Bethea (12) and of Ward and Elliott (13) constitute a useful basis for testing the accuracy of theoretical methods for calculating the third-order hyperpolarizabilities of organic molecules.

At this stage, around 1975, it became already quite clear that the EFISH effect could be applied to the design of modulator or amplifier devices only if we could discover and identify materials with unusually large β and γ values. In such a search theory could potentially play a useful role. If theoretical considerations could indicate which types of materials might have unusually large β or γ values then theory would make a useful contribution even if its numerical predictions had only limited accuracy. We should add that the calculation of higher order electric susceptibilities constitutes an interesting theoretical problem even if we disregard the potentially useful applications.

During the past twenty years an enormous amount of experimental and theoretical work on second-order and third-order nonlinear susceptibilities and on related topics have been reported in the literature. The field was extensively reviewed in a book, edited by Chemla and Zyss in 1987 (14) and in a book by Prasad and Williams (15). Recently, André and Delhalle (16) and Brédas, Adant, Tackx, Persoons and Pierce (17) published two very extensive review articles quoting almost five hundred references. However, the bulk of the work quoted in these two reviews (16, 17) deals with the second-order nonlinear susceptibility β and only a small part describes work on the third-order susceptibility γ.

For example, we quote from reference 16: "Definite successful theoretical and experimental achievements have been made in the area of β. The same cannot be claimed for the nonlinear hyperpolarizability γ where both theory and experiment are at a very early stage of the understanding of the underlying processes in second-order effects." We also quote from reference 17: "Only the pure electronic third-order polarizability of centrosymmetric molecules is directly accessible by the EFISH technique. It is clear that for this reason it is a limited technique".

In the present paper we plan to limit ourselves to calculations of the third-order nonlinear susceptibility γ and we focus on some general and fundamental aspects of the theory only. We hope that the limited scope of our efforts supplements the recent review articles (15, 16, 12) rather than competes with them.

II. General Theory

From a quantum chemical point of view we find it convenient to relate the third-order nonlinear susceptibility to the fourth-order energy perturbation of a molecule due to a homogeneous electric field **F**. We realize that the variety of nonlinear optical effects are related to time-dependent susceptibilities where the perturbing electric fields are time-dependent. However the extension of the perturbation procedure from static to time-dependent field is fairly straightforward and it is convenient to focus on the static third-order susceptibilities first.

If we want to consider the perturbation of a freely rotating molecule in a gas or liquid we should differentiate between the space-fixed coordinate system (x,y,z) and a coordinate system (X,Y, Z) attached to the molecule. The transformation between the two sets of coordinate systems is described by a set of Eulerian angles (18). First we consider the perturbation of the molecule, described by a Hamiltonian H_o, due to a homogeneous electric field F_x relative to the space fixed coordinate system. We write the perturbed Hamiltonian H as

$$H = H_o + \lambda V = H_o - \mu_x F_x \tag{2}$$

The perturbed energy E_o and eigenfunction Ψ_o of the Hamiltonian H are defined as a power series in λ:

$$\begin{aligned} E_o &= \varepsilon_o + \lambda E_{o,1} + \lambda^2 E_{o,2} + \lambda^3 E_{o,3} + \lambda^4 E_{o,4} + \ldots\ldots. \\ \Psi_o &= \Psi_o + \lambda \Psi_{o,1} + \lambda^2 \Psi_{o,2} + \lambda^3 \Psi_{o,3} + \ldots\ldots \end{aligned} \tag{3}$$

where ε_o and α_o are the unperturbed eigenvalue and eigenfunction of the Hamiltonian H_o. The energy perturbations $E_{o,3}$ and $E_{o,4}$ are expressed in terms of the eigenfunction perturbations as follows (19)

$$\begin{aligned} E_{o,3} &= \langle \Psi_{o,1} \mid V - E_{o,1} \mid \Psi_{o,1} \rangle \\ E_{o,4} &= \langle \Psi_{o,1} \mid V - E_{o,1} \mid \Psi_{o,2} \rangle - E_{o,2} \langle \Psi_{o;1} \mid \Psi_{o,1} \rangle \end{aligned} \tag{4}$$

It is important to note that the third-order energy perturbation $E_{o,3}$ and the hyperpolarizabilities β may be derived from the first-order perturbed wave function $\Psi_{o,1}$ but that the fourth-order energy perturbation depends on the second-order perturbed wave function $\Psi_{o,2}$. Consequently, the calculation of $E_{o,4}$ and γ is an order of magnitude more complex than the calculation of $E_{o,3}$ and β. The calculation of γ for even small systems such as the helium atom or the hydrogen atom requires a considerable computational effort. Computations of γ for molecules of practical interest such as benzene or conjugated hydrocarbons require some drastic approximations and it is difficult to make a priori predictions about the accuracy of the computed results.

Let us now return to the transformation between the space-fixed coordinate system (x,y,z) and the molecular coordinate system (X,Y,Z). We define the energy perturbation E_o (x,x,x,x) as the fourth order energy perturbation due to the perturbation

$-\mu_X F_X$, the experimental hyperpolarizability γ is given by

$$\gamma = \chi^{(3)}_{xxxx} = -4E_o\,(x,x,x,x) = -4E_o\,(x^4) \tag{5}$$

In order to compute this quantity we must transform it to the molecular coordinate system (X,Y,Z). By introducing Eulerian angles (18) and averaging over molecular orientations (20) we obtain

$$E_o\,(x^4) = (1/5)\,[E_o\,(X^4) + E_o\,(Y^4) + E_o\,(Z^4)] + (1/15)\,[E_o\,(X^2Y^2) + E_o\,(X^2Z^2) + E_o\,(Y^2Z^2)] \tag{6}$$

For larger molecules it is preferable to expand the perturbation energies in terms of the unperturbed eigenvalues E_K and eigenfunctions Ψ_K of the molecule. The corresponding expression is

$$E_o\,(X^4) = -\Sigma_K\,\Sigma_L\,\Sigma_M\,(E_K - E_o)^{-1}\,(E_L - E_o)^{-1}\,(E_M - E_o)^{-1}\,V_{OK}\,V_{KL}\,V_{LM}\,V_{MD} + [\Sigma_K\,(E_K - E_o)^{-1}\,V_{OK}\,V_{KO}]\,[\Sigma_L\,(E_L - E_o)^{-2}\,V_{OL}\,V_{LO}] \tag{7}$$

where the matrix elements are defined as

$$V_{KL} = <\Psi_K\,|\,\mu_X\,|\,\Psi_L> \text{etc.} \tag{8}$$

The other contributions to Eq (6) are double perturbation terms, their expansions are given by

$$\begin{aligned}E_o\,(X^2Y^2) = &-\Sigma_K\,\Sigma_L\,\Sigma_M\,(E_K - E_o)^{-1}\,(E_L - E_o)^{-1}\,(E_M - E_o)^{-1}\\ &x[V_{OK}\,V_{KL}\,W_{LM}\,W_{MO} + V_{OK}\,W_{KL}\,V_{LM}\,W_{MO} + V_{OK}\,W_{KL}\,W_{LM}\,V_{MO}\\ &+ W_{OK}\,V_{KL}\,V_{LM}\,W_{MO} + W_{OK}\,V_{KL}\,W_{LM}\,V_{MO} + W_{OK}\,W_{KL}\,V_{LM}\,V_{MO}]\\ &+ \Sigma_K\,\Sigma_L\,(E_o - E_K)^{-1}\,(E_o - E_L)^{-2}\,[V_{OK}\,W_{KO} + W_{OK}\,V_{KO}][V_{OL}\,W_{LO} + W_{OL}\,V_{LO}]\\ &+ \Sigma_K\,\Sigma_L\,(E_o - E_K)^{-1}\,(E_o - E_L)^{-2}\,[V_{OK}\,V_{KO}\,W_{OL}\,W_{LO} + W_{OK}\,W_{KO}\,V_{OL}\,V_{LO}]\end{aligned} \tag{9}$$

with

$$V_{KL} = <\Psi_K\,|\,\mu_X\,|\,\Psi_L> \quad W_{KL} = <\Psi_K\,|\,\mu_Y\,|\,\Psi_L> \text{etc.} \tag{10}$$

It may seem that the above expressions are unduly complicated but they are easily programmed.

The general perturbation expressions that we have presented outline two alternative approaches to the calculation of γ. The first approach is based on the perturbation expansions (7) and (9). The second approach is based on Eq. (4), here a procedure must be developed to solve two successive inhomogeneous differential equations in order to obtain the functions $\Psi_{o,1}$ and $\Psi_{o,2}$. There is a third approach to compute γ, this involves a numerical procedure in solving the molecular Schrödinger equation in the presence of an electric field. The latter method was first proposed by Cohen and Roothaan (21).

It is of course necessary to reduce these general theoretical approaches to procedures that may be applied to reasonably sized molecules with acceptable accuracies. Most of these procedures are related to the use of the Hartree-Fock method. Much attention has been focused on conjugated hydrocarbon chains, either finite length chains or polymers and the latter applications have specific theoretical requirements.

We limit ourselves to a brief outline of the numerical approaches. The original suggestion by Cohen and Roothaan (21) was based on the use of the Hartree-Fock procedure. The method has been applied at the ab initio level (22) and it has also been interfaced with the MOPAC semiempirical program (23) by Kurtz, Stewart and Dieter (24). The latter obtained a reasonably good result for benzene ($\gamma = 0.71 \times 10^{-36}$ esu) by using the parameters that had been introduced for linear susceptibilities. The numerical approach has also been used by Matsuzawa and Dixon (25, 26). Matsuzawa and Dixon's work is based on the use of semiempirical computer programs, either MNDO or MOPAC (23). Their theoretical γ value for benzene is 0.5×10^{-36} esu, about half the experimental value.

III. Applications to Polyene Chains

We mentioned that theoretical work on calculations of γ are directed towards materials that might have unusually larger γ values. Consequently the theory has been focused on conjugated organic molecules. Conjugated hydrocarbon chains (polyene chains) have especially attracted attention and we discuss them in this section, other conjugated molecules and especially benzene are considered in section IV and V.

One of the earliest measurements on third order nonlinear optical efforts was the work by Hermann, Richard and Ducuing (10) on β-carotene. This was followed by measurements on some other molecules with long hydrocarbon chains that occur in the human retina (11). Meanwhile Rustagi and Ducuing (27) had calculated the third-order electric susceptibilities of linear hydrocarbon chains based on the free-electron model.

We decided at the time that the accuracy of this work could be improved by making use of the Hückel method (28) especially since we discovered that the Hückel equations of a linear conjugated hydrocarbon chain in a homogeneous electric field can be solved analytically (29). This was a fairly crude calculation neglecting bond alternation and correlation effects. One important conclusion describes the relation between γ and the chain length N, which was found to be

$$\gamma = A\, N^{5.3}$$

Our calculation (29) seemed to be based on a reasonable set of assumptions but subsequent investigations revealed that many of these assumptions were not justified. First we discovered that the third-order susceptibilities of conjugated hydrocarbon chains could not be approximated by a one-dimensional model. Inclusion of the terms $E_o\,(Y^4)$ and $E_o\,(X^2Y^2)$, in addition to $E_o\,(X^4)$ caused a significant change in the theoretical γ values (30).

In a subsequent calculation (31) we abandoned the analytical approach altogether and we used the perturbation expressions of Eqs. (7) and (9) instead. We assumed that the ground state Hartree-Fock wave function could be approximated by a single-determinant SCF function. The excited states K and M in the perturbation expansions (7) and (9) may then be represented by all possible single-excitation molecular wave functions. In the case of the excited states L nonzero contributions are obtained from both single-excitation and double-excitation excited states, we presented a detailed analysis of all possibilities (31).

The theoretical results for the nonlinear susceptibilities γ of conjugated hydrocarbon chains show a surprising trend. The γ values are negative for small values, N = 4 to N = 12, and they become positive for N = 14 and higher. It is therefore not feasible to represent γ by a single power of N which is valid for all values of N. The numerical γ values of the perturbation calculation (31) are quite different from the earlier analytical theory (29). It should therefore be concluded that the results of the earlier calculation (29) are not meaningful even though the underlying approximations seemed quite reasonable at the time.

There is a good reason for the erratic behavior of the theoretical γ values. They are obtained as small differences between relatively large positive and negative contributions and relatively small changes in these contributions give rise to large changes in the final theoretical predictions for γ.

During the past two decades many experimental and theoretical studies of polyacetylene chains and related polymers have been reported (16). One aspect of these efforts relates to the calculations of the linear polarizabilities α. These are much easier to calculate than the third order polarizabilities γ and the various theoretical procedures have a much higher degree of sophistication (32). We briefly discuss some of the procedures for calculating α values since they allow us to evaluate the accuracy of the approach. Some of the Hückel calculations were reviewed by André and Champagne (33).

Risser et al (34) computed α from the Hückel equations using conventional perturbation theory, it appears that these authors were not aware that these equations could be solved analytically. Flytzaniz (35) solved the perturbed Hückel equations in Terms of Bessel functions, he was also unaware of our earlier work.

It should be noted that we recalculated the polarizabilities of polyacetylene chains by using the Pariser-Parr-Pople method (36, 37, 38), we found that the PPP values are about one fourth the Hückel values (39). Soos and Hayden (40) found an even larger decrease in value when they derived the static polarizabilities α of the II electrons in conjugated polymers from exact solutions of the PPP Hamiltonian rather than from the Hückel methodg. The linear polarizabilities of finite polyacetylene chains have also been computed from ab initio procedures, either by using the finite-field method (21, 41) or by using the coupled perturbed Hartree-Fock method (42, 43). We should also mention the work by de Delo and Silbey (44).

Since Ward and Elliott (13) reported experimental γ values for the smallest polyenes, ethylene, 1,2-butadiene and trans,-1,3,5-hexatriene some effort was made to calculate these quantities with a greater degree of accuracy. We calculated the static γ values (45) with the Pariser-Parr-Pople method (36, 37, 38), the results are reported in Table I. Karna et al (46) calculated both the static and frequency-dependent values with the ab initio time-dependent Harfree-Fock method. We list the experimental Ward and Elliott values at $\lambda = 0.69$ μm (13), our static values (45), Karna et al's static values (46) and their $\lambda = 0.69$ values all in Table I. It may be seen that Karna's theoretical results agree remarkably well with the experimental values. Our static results should not be directly compared with the frequency dependent values but they agree within an order of magnitude. Nevertheless Karna's values are closer to experiment than ours.

Our survey of the various theories of linear susceptibilities is far from complete but our brief overview confirms the idea (32) that the PPP level of approximation leads to satisfactory results for linear susceptibilities irrespective of the specific theoretical approach. At the present time the PPP level of approximation is the best that is compartible with the present computer capabilities for polymers and large molecules.

Finally we should mention that Yaron and Silbey (47) have studied the vibrational contributions to third-order nonlinear susceptibilities and that Heflin, Wong, Zamani-Khamiri and Garito (48) performed a CNDO calculation including configuration interaction. These authors find that the space averaged susceptibility γ is essentially determined from the longitudinal energy perturbation E_0 (X^4), a conclusion which is contradictary to our previous findings (30).

Table I Experimental and theoretical γ values for ethylene, 1,3-butadiene and 1,3,5-hexatriene

	Ethylene	Butadiene	Hexatriene
exp (ref 13)	0.758	2.30	7.53
2MH (ref 45)	–0.028	1.345	9.92
KP (ref 46)	0.500	1.41	3.19
KP ($\lambda = 0.69$)	1.129	2.46	7.36

Table II Fourth-order energy perturbations of benzene according to the extended Hückel method

	π	σ	**Total**
$E_0\ (X^4)$	– 0.3295	– 0.0320	– 0.3615
$E_0\ (Y^4)$	– 0.3305	– 0.0312	– 0.3617
$E_0\ (Z^4)$	0	0	0.0064
$E_0\ (X^2Y^2)$	– 0.6555	– 0.0715	– 0.7270
$E_0\ (X^2Z^2)$	0	0	– 0.1696
$E_0\ (Y^4Z^2)$	0	0	– 0.1699
$E_0\ (x^4)$	– 0.1757	– 0.0174	– 0.21445
γ	0.7028	0.0696	0.8578

IV. Applications to Benzene

There are good reasons why much of the subsequent theoretical studies of third-order susceptibilities dealt with the benzene molecule. During many years this was one of the few systems for which accurate experimental information was available and this was one of the few larger molecules where approximate theories could be verified.

We already mentioned the experimental work by Levine and Bethea (12) and subsequent more precise measurements by Ward and Elliott (13). Earlier results by Hermann (49) are not consistent with these authors (12, 13) and we will not consider them. Both Levine and Bethea (12) and Ward and Elliott (13) measured frequency dependent γ values. More recently Shelton (50) measured γ at a few more frequencies and he extrapolated his results to zero frequency in order to predict the static polarizability $\gamma = \chi_{xxxx}$. Shelton reports a value $\gamma = 0.73 \pm 0.09$ (10^{-60} C^4 m^4 J^{-3}). We use esu units and we convert Shelton's result to $\gamma = 0.98 \pm 0.12$ (10^{-36} esu).

Schweig (51) had already calculated γ values for a group of conjugated molecules using a variation method. He reports a value of 1.4 x 10^{-36} esu units for the molecular parameters χ_{xxxx} and χ_{yyyy}. We are not quite sure how to interpret his result but if we use Eq. (6) and if we assume the double perturbation term to be zero we obtain γ = 0.56 x 10^{-36} esu, which is not a bad value.

Our first effort to calculate γ for benzene (52) was based on the rather crude CNDO method and it produced a value that was much too high, γ = 6.15 x 10^{-36} esu. A subsequent effort (53) based on semiempirical Hückel theory and limited to π electrons only, yielded an equally poor result γ = 0.07 x 10^{-36} esu.

It seemed to us that a more accurate approach to the problem is needed in order to get reliable results. In a subsequent effort (54) we made use of the extended Hückel method (55) and the Pariser, Parr Pople method (36, 37, 38).

We list some of the details and intermediate results of our calculations in Table II and III because we hope that they may be helpful in evaluating the accuracy of the results. Table II contains our results of the extended Hückel calculations. First they show the relative magnitude of the π and σ electron contributions. It should be noted first that the total effect contains both the π and σ contributions and also a contribution of the π - σ interactions so that it is not simply the sum of the π and σ terms only. Nevertheless, the bulk of the effect, roughly 80%, is due to the π electrons only. It should be noted also that the X^4 and Y^4 terms are slightly different, violating symmetry conditions. The small differences are due to truncation of the set of wave functions that were used in the expansion because the X^4 and Y^4 terms would have been exactly equal if we had used a complete set of functions. Nevertheless, the differences are relatively small and this indicates that the truncations do not lead to significant errors. We should point out that the symmetry condition was not satisfied in an earlier paper (52) where the limited size of the set of eigenstates in the expansion did not produce a satisfactory theoretical result. Finally, it should be noted that the double perturbation terms X^2 Y^2 etc. produce about a third of the total effect. It is a matter of speculation wether inclusion of these terms in Schweig's work (51) might give rise to a significant improvement in the accuracy of his results.

In computing the energy perturbations E_0 (X^4) and E_0 (X^2 Y^2) from Eqs. (7) and (9) we separate the expansions into three parts which we denote by A, B and C. It is assumed here that we make use of the Hartree Fock SCF approximations so that the ground state O corresponds to a closed shell single-determinant eigenstate. In the triple sums of Eqs. (7) or (9) the eigenstates K and M correspond to single-excitation eigenstates whereas the eigenstates L correspond to either singly or doubly excited eigenstates (31). The contributions to the triple sum due to singly excited L states is denoted as the A term and the contributions to the triple sum due to doubly excited L states is denoted as the B term. The other part of each perturbation expansion, the double sum, is called the C term. It follows from Eq. (4) that the A and B terms are both negative and produce a positive contribution to γ whereas the C terms is positive and produces a negative contribution to γ. In principle, either negative or positive theoretical γ values are possible depending on the relative magnitude of the A, B and C terms.

It may be seen from Table III that in the benzene case the final γ value is a relatively small difference between the positive and A and B contributions and the negative C term and that our theoretical result is quite sensitive to small changes in the

Table III Fourth-order energy perturbations of the benzene π electron system according to PPP theory

	A term	B term	C term	Total
$E_0\ (X^4)$	– 0.6423	– 1.3226	1.5461	– 0.4188
$E_0\ (Y^4)$	– 0.6423	– 1.3226	1.5461	– 0.4188
$E_0\ (X^2Y^2)$	– 1.2855	– 2.9631	3.0921	– 1.1555
$E_0\ (x^4)$	– 0.34255	– 0.72658	0.82458	– 0.24455
γ	1.3702	2.9063	– 3.2983	0.9782

Table IV Theoretical γ values for Benzene (in terms of 10^{-36} esu) from various authors

γ	Authors	Ref
0.71	Kurtz et al (1990)	24
0.56	Schweig (1967)	51
1.133	Zamani et al (1980)	54
1.243	Waite et al (1982)	56
1.279	Perrin et al, SCF (1989)	57
1.703	Perrin et al, MPZ (1989)	57
1.279	Karna et al (1991)	59
2.12	Pierce (1989)	60

various contributions. It is therefore much harder to calculate reliable γ values than, for example, linear polarizabilities α or even nonlinear second-order polarizabilities β.

It is well known that the actual excited molecular eigenstates K, L, M occurring in the perturbation expansions are usually linear combinations of single-determinant SCF singly and doubly excited eigenstates. Reliable theoretical γ values require therefore configuration interaction procedures in order to predict the correct molecular eigenstates K, L and M. Since the PPP eigenstates are obtained by means of a CI procedure the method should lead to more accurate γ values than the extended Hückel method even though the PPP method depends on some semiempirical parameters.

At first sight it may seem that our theoretical γ value reported in Table III is exactly equal to Shelton's experimental value $\gamma = 0.98 \times 10^{-36}$ esu (50) but this is not the case since our result refers to the π electron contribution only while Shelton's result refers to the total value. We should add the σ and the σ-π interaction to our theoretical values. If we take the extended Hückel value of 0.155×10^{-36} esu for the sum of the two contributions then we predict a total γ value of $\gamma = 1.133 \times 10^{-36}$ esu by combining the PPP result for the π contribution and the extended Hückel result for the σ and σ-π interaction contributions. It is interesting to note that Shelton's experimental value $\gamma = 0.98 \pm 0.12$ is just between the extended Hückel value $\gamma = 0.86$ and the PPP value $\gamma = 1.13$.

It may be concluded from the numerical values in Table II that it is not particularly easy to calculate γ since it involves a large number of singly and doubly excited states and since it is a difference of various contributions. We list a variety of recent theoretical results in Table III and the accuracy of the various results is surprisingly good.

Three of the results in Table IV were derived by means of numerical methods where finite electric fields are introduced and the various polarizabilities are derived by numerical differentiation of the energy or the induced dipole moment. We have reduced all results to the same units, 10^{-36} esu. The first calculation by Schweig (51) does not mention any result for E_o (X^2 Y^2) but his reported value for E_o (X^4) = E_o (Y^4) is in rough agreement with our results. The work by Kurtz, Stewart and Dieter (24) is based on a numerical extension of the semiempirical MOPAC program (23). Another calculation, reported by Waite, Papadopolis and Nicolaides (56) is also based on a semiempirical approach, a CNDO/2 method enhanced by some empirical parameters. A more recent calculation by Perrin, Prasad, Mougenot and Dupuis (57) reports two γ values, $\gamma = 1.279$ derived from an SCF calculation and a second value $\gamma = 1.703$ derived from an MP2 computation (58).

Two recent studies attempt to predict the frequency dependence of the γ in benzene order to compare it with experimental results. Both also report the static value for zero frequency. The value $\gamma = 1.279$ reported by Karna, Talapatra and Prasad (59) is consistent with the previously reported SCF value (57). Pierce (60) reports a static polarizability value $\gamma = 2.12$ which is significantly higher than any of the others. Pierce's π electron contribution $\gamma = 0.74$ is not very different from our extended Hückel value $\gamma_\pi = 0.70$ (53) but Pierce's σ electron contribution is much higher than anybody else's and this causes his total value to be quite high. It should be noted also that Pierce does not report a π-σ interaction contribution.

We mentioned already that the various authors have used a variety of units in reporting γ values, we converted all values to 10^{-36} esu units. It should be noted that Waite (56) misquoted our PPP result (54) in his paper.

We concluded from Table III that most recent theoretical results agree reasonably well with experiment. The various numerical approaches give satisfactory results. We prefer the PPP approach because it contains configuration interaction and it allows us to substitute the proper molecular eigenstates with accurate excitation energies into the perturbation expansions.

V. Applications to Other Aromatic Molecules

The theoretical work on benzene was motivated by the desire to derive a suitable and practical theoretical procedure for γ that yields results of acceptable accuracy. Both the theoretical and experimental work on other materials is focussed on the search for materials with substantially larger γ values, which have the potential to be used in photonic devices. In this quest the interaction between theory and experiment is of primary importance. In an ideal case theoretical studies indicate the direction of experimental investigations and experimental results may stimulate more precise theoretical studies of promising molecules.

It follows from general considerations that the presense of low-lying excited states gives rise to enhanced β and γ values. Consequently there was reason to believe that large aromatic molecules might be worth investigating. It was also believed that the presence of amino, nitro or cyano substituents could enhance the β or γ values.

In 1977 Oudar, Chemla and Batifol (61) presented a systematic study of the β and γ values of a group of mono- and disubstituted benzenes by simultaneous use of static-field induced second harmonic generation and tunable four-wave mixing. The two experimental procedures measure slightly different quantities and their results are not quite consistent. Nevertheless the experiments clearly indicate that paranitroanaline is the molecule with the largest β or γ values.

Subsequently, Meredith, Buchalter and Hanzlik (62, 63) measured the third-order nonlinear susceptibilities of liquids using third-harmonic generation. These measurements again indicated that paranitroanaline deserves special attention. Subsequent work on paranitroanaline only (64) showed that the results for β or γ are strongly solvent dependent.

It is not surprising that paranitroanaline has been the subject of theoretical studies also. Karna and Prasad (65) presented an ab initio time-dependent coupled perturbed Hartree-Fock study of p-nitroanaline. The result of the calculation was somewhat disappointing, whereas Oudar and Chemla (66) had previously reported a value of $\gamma = 48 \times 10^{-36}$ esu and Perrin and Prasad (67) had measured a value of $\gamma = 18 \times 10^{-36}$, the calculations predicted a much lower value $\gamma = 1.48 \times 10^{-36}$ esu.

We should note that we have had similar experiences. We investigated a group of about forty nitrogen-containing aromatics with the relatively crude Hückel method (68). Our calculations indicated that two of the molecules we investigated might have exceptionally larger γ values, namely 7, 7, 8, 8-tetracyanoquinodimethane (TCNQ) and 2, 2, 6, 6,-tetracyanonaphtoquinodimethane. Our theoretical predictions led Meredith and Buchalter (69) to measure γ for TCNQ. They found that the γ value for TCNQ is less than 26 times the benzene value whereas we had predicted a ratio in excess of 1000. This experience shows that theory has its limitations in predicting γ values. However, even though theory's numerical prediction for large molecules are not particularly accurate, theory can offer some guidelines for the experimentalists. Our numerical predictions for TCNQ were off by a factor 40 but our prediction that the γ value for TCNQ was much larger than for other molecules was still valid.

Another group of molecules that has recently attracted attention are thiophene, its derivatives and its polymers. Experimental results on various thiophene chains from monomers to hexamers were presented by Zhao, Singh and Prasad (72). The linear polarizabilities of these molecules were computed by Champagne, Mosley and André (71). The nonlinear susceptibilities γ were evaluated by Karna et al (72), again with the ab initio time-dependent Hartree-Forck method.

We have encountered a number of studies on smaller molecules, for example water (73) and carbon dioxide (74) but we do not intend to review this aspect of the theory.

VI. Concluding Remarks

We may conclude that the theoretical studies of the third-order nonlinear susceptibilities γ of conjugated and aromatic molecules were motivated by two different reasons. First, it was hoped that the reseach might lead to the identification of materials with enhanced γ values which might find applications in light wave communication systems. Initially it appeared that the EFISH effect might find applications in the construction of modulator devices. Second, the calculation of values for large molecules constituted an interesting and challenging theoretical problem in quantum chemistry. For the first time in history a branch of quantum chemistry depended on the evaluation of a fourth-order energy perturbation.

It is unfortunate that practical applications of the EFISH effect never materialized in a major way. The extensive search never led to the identification of materials with γ values sufficiently large to be of practical use. Also, alternative modulator devices, based on different mechanisms were developed. Of course, the evaluation of γ still remains an interesting theoretical problem which awaits new developments beyond the PPP level of approximation. We anticipate that future efforts to improve the accuracy of the theoretical results will attempt to incorporate configuration interaction methods in the theoretical approaches. However, the size-inconsistency of CI beyond the singles level is a serious problem which may prevent the incorporation of CI for some time to come. This is especially true since materials with very large γ will likely be larger molecules such as polymers where size-inconsistency is especially problematic. We can only speculate about the directions of future improvements in the theory but is is safe to assume that they will materialize one way or another.

References

1. Franken, P.A.; Ward, J.F. *Revs. Mod. Phys.*, **1963**, *35*, 23-39.
2. Ward, J.F. *Revs. Mod. Phys.* **1965**, *37*, 1-18.
3. Bloembergen, N. *Nonlinear Optics*, Benjamin, New York, **1965**.
4. Hameka, H.F. *J. Chem. Phys.*, **1977**, *67*, 2935-2942.
5. Terhune, R.W; Maker, P.D.; Savage, C.M. *Phys. Rev. Lett.*, **1962**, *8*, 404.
6. Minck, R.W.; Terhune, R.W.; Wang, C.C. *Appl. Opt.*, **1966**, *5*, 1595.
7. Mayer, G. *Compt. Rend.*, **1968**, *267 B*, 54.
8. Kielich, S. *Chem. Phys. Lett..*, **1968**, *2*, 569-572.
9. Hameka, H.F. *Can. J. Chem..*, **1971**, *49*, 1823-1829.
10. Hermann, J.P.; Richard, D.; Ducuing, J. *J. Appl. Phys. Lett.*, **1973**, *23*, 178.
11. Hermann, J.P.; Ducuing, J. *J. Appl. Phys.*, **1974**, *45*, 5100.
12. Levine, B.F.; Bethea, C.G. *J. Chem. Phys.*, **1975**, *63*, 2666-2682.
13. Ward, J.F.; Elliott, D.S. *J. Chem. Phys.*, **1978**, *69*, 5438-5440.
14. Chemla, D.S.; Zyss, J. *Eds. Nonlinear Optical Properties of Organic Molecules and Crystals*; Academic Press, New York, NY., **1984**; Vols. 1 and 2.

15. Prasad, P.N.; Williams, D.J. *Introduction to Nonlinear Optical Effects in Molecules and Polymers*; Wiley, New York, N.Y., **1991**.
16. André, J.M.; Delhalle, J. *Chem. Rev.*, **1991**, *91*, 843-865.
17. Brédas, J.L.; Adant, C.; Tackx, P.; Persoons, A.; Pierce, B.M. *Chem. Rev.*, **1994**, *94*, 243-278.
18. Wigner, E.P. Gruppentheorie and ihre Anwendung auf die Quanten mechanik der Atomspektren, Friedr. Vieweg & John, Akt. Ges., Braunschweig, **1931**.
19. Hameka, H.F. *Quantum Mechanics*, Wiley-Interscience, New York, N.Y., **1981** (page 226)
20. McJntyre, E.F.; Hameka, H.F. *J. Chem. Phys.*, **1978**, *68*, 5534-5537.
21. Cohen, H.D.; Roothaan, C.C.J. *J. Chem. Phys.* **1965**, *43*, 534.
22. André, J.M.; Pelhalle, J.; Brédas J.L. *Quantum Chemistry Aided Design of Organic Polymers, An Introduction to the Quantum Chemistry of Polymers and its Applications*, World Scientific, Singapore, **1991**.
23. Stewart, J.J.P. *QCPE Program 455*, **1983**; vission 3.1 (1986).
24. Kurtz, H.A.; Stewart, J.J.P.; Dieter, K.M. *J. Comp. Chem..*, **1990**, *11*, 82-87.
25. Matsuzawa, N and Dixon, D.A. *J. Phys. Chem..* **1992**, *96*, 6232-6241.
26. Matsuzawa, N and Dixon, D.A. *Int. J. Quantum Chem..*, **1992**, *44*, 497-515.
27. Rustagi, K.C.; Ducuing, J. *Opt. Comm..*, **1974**, *10*, 58.
28. Hameka, H.F. *Quantum Theory of the Chemical Bond*, Hafner Press, New York, **1975**.
29. Hameka, H.F. *J. Chem. Phys.*, **1977**, *67*, 2935-2942.
30. McIntyre, E.F. and Hameka, H.F. *J. Chem. Phys.*, **1978**, *68*, 3481-3484.
31. McIntyre, E.F. and Hameka, H.F. *J. Chem. Phys.*, **1978**, *69*, 4814-4820.
32. Champagne, B.; Fripiat, J.G.; André, J.M. *J. Chem. Phys.*, **1992**, *96*, 8330-8337.
33. André, J.M.; Champagne, B. *Organic Molecules for Nonlinear Optics and Photonics*, (J. Messier, et al eds.), Kluwer Academic Publishers, Dordrecht, Netherlands, **1991**.
34. Risser, S.; Klemm, S.; Allender, D.W.; Lee, M.A. *Mol. Cryst. Liq. Cryst.*, **1987**, *150b*, 631.
35. C. Flytzanis, reference 14, vol. 2, page 121.
36. Pariser, R.; Parr, R.G. *J. Chem. Phys.*, **1953**, *21*, 466, 767.
37. Pople, J.A. *Trans. Far. Soc.*, **1953**, *49*, 1375.
38. Pariser, R. *J. Chem. Phys.*, **1956**, *24*, 250.
39. Zamani-Khamiri, O.; Hameka, H.F. *J. Chem. Phys.*, **1979**, *71*, 1607-1610.
40. Soos, Z.G.; Hayden, G.W. *Phys. Rev.*, **1989**, *B40*, 3081.
41. Bodart, V.P.; Delhalle, J.; André, J.M.; Zyss, J. *Can. J. Chem.*, **1985**, *63*, 1631.
42. Dalgarna, A. *Adv. Phys.*, **1962**, *11*, 281.
43. Hurst, G.J.B.; Dupuis, M.; Clementi, E. *J. Chem. Phys.*, **1988**, *89*, 385-395.
44. de Melo, C.P.; Silbey, R. *J. Chem. Phys.*, **1988**, *88*, 2558-2566.
45. Zamani-Khamiri, O.; McIntyre, E.F.; Hameka, H.F. *J. Chem. Phys.*, **1980**, *72*, 5906-5908.
46. Karna, S.P.; Talapatra, G.B.; Wijekoon, M.K.P.; Prasad, P.N. *Phys. Rev.*, **1992**, *A45*, 2763-2769.
47. Yaron, D.; Silbey, R. *J. Chem. Phys.*, **1991**, *95*, 563-568.
48. Heflin, J.R.; Wong, K.Y.; Zamani-Khamiri, D.; Garito, A.F. *Phys. Rev. B*, **1988**, *38*, 1573-1576.
49. Hermann, J.P. *Opt. Commun.*, **1973**, *9*, 74.
50. Shelton, D.P. *J. Opt. Soc. Am.*, **1985**, *B2*, 1880-1882.
51. Schweig, A. *Chem. Phys. Lett.*, **1967**, *1*, 195-199.
52. Svendsen, E.N.; Stroyer-Hansen, T.; Hameka, H.F. *Chem. Phys. Lett.*, **1978**, *54*, 217-219.
53. McIntyre, E.F.; Hameka, H.F. *J. Chem. Phys.*, **1978**, *69*, 4814-4820.
54. Zamani-Khamiri, O.; McIntyre, E.F.; Hameka, H.F. *J. Chem. Phys.*, **1980**, *72*, 1280-1284.

55. Hoffmann, R. *J. Chem. Phys.*, **1963**, *39*, 1397.
56. Waite, J.; Papadopoulos, M.G.; Nicolqides, C.A. *J. Chem. Phys.*, **1982**, *77*, 2536-2539.
57. Perrin, E.; Prasad, P.N.; Mougenot, P.; Dupuis, M. *J. Chem. Phys.*, **1989**, *91*, 4728-4732.
58. Møller, C.; Plesset, M.S. *Phys. Rev.*, **1934**, *46*, 618.
59. Karna, S.P.; Talapatra, G.B.; Prasad, P.N. *J. Chem. Phys.*, **1991**, *95*, 5873-5881.
60. Pierce, B.M.; *J. Chem. Phys.*, **1989**, *91*, 791-811.
61. Oudar, J.L.; Chemla, D.S.; Batifol, E. *J. Chem. Phys.*, **1977**, *67*, 1626-1636.
62. Meredith, G.R.; Buchalter, B.; Hanzlik, C. *J. Chem. Phys.*, **1983**, *78*, 1533-1542.
63. Meredith, G.R.; Buchalter, B.; Hanzlik C. *J. Chem. Phys.*, **1983**, *78*, 1543-1551.
64. Meredith, G.R. and Buchalter, B. *J. Chem. Phys.*, **1983**, *78*, 1938-1945.
65. Karna, S.P.; Prasad, P.N. *J. Chem. Phys.*, **1991**, *94*, 1171-1181.
66. Oudar, J.L.; Chemla, D.S. *J. Chem. Phys.*, **1977**, *66*, 2664.
67. Perrin, E.; Prasad, P.N. (unpublished results).
68. McIntyre, E.F.; Hameka, H.F. *J. Chem. Phys.*, **1979**, *70*, 2215-2219.
69. Meredith, G.R.; Buchalter, B. *J. Chem. Phys.*, **1983**, *78*, 1615-1616.
70. Zhao, M.T., Singh, B.P.; Prasad, P.N. *J. Chem. Phys.*, **1988**, *89*, 5535-5541.
71. Champagne, B.; Mosley, D.H.; André, J.M. *J. Chem. Phys.*, **1994**, *100*, 2034-2043.
72. Karna, S.P.; Zhang, Y.; Samoc, M.; Prasad, P.N.; Reinhardt, B.A.; Dillard, A.G. *J. Chem. Phys.*, **1993**, *99*, 9984-9993.
73. Maroulis, G. *J. Chem. Phys.*, **1991**, *94*, 1182-1189.
74. Maroulis, G.; Thakkar, A.J. J. *Chem. Phys.*, **1990**, *93*, 4164-4171.

RECEIVED September 29, 1995

Chapter 6

Effect of Higher Excited Configurations on the Linear and Nonlinear Optical Properties of Organic Molecules

Israel D. L. Albert, Tobin J. Marks, and Mark A. Ratner

Department of Chemistry and Materials Research Center, Northwestern University, Evanston, IL 60208–3113

Using the INDO/1 Hamiltonian and a set of configuration interaction (CI) calculations the order of which ranges from singly excited (SCI) to full CI, we demonstrate the importance of higher excited configurations on the linear and nonlinear optical (NLO) properties of two archetypical organic π-conjugated systems. The inclusion of higher excited configurations affects, both qualitatively and quantitatively, the electronic spectrum and the NLO response properties of these systems. It is shown that configurations at least at the level of singles and doubles are essential in obtaining reliable electronic and NLO properties of these chromophores. However, as the standard INDO/1 parametrization is based on a SCI level, the parameters used in higher order CI must be modified to obtain reliable properties. In this study we have found that by using a reduced value of the two-electron repulsion integral one can obtain reliable values of the linear and NLO response properties.

I. Introduction

Organic molecules and macromolecules having extended π-conjugation are known to exhibit large nonlinear optical (NLO) responses[1-5] since the delocalized π-electrons in these systems are readily polarized (distorted from their equilibrium position) by the application of an electric field. In addition to the large NLO responses, these structures also exhibit a wide variety of interesting properties that have attracted the attention of chemists, physicists, and biologists. For example, on excitation these molecules undergo a variety of photochemical reactions that are of importance in biological systems. As examples, we cite the mechanisms of photosynthesis in plants and vision in animals[6]. Such molecules and macromolecules are also known to exhibit electrical conductivity properties that rival those of Cu[7]. Although there are no current examples of molecules exhibiting room temperature superconducting properties, π-conjugated (BETD-

0097–6156/96/0628–0116$15.00/0

TTF)$_2$X salts are known to exhibit superconducting properties at lower temperatures[8]. The possibility of a ferromagnetic ordering in such compounds has also been of recent interest[9].

For a satisfactory understanding of the electronic and optical properties of organic π-conjugated systems, a detailed knowledge of the electronic states is a prerequisite. There have been a number of reports on the quantum chemical description of the electronic states of these structures and their relation to chemical and physical properties[10-12]. The major outcome of these studies is the clear demonstration of strong electron correlations in these systems. Some of the manifest properties are the existence of optically forbidden states below the optically allowed states in finite polyenes[13], negative spin densities at even numbered carbon sites in neutral polyene radicals, etc[14]. It is clear then that correlations must be properly accounted for in modelling chromophores for NLO applications and for understanding the origin of large NLO responses frequently observed in these systems.

In any semi-empirical quantum theory[15], such as the Pariser-Parr-Pople (PPP), complete-neglect of differential overlap (CNDO), or intermediate neglect of differential overlap (INDO) approximations, the parameters describing the model can be tuned to reproduce experimental observables such as heat of formation, excitation energy etc.. This adjustment of the parameters is equivalent to introducing some electron correlation in the model. In order to exemplify this point we describe below the PPP hamiltonian in the second-quantized notation,

$$H_{PPP} = \sum_i \varepsilon_i n_i + \sum_{i\sigma} t_i (a_{i\sigma}^{+} a_{i+1\sigma} + h.c.) + \sum_i U_i n_i (n_i - 1)/2 + \sum_{i,j} V(R_{ij})(n_i - z_i)(n_j - z_j)/2 \qquad (1)$$

where $a_{i\sigma}^{+}(a_{i\sigma})$ creates (annihilates) an electron with spin σ in the ith atomic orbital, ϕ_i, ε_i is the site energy (Huckel α) of the ith site, $t_{i,i+1}$ is the nearest neighbor transfer integral (Huckel β), U_i (V_0) is the on-site, or Hubbard, repulsion associated with doubly filled ϕ_i and V(R), the coulomb interactions of electrons at two different sites, ϕ_i and ϕ_j, is usually interpolated between U at R=0 and e^2/R at $R \rightarrow \infty$. In the above Hamiltonian if the on-site repulsion integral, U, and the intersite repulsion integral, V, are set to zero one obtains an uncorrelated Huckel Hamiltonian. On-site correlations can be included by tuning the on-site repulsion integral, U to obtain the Hubbard Hamiltonian; when intersite repulsion is also introduced one obtains the PPP Hamiltonian. Thus by tuning the parameters of the PPP model Hamiltonian one can introduce some amount of correlations. However it should be noted that in the above process we have not taken into account the dynamic electron correlations which can be introduced explicitly using one of the methods described in the next section. A treatment that introduces correlations explicitly through configuration interaction is found to be necessary to account for the properties of the excited states such as those described in the previous paragraph[10-12]. Models including electron correlations produce results that are markedly different from those that do not. For example, the sign of the THG coefficient of unsubstituted polyenes computed from an uncorrelated model is opposite to that computed from a correlated model[10-12,16]. The ordering of the electronic low-lying electronic states from a correlated model is opposite to that from an uncorrelated model. Thus, a proper account of electron correlations is an absolute prerequisite for accurately computing the NLO properties of organic conjugated systems.

In this chapter we propose a method of including correlations through CI calculations that include configurations higher than singly excited configurations and we analyze the effect of these higher excited configurations on the linear and NLO response properties of two typical organic chromophores, namely *para*-nitroaniline (*p*NA) and 1,3,5,7-*trans*-octatetraene (*trans*-octatetraene). These two molecules are representative of broad classes of chromophores which exhibit large second and third order nonlinearities. These two chromophores are also chosen because there is a substantial database of experimental and theoretical results on electronic and optical properties available for comparison. In the next section, we briefly outline the method of performing CI calculations that include configurations of any arbitrary level of excitation using the INDO/1 Hamiltonian. In Section III we present the results of our calculations on the linear and NLO properties of these two archetypical chromophoric systems.

II. Computational Methodology

Major approaches for treating electron correlations include: (1) the perturbation theoretical approach; (2) the configuration interaction approach; and (3) the coupled-cluster method. A number of excellent treatises can be found in literature which describe these three approaches. We briefly outline the first two approaches and give more details on the proposed method[17]. Most of the perturbation theory that is utilized today for electronic structure calculations is of the Møller-Plesset (MPPT) type or of the Nesbet-Epstein type. The former is characterized by a zeroth-order Hamiltonian which is the sum of effective one-particle operators, such as the Fock operator, and from Brillouin's theorem the perturbation contains only bielectronic coulombic and exchange integrals. The Nesbet-Epstein theory is based on the partitioning of the CI matrix. The zeroth-order Hamiltonian in this case is characterized by the diagonal of the CI matrix. Although the MPPT approach is size-consistent in every order, it is not convergent for all values of the model parameters and hence, the infinite summation is usually truncated to any arbitrary order. The most commonly used procedures are the second (MP2) and the fourth (MP4) order perturbation corrections. Despite the size-consistency of the MPPT approach, the CI method is more commonly used to treat electron correlations, especially in semi-empirical methods, as it is variational and conceptually simple. In the CI method, N-electron states are expanded in terms of a basis set of all possible N-electron Slater determinants of all possible excitation orders, and the coefficients of the expansion are determined variationally. This procedure is called the full CI calculation and a CI calculation is size consistent only when such a calculation is performed. However, FCI calculations are not feasible even for moderately sized molecular systems as the number of configurations is enormous, and usually the CI expansion is truncated to an arbitrary level of excitation; singly excited (SCI) and singly and doubly excited (SDCI) CI calculations being the most common. Both these methods have been applied to estimate the NLO responses of organic π-conjugated systems, the former being successful only in estimating the first hyperpolarizability of donor-acceptor systems[18,19]. Whatever the order of the CI calculation, it is important to obtain a spin-symmetry adapted linear combination of the configurations so as to conserve the total spin and hence utilize the spin invariance of the Hamiltonian. This exploitation of the spin invariance reduces both the cpu times and the storage requirements in large scale computations and gives more easily interpreted results. Table I shows the advantage of a spin-symmetry adapted basis over the N-electronic Slater determinant in a full CI calculation.

A number of methods exist in the literature for generating spin-symmetry adapted basis sets[20]. These include, for example, the explicit diagonalization of the $\mathbf{S}^2$ (total spin) operator[20], the Löwdin projection operator technique[20], the symmetry group[21] and the unitary group approaches[22] and the Rumer spin pairing approach[23]. Of all the above procedures, we have chosen the Rumer spin pairing method or valence bond (VB) method as it is chemically intuitive and has been extensively used in many quantum chemical calculations. The next subsection briefly outlines the Rumer spin pairing method for generating spin-symmetry adapted basis.

A. Rumer Spin Pairing Method for Spin Symmetry Adaptation

Given a number of orbitals N (atomic or molecular), electrons N_e, and the spin S, one can generate all possible linearly independent Rumer diagrams $|i\rangle$ following the Rumer-Pauling rules[23]. In a Rumer diagram a "X" represents a doubly occupied orbital, a "." represents an empty orbital and a line between orbitals *i* and *j* represents a covalent bond between the two orbitals. The diagrams can be represented by an integer by associating two bits per orbital. For example a "X" is represented by the binary number "11", a "." by "00", a line beginning by "10" and a line ending by "01". Thus each of the Rumer diagrams can be generated and stored as positive integers $I_k < 2^{2N}$. In order to generate the linearly independent set of Rumer diagrams, we start with the lowest integer diagram, a diagram in which all the $N_e/2$ orbitals to the right are doubly occupied (the HF ground state in which all orbitals below the HOMO are doubly occupied), and systematically shift the bits to the left and check the validity of the diagram thus formed. The bit shifting is continued until the highest integer diagram, a diagram in which all the $(N-N_e/2)$ orbitals to the left are doubly occupied (all highest-energy orbitals) is reached. This procedure is extremely rapid. For example, the generation 226512 singlet diagrams in a FCI with 12 electrons and 12 orbitals requires less than a minute on a IBM RISC/6000-560 machine.

One of the drawbacks of Rumer CI procedure is that it is a non-orthogonal basis set and one has to obtain the overlap between the Rumer functions. This is done using the Pauling Island counting scheme[23] which uses the charge orthogonality of the Rumer functions. In this procedure, a nonvanishing S_{ij} requires two Rumer diagrams to have an identical occupancy in each orbital. For example, Rumer diagrams with one electron in all the orbitals (a purely covalent diagram in the VB representation) are orthogonal to any diagram which has more than one electron in any of the orbitals, and diagrams with all the orbitals doubly

Table. I. Number of Slater determinants and spin symmetry adapted basis (SSAB) in a full CI calculation involving N orbitals and N electrons

N	Slater Determinants	SSAB
4	36	20
6	400	175
8	4 900	1 764
10	63 504	19 404
12	853 776	226 512
14	11 778 624	2 760 615

occupied or those with only two orbitals singly occupied are orthogonal to all others. Thus by this procedure, instead of calculating a 226512 dimensional overlap matrix for a FCI calculation involving 12 orbitals and 12 electrons, we need to calculate only one 132, one 42, one 14, one 5 and one 2 dimensional submatrices.

Once a set of orthonormal basis is obtained, the CI matrix can be set up using the formulae for matrix elements of spin-independent operators between Rumer diagrams given by Cooper and Mc Weeny[24] and by Sutcliffe[25]. However, as mentioned earlier, a FCI calculation is feasible only for smaller systems with minimal basis set. One way to overcome this limitation is by choosing an active subspace of a particular symmetry and to perform a FCI calculation within that subspace. Since most molecules used in NLO applications are conjugated π-systems, one is prompted to choose an active space formed by orbitals with π-symmetry. This is further supported by the success of π-electron theories such as the Pariser-Parr-Pople models in describing the linear and NLO properties of various π-conjugated systems. In our calculations we first carry out a Hartree-Fock calculation using the INDO/1 Hamiltonian, and from the INDO/1 ground state, the orbitals of π-symmetry are selected and CI calculations are performed within this subspace. The level of CI, ranging from singles only (SCI) to a quadruply excited (SDTQCI) calculation, is systematically varied to demonstrate the importance of higher excited configurations. For all CI levels beyond SCI the ground state is explicitly correlated.

One of the major concerns of including higher excited configurations in a CI calculation within the framework of a semi-empirical method is the transferability of the parameters. The parameters in most semiempirical models are chosen based on an SCI level to match experiment. The ground state is not correlated in a SCI calculation. Thus by tuning the values of the parameters to reproduce an experimentally observed variable, usually the optical absorption maximum, one introduces some amount of electron correlation into the model Hamiltonian itself. Now when these parameters are used in an higher level study where correlations are explicitly included through CI, it would mean that electron correlations are in a sense double counted. In standard semiempirical models, the only parameters that can include electron correlations are the two-electron repulsion integrals. To remove the inherently built-in electron correlations used in the model one has to modify this two-electron integral. One might choose this to be equal to the value used in *ab initio* calculations. However, from our own previous experience and from other studies[16] we have chosen to use a value of 10.33 eV instead of 11.11 eV used in the original INDO/1 Hamiltonian. It is interesting to note that a reduced value of 10.33 eV is also the value of the two-eletron repulsion integral calculated by the uniformly-charged-sphere approximation with the assumption that the two electrons occupy different lobes of the p atomic orbital and their mutual repulsion is then given by $(e^2/4R^2)$[15]. An even lower value of 9.87 eV was suggested by Julg[26] who used the original approach of Pariser in which the two-eletron repulsion integral is calculated from the ionization potential and the electron affinity. In addition, he also made allowance for the changes in the size of the $2p_z$ orbital with changes in the number of electrons in it and for the polarization of the core. A good description of the methods of calculating the two electron repulsion integral can be found in the book of Dewar[15]. The effect of this reduced two-electron repulsion integral on the linear and NLO properties will also be studied.

B. Methods of Computing NLO Properties

There are a number of procedures to compute the NLO response properties of organic molecules. Some of the commonly used methods are the finite-field self-consistent field (FF-SCF) or coupled perturbed Hartree Fock (CPHF) method[27], the sum-over-states (SOS) methods[28], the time dependent Hartree Fock (TDHF) method[29], and the Correction Vector approach[11,16,30]. In the FFSCF method, an additional term equal to **μ.E**, describing the interaction between an external electric field and the elementary charges constituting the molecule is added to the molecular Hamiltonian. At the restricted Hartree-Fock (RHF) level, the one-electron orbitals are self-consistent eigenfunctions of a one-electron field-dependent Fock operator, consisting of the field-free Fock operator $h_0(r)$ and a field term:

$$h(r) = h_0(r) - e\boldsymbol{r}\cdot\boldsymbol{E} \tag{2}$$

In the matrix representation of h(r), besides the usual integrals, one-electron moment integrals <p|r|q>, the calculation of which is fairly standard, also appear. The solution of h(r) yields the field-dependent density matrices, energies, and dipole moments. The NLO properties are evaluated by taking the derivatives of the field-dependent dipole moments with respect to the applied electric field. This method has been used in computing the NLO response properties of molecules of widely varying size[31]. It should be noted that in this procedure, the entire SCF procedure and the CI calculation[32] must be performed at each value of the external electric field required in the differentiation scheme. This becomes computationally tedious and the differentiation procedure can be numerically unstable for second hyperpolarizability calculations[32]. Moreover, the frequency dependence of the NLO coefficients cannot be taken into consideration in this procedure.

The SOS method relies on a direct summation of the perturbation expression for the NLO properties (detailed expressions for the NLO coefficients to the fourth order can be found elsewhere)[28]. In the general procedure, used mostly in the context of semi-empirical methods, one computes the approximate many-body electronic ground state (usually the HF ground state) of the molecule with an antisymmetrized product of one-electron eigenfunctions of the valence electrons. Correlations are accounted for within a limited CI scheme, usually SCI, that includes determinants formed by exciting electrons from occupied valence orbitals to unoccupied virtual orbitals. From the CI eigenfunctions and the eigenvalues, molecular NLO coefficients are evaluated using the SOS expressions. This is the most widely used method for computing frequency-dependent NLO responses and for analyzing the molecular origin of the large NLO responses of organic and organometallic chromophores within semi empirical approaches[33].

This procedure is usually very time consuming as one has to compute the transition moments between the various CI eigenstates, and the summation in the SOS expression must be carried out over all the excited states. The summation is often truncated at an arbitrary number of low-lying excited states when a desired convergence in the computed NLO property is obtained. This procedure, however, is useful in determining the states contributing to the final value of the NLO coefficients and in interpreting experimentally observed values.

In the correction vector approach, one computes the first and second order correction vectors, $\phi^{(1)}$, $\phi^{(2)}$, defined by the equations (for details see refs. 16 and 30)

$$(H - E_G + \hbar\omega_1 + i\Gamma)\,\phi_i^{(1)}(\omega_1) = \mu_i|G\rangle \tag{3}$$

$$(H - E_G + \hbar\omega_2 + i\Gamma)\phi_{ij}^{(2)}(\omega_1,\omega_2) = \mu_j\phi_i^{(1)}(\omega_1) \tag{4}$$

where E_G is the ground state energy after CI, the ω's are the excitation frequencies, μ is the dipole displacement operator and Γ is the average lifetime of the excited states. Equations 2 and 3 can be solved by expanding the correction vectors in the basis of the configuration functions. Since the CI, the dipole, the overlap matrices, and the CI wavefunctions are also constructed in the basis of the configuration functions, the operator equations (2) and (3) can be cast into the following set of matrix equations by matching coefficients,

$$\sum_j [H_{ij} - (E_G - \hbar\omega_1 - i\Gamma)S_{ij}]C_{j\alpha} = \sum_j (\mu_{\alpha,ij} - \langle G|\mu_\alpha|G\rangle S_{ij})g_j \tag{5}$$

where c_{j1} and g_j are the expansion coefficients of $\phi^{(1)}$ and |G>, H_{ij} and S_{ij} are the CI and overlap matrices, and α is the coordinate axis of the dipole operator. Similar equations for the second order correction vector $\phi^{(2)}$ can also be generated. The above set of linear inhomogeneous equations is solved using the small matrix algorithm proposed by Ramasesha[34]. Once the first and second order correction vectors are known, NLO coefficients up to third order can be readily written as

$$\alpha_{ij} = \langle\phi_i^{(1)}(\omega)|\mu_j|G\rangle + \langle G|\mu_i|\phi_j^{(1)}(-\omega)\rangle \tag{6}$$

$$\beta_{ijk} = (8)^{-1}\,\wp\langle\phi_i^{(1)}(-2\omega)|\mu_j|\phi_k^{(1)}(-\omega)\rangle/8 \tag{7}$$

where ℘ is the permutation operator implying the additional terms which are obtained when the coordinates and the frequencies are permuted. The 2nd hyperpolarizability γ, in terms of the first and second order correction vectors $\phi_i^{(1)}$ and $\phi_{ij}^{(2)}$, is written as

Table II. Linear and nonlinear optical properties of *trans*-octatetraene at an excitation energy of 0.65 eV

CI Order	ΔE $(2^1A_g)^*$ (eV)		ΔE $(1^1B_u)^{**}$ (eV)		α_{xx} (10^{-23} esu)$^+$		γ_{xxxx} (10^{-36} esu)$^@$	
	Γ_C^a	Γ_C^b	Γ_C^a	Γ_C^b	Γ_C^a	Γ_C^b	Γ_C^a	Γ_C^b
SCI	5.32	5.22	4.29	4.57	33.64	35.45	22.86	22.19
SDCI	4.62	4.58	5.37	4.99	16.07	18.72	33.62	42.89
SDCTCI	4.46	4.42	5.12	4.79	17.50	20.51	28.63	36.53
SDCTQCI	4.37	4.34	5.15	4.81	16.93	19.57	29.43	37.60
FCI	4.36	4.34	5.15	4.81	16.93	19.57	29.36	37.54

* Experimental value of ΔE (2^1A_g) is 3.97 eV (Ref. 13).

** Experimental value of ΔE (1^1B_u) is 4.42 eV (Ref. 13).

[a] The two electron repulsion integral, Γ_C, is taken to be 11.11 eV.

[b] A reduced value (10.33 eV) is assumed for the two electron integral. See text for details.

[+] The 4-31G *ab initio* value of α_{xx} at an excitation energy of 0.65eV is 38.90 X 10^{-23} esu[38]

[@] The 4-31G *ab initio* value of γ_{xxxx} at an excitation energy of 0.65eV is 39.49 X 10^{-36} esu[38]

$$\gamma_{xxxx} = (8)^{-1}[\langle\phi_x^{(1)}(-3\omega)|\mu_x|\phi_{xx}^{(2)}(-2\omega,-\omega)\rangle + \langle\phi_{xx}^{(2)}(2\omega,\omega)|\mu_x|\phi_x^{(1)}(-\omega)\rangle + \omega \rightarrow -\omega] \quad (8)$$

where $\omega \rightarrow -\omega$ indicates the same matrix elements with new arguments. This procedure of computing the NLO coefficients is extremely efficient. For example, the entire calculation of the polarizability, SHG, and THG coefficients, at two different excitation energies of *p*NA, with all singles and doubles generated from an active space of 10 π-orbitals with 12 π-electrons spanning a 325 dimensional Hilbert space, requires less than 5 minutes on an IBM RISC/6000-560 workstation.

III. Results and Discussion

The geometries of the molecules studied here were optimized using the AM1 Hamiltonian in the MOPAC software package[35]. The molecules are planar, and the coordinate system has been chosen so that the molecules lie on the x-y plane with the x-axis directed along the molecular backbone (long axis). In this configuration, the dominant components are α_{xx} for the polarizability, β_{xxx} for the SHG coefficient, and γ_{xxxx} for the THG coefficient. The electronic structure calculation involves a set of CI calculations the order of which ranges from SCI to FCI between a set of chosen active orbitals. In the case of SCI and SDCI calculations, all singly and doubly excited configurations between all the π-molecular orbitals, 8 orbitals for *trans*-octatetraene, and 10 in the case of *p*NA, were used in the CI calculation. In the case of higher order CI, the singles and doubles were the same as used in the SCI and SDCI calculations, and the higher excited configurations were generated between three HOMO's and three LUMO's. The two-electron integrals were parametrized using the Mataga-Nishimoto approximation[36] for the SCI calculations and the Ohno-Klopman approximation[37] for all other CI calculation. This choice of parametrization was based on our previous calculations of the NLO properties of organic molecules[16].

Tables II and III present the results of our calculations on the electronic and optical properties of *trans*-octatetraene and *p*NA, respectively. The Tables include the computed values of the excitation energies of the lowest dipole allowed states, the dominant components of the frequency dependent polarizability, as well as SHG and THG coefficients at an excitation energy of 0.65 eV. In the case of *trans*-octatetraene, we also report the excitation energy of the two photon 2^1A_g state. All the above properties are examined as a function of increasing order of CI used in the calculation and at two different values of the two-electron repulsion integral. In the following paragraphs the results for the individual molecules are discussed.

A. *Trans*-Octatetraene

It can be seen from Table II that electron correlation at least at the level of SDCI is required to reproduce the experimentally observed order of the excited states, as concluded by several other studies[10-12,16]. This is true for both values of the two-eletron repulsion integral. It is interesting to note that the excitation energy of the 1^1B_u state from the SCI calculation (4.29 and 4.57 eV from the two parametrizations) is remarkably close to the experimentally observed value of 4.40

Table III. Linear and nonlinear optical properties of *para*-nitroaniline at an excitation energy of 0.65 eV

CI Order	$\Delta E\ (^1B_2)^a$ (eV)		α_{xx} (10^{-23} esu)[b]		β_x (10^{-30} esu)[c]		γ_{xxxx} (10^{-36} esu)[d]	
	Γ_C^*	Γ_C^{**}	Γ_C^*	Γ_C^{**}	Γ_C^*	Γ_C^{**}	Γ_C^*	Γ_C^*
SCI	3.93		10.42	-	11.57	-	46.12	-
SDCI	4.68	4.59	5.97	6.29	4.58	4.54	16.46	16.26
SDCTCI	4.59	4.52	6.31	6.61	5.54	5.42	18.12	17.42
SDCTQCI	4.59	4.52	6.24	6.55	5.38	5.29	18.09	17.43
FCI	4.59	4.52	6.25	6.54	5.36	5.26	18.07	17.42

[a] Experimental value of $\Delta E\ (^1B_2)$ is 3.40 eV in acetone (Ref. 40).
[b] Experimental value of α_{av} in acetone is 1.7×10^{-23} esu (Ref. 40). Double zeta *ab initio* value is 20.54×10^{-23} esu[42a].
[c] Experimental value of β_x in acetone is 9.2×10^{-30} esu (Ref. 40), but this should[41] be 5.34×10^{-30} esu. Double zeta *ab initio* values are 4.84[42a] and 5.01×10^{30} esu[42b].
[d] Experimental value of γ_{scalar} in acetone is 15×10^{-36} esu (Ref. 40). Double zeta *ab initio* value is 8.45×10^{-36} esu[42a]
* The two electron one center integral, Γ_C, is taken to be 11.11 eV.
** A reduced value of 10.33 eV for the two electron one centre integral, Γ_C, is taken. See text for details.

eV[13]. This is because the INDO/1 Hamiltonian has been parametrized at the SCI level to reproduce the optical absorption spectra of organic molecules. The SCI calculation, does not however, reproduce the excitation energy of the two photon (2^1A_g) state. This is because, while the contribution of the singly excited configuration (HOMO->LUMO) to the 1^1B_u state is 95%, the contribution of the singly excited configuration (HOMO-1 -> LUMO and HOMO -> LUMO+1) to the 2^1A_g is only 33%, and the major contribution (60%) comes from a doubly excited configuration (HOMO,HOMO -> LUMO, LUMO). This is essentially why none of the SCI calculations reproduce the correct ordering of the 2^1A_g and 1^1B_u states[10,16].

Even though the SDCI calculation reproduces the correct ordering of the electronic states in *trans*-octatetraene, the inclusion of higher excited configurations does quantitatively affect the linear and NLO response properties. We see that configurations at least at the level of quadruples, beyond which there are no significant changes in the linear and NLO properties, are necessary to attain stability in the electronic and optical properties. This is true in both cases of parametrization used in this study. The effect of a reduced two-electron integral is to remove some amount of electron correlation that is inherently built into the INDO/1 Hamiltonian as described above. This reduction in the two-electron integral provides much better agreement with the experimentally observed excitation energies for higher order CI, as expected. Although there are no available experimental values of the frequency-dependent polarizability and THG coefficient of *trans*octatetraene, results from *ab initio* calculation using 4-31G and semi-diffuse p and d basis function are available[38]. The dominant component of the frequency-dependent polarizability and the THG coefficient at an excitation energy of 0.65 eV are given in Table II for comparison. A direct comparison of the results from *ab initio* and the correction vector calculations is impeded by the different conventions used in literature in defining the NLO coefficients[39]. The difference in the conventions used in the perturbative correction vector approach and in the Taylor series based CPHF method can be easily understood by writing the energy expansion from the two methods. The Taylor series expansion of the energy in the presence of a electric field can be written as

$$E_{Taylor} = E_0 + \mu_i F_i + 1/2!\alpha_{ij} F_i F_j + 1/3!\beta_{ijk} F_j F_j F_k + 1/4!\gamma_{ijkl} F_i F_j F_k F_l + \dots\dots\dots \quad \textbf{(9)}$$

and the corresponding perturbation expansion of the energy can be written as

$$E = E_0 + \mu F + \alpha_{ij} F_i F_j + \beta_{ijk} F_i F_j F_k + \gamma_{ijkl} F_i F_j F_k F_l + \dots\dots \quad \textbf{(10)}$$

Comparing the two equations we see that the NLO coefficients from the correction vector method differs from those from the *ab initio results* in the Taylor expansion coefficients. Thus for example the polarizability from the correction vector approach is half that from the *ab initio* calculated polarizability and the THG coefficient differs by a factor of 24. A comparison of the two results, after taking into account the difference in the convension, shows that while the polarizability from the two approaches is in excellent agreement the THG coefficients differs by a larger extent. Such a difference in the THG coefficients computed from semi-empirical and *ab initio* methods have been reported earlier on the smaller homologue 1,3,5-hexatriene. However the available experimental results for the smaller homologue was supportive of the results obtained from the semi-empirical CNDO method. It is also known that the higher order NLO coefficients from *ab initio*-TDHF or CPHF calculations are always too small compared to the experimental value.

B. *Para*-Nitroaniline

In contrast to *trans*-octatetraene, we see that the change in electronic and optical properties are minimal after the SDCI level, although some changes in the SHG and THG coefficients are observed when going from the SDCI to the SDTCI level. The overall agreement of the computed linear and NLO properties of *p*NA at the SDTCI level with the experimentally observed properties is good (Table III). It should be noted that even though the literature experimental value[40] of the vector component of the SHG coefficient is 9.2 10^{-30} esu, it must be multiplied by 0.58 to account for the recent change in the d_{33} value of the quartz[41]. This yields a value 5.34 10^{-30} esu which is in excellent agreement with our computed value of 5.42 10^{-30} esu at the SDTCI level. There are also a few *ab initio* calculated NLO response properties of *p*NA available in literature[42]. A comparison, after accounting for the difference in the two convensions, of the INDO NLO response properties of *p*NA with the *ab initio* results is in general poor, despite the good agreement between the INDO and experimental results. However it is known that the *ab initio*-TDHF results of β and γ are always too small compared to experiment[43]. This has been attributed to the neglect of electron correlations in the *ab initio* calculations. This is further supported by the fact the MP2 calculated β_{vec} is about twice that from the HF value[44]. Our polarizability value, however appears to be somewhat overestimated. Again we see that a reduced value of the two-electron integral yields much better agreement with the experimentally observed linear and NLO properties for higher order CI calculations.

III Conclusions

Using a set of CI calculations that include configurations of varying levels of excitations, ranging from singles to FCI, in this chapter, we have demonstrated the importance of the higher excited configurations in describing the linear and NLO properties of two archetypical π-conjugated chromophores. We find that configurations at least at the level of double excitation are important in accurately reproducing the electronic and optical properties of organic π-conjugated molecules, both qualitatively and quantitatively. However, when correlations are explicitly included through CI calculations, the two-electron integrals, which are parametrized at the SCI level, must be reparametrized to remove the inherently built-in electron correlations in a semi-empirical model Hamiltonian. We find that a reduced value of the two-eletron repulsion integral gives good agreement with experimentally observed linear and NLO response properties. A similar conclusion was arrived at by Prasad and Karna from a INDO/SCI and INDO/SDCI calculation of the NLO response properties of NTE (1-nitro-2-thiophene-ethylene) and NTB (1-nitro-4-thiophene-butadiene)[43]. These authors have concluded that, an extended set of basis functions capable of describing the valence and diffuse states in the semi-empirical formalism and also including larger CI space with higher excited configurations, which were absent till very recently, are essential.

We and a number of other groups have studied the β-response properties of organic and organometallic chromophores as well as molecular clusters using appropriate PPP or INDO/S Hamiltonians, with the SOS method including only singly-excited configurations. For a wide variety of chromophores exhibiting fairly large β-responses this method works well and it serves as a very useful correlative, interpretive, and predictive tool; for molecules with only weak NLO response, or for the weaker tensor components of β, it fails badly. This is the expected behavior since accurate representation of weak response requires both basis set and correlation treatment well beyond the SOS-SCI semi-empirical level.

In the present contribution we have shown that when more complete treatments of CI are included, appropriate semi-empirical models can yield linear and NLO responses that are essentially quantitative (at least in the favorable cases analyzed here). Thus appropriate semi-empirical electronic structure models treated with suitable levels of correlations can yield useful interpretations of the response of π-electron species at either a semi-quantitative (SCI) or quantitative (SDTCI) level.

Acknowledgements

This research was sponsored by the NSF through the Northwestern Materials Research Center (Grant DMR-9120521) and by the AFOSR (Contract 94-1-0169).

We thank Professor M. C. Zerner for the original ZINDO program. We thank Professor S. Ramasesha for the modifications to the ZINDO program required in a correction vector calculation of the NLO coefficient using the Rumer CI procedure.

References

1. *Nonlinear Optical Properties of Organic Molecules and Crystals*, Edited by J. Zyss and D. S. Chemla (Academic Press, New York 1987), Vols. 1 and 2.
2. P. N. Prasad and D. J. Williams, *Introduction to Nonlinear Optical Effects in Molecules and Polymers* (Wiley Interscience, New York, 1990).
3. *Polymers for Second-Order Nonlinear Optics*, Edited by G. A. Lindsay and K. D. Singer, ACS Symposium Series 601, (American Chemical Society, Washington DC, 1995).
4. *Molecular Nonlinear Optics: Materials, Physics and Devices*, Edited by J. Zyss, Quantum Electronics Principles and Application Series, (Academic Press, New York, 1994).
5. *Nonlinear Optical and Electroactive Polymers*, Edited by P. N. Prasad and D. Ulrich, (Plenum Press, New York, 1988).
6. J. E. Darnell, H. F. Lodish, and D. Baltimore, *Molecular Cell Biology*, (Scientific American Books, W. H. Freeman, New York, 1990).

7. *Handbook of Conducting Polymers*, Edited by T. A. Skotheim (Marcel Dekker, New York, 1986) Vol. 1 and 2.
8. W. A. Little, *Phys. Rev. A* **134**, 1416, (1964).
9. (a) J. S. Miller, A. J. Epstein, and W. M. Reiff, *Chem. Rev.*, **88**, 201, (1988)(b). J. S. Miller and A. J. Epstein, *Angew. Chem. Int. Ed. Engl.*, **33**, 385, (1994). (c) O. Kahn, *Molecular Magnetism* (VCH, New York, 1993).
10. P. Tavan and K. Schulten, *J. Chem. Phys.*, **85**, 6602, (1986); *Phys. Rev. B*, **36**, 4337, (1987); K. Schulten, I. Ohmine, and M. Karplus, *J. Chem. Phys.*, **64** 4422, (1976).
11. Z. G. Soos and S. Ramasesha, *Phys. Rev. Lett.*, **52**, 2374, (1983); S. Ramasesha and Z. G. Soos, *J. Chem. Phys.*, **80**, 3278, (1986); **90**, 1067, (1989).
12. J. R. Heflin, K. Y. Wong, O. Zamini-Kamiri, and A. F. Garito, *Phys. Rev. B*, **38**, 1573, (1988).
13. B. S. Hudson, B. E. Kohler, and K. Schulten, *Excited States*, Edited by E. C. Lim (Academic Press, New York, 1982) Vol. 6 p. 1. M. F. Granville, G. R. Holtom and B. E. Kohler, *J. Chem. Phys.*, **72**, 4671, (1980). K. L. D' Amico, C. Manos, and R. L. Christiansen, *J. Amer. Chem. Soc.*, **102**, 1777, (1980).
14. H. Thoman, L. R. Dalton, Y. Tomkeiwicz, N. J. Shiren, T. C. Clarke, *Phys. Rev. Lett.*, **50**, 553, (1983).
15. Instead giving the original references we have given the reference some excellent monographs on these semi-empirical methods. The original references can be found in these texts. J. A. Pople and D. L. Beveridge, *Approximate Molecular Orbital Theory*, (McGraw-Hill, New York, 1970). G. A. Segal, *Semiempirical Methods of Electronic Structure Calculation*, (Plenum Press, New York, 1977). J. Sadlej, *Semi-Empirical Methods of Quantum Chemistry*, (Halsted Press, New York, 1985). M. J. S. Dewar, *The Molecular Orbital Theory of Organic Chemistry* (Mc Graw Hill, 1969). Also see M. A. Ratner, in *Structure and Dynamics of Atoms and Molecules: Conceptual Trends*, ed. J. L. Calais and E. S. Kryachko (Kluwer Academic, 1995).
16. I. D. L. Albert, J. O. Morley and D. Pugh *J. Chem. Phys.*, **99**, 5197, (1993), **102**, 237, (1995). B. Pierce, *J. Chem. Phys.*, **91**, 791, (1989).
17. A. Szabo and N. S. Ostlund, *Modern Quantum Chemistry: An Introduction to Advanced Electronic Structure Theory* (McMillan, New York, 1982). I. Shavitt, in *Methods of Electronic Structure Theory*, Edited by H. F. Schaefer III (Plenum, New York, 1977) Ch. 6.
18. D. Li, T. J. Marks, and M. A. Ratner, *Chem. Phys. Lett.*, **131**, 370, (1986). D. Li, M. A. Ratner, and T. J. Marks, *J. Amer. Chem. Soc.*, **110**,1704,(1988). S. J. Lalama and A. F. Garito, *Phys. Rev. A*, **20**, 1179, (1979). C. W. Dirk, R. J. Tweig, and G. Wagniere, *J. Amer. Chem. Soc.*, **108**, 5387, (1986). V. J. Docherty, D. Pugh, and J. O. Morley, *J. Chem. Soc. Faraday Trans. 2*, **81**, 1179, (1985). D. R. Kanis, M. A. Ratner, and T. J. Marks, *Chem. Rev.*, **94**, 195, (1994).

19. I. D. L. Albert, J. O. Morley, and D. Pugh, *J. Chem. Soc. Faraday Trans. 2*, **90**, 2617, (1994). *J. Phys. Chem.*, **99**, 8024, (1995).
20. R. Puancz, *Spin Eigen Functions: Construction and Use* (Plenum, New York, 1979).
21. W. Duch and J. Karwowski, *Comput. Phys. Rep*, **2**, 93, (1985).
22. J. Paldus, *J. Chem. Phys.*, **61**, 5321, (1974). I. Shavitt, C. F. Bender, A. Pipano, and R. P. Hosteny, *J. Comput. Phys.*, **11**, 90, (1973). M. A. Robb and U. Niazi, *Comput. Phys. Rep.*, **1**, 127, (1984).
23. G. Rumer, *Gottingen Nach.*, 337, (1932). L. Pauling and J. Sherman *J. Chem. Phys.*, **1**, 606, (1933). L. Pauling and G. W. Wheland *J. Chem. Phys.*, **1**, 362, (1933). S. Ramasesha and Z. G. Soos *Int. J. Quantum Chem.*, **25**, 1004, (1984).
24. I. L. Cooper and R. McWeeny *J. Chem. Phys.*, **45**, 226, (1966).
25. B. T. Sutcliffe *J. Chem. Phys.*, **45**, 235, (1966).
26. A. Julg, *J. Chim. Phys.*, **55**, 413, (1958).
27. H. D. Cohen and C. C. J. Roothaan *J. Chem. Phys.*, **43**, S34, (1965). A. Schweig, *Chem. Phys. Lett.*, **1**, 163, (1967). J. Zyss, *J. Chem. Phys.*, **70**, 3333, 3341, (1979).
28. J. F. Ward *Rev. Mod. Phys.*, **37**, 1, (1965). B. J. Orr and J. F. Ward *Mol. Phys.*, **20**, 513, (1971).
29. H. Sekino and R. J. Bartlett, *Int. J. Quantum Chem.*, **43**, 119, (1992). *J. Chem. Phys.*, **85**, 976, (1986). J. E. Rice, R. D. Amos, S. M. Colwell, N. C. Handy, and J. Sanz *J. Chem. Phys.*, **93**, 8828, (1990). J. E. Rice and N. C. Handy, *Int. J. Quantum Chem.* **43**, 91, (1992). S. P. Karna and M. Dupius *J. Comput. Chem.*, **12**, 487, (1991). *Chem. Phys. Lett.* **171**, 201, (1990). J. Yu and M. C. Zerner *J. Chem. Phys.*, **100**, 7487, (1994).
30. S. Ramasesha, Z. Shuai and J. L. Bredas, *Chem. Phys. Lett.* (in press). E. N. Svendsen, C. S. Willand, and A. C. Albrecht *J. Chem. Phys.*, **98**, 5760, (1985).
31. F. Meyers, C. Adant, and J. L. Bredas, *J. Am. Chem. Soc.* **113** 3715, (1991). F. Meyers and J. L. Bredas, *Int. J. Quantum Chem.*, **42**, 1595, (1992). J. L. Bredas, F. Meyers, B. M. Pierce, and J. Zyss, *J. Am. Chem. Soc.*, **114**, 4928, (1992) T. Tsunekawa and K. Yamaguchi, *J. Phys. Chem.* **96** 10268, (1992).
32. I. D. L. Albert and S. Ramasesha, *Mol. Cryst. Liq. Cryst.*, **168**, 95, (1989). S. Ramasesha and I. D. L. Albert, *Chem. Phys. Lett.*, **154**, 501 (1989). S. Ramasesha and I. D. L. Albert, *Phys. Rev. B*, **42**, 8587 (1990).
33. A. Ulman, C. S. Willand, W. Kohler, D. R. Robello, D. J. Williams, and L. Handley, *J. Am. Chem. Soc.*, **112**, 7083, (1990). D. Li, M. A. Ratner, and T. J. Marks, *J. Phys. Chem.*, **96**, 4325, (1992). D. R. Kanis, M. A. Ratner, and T. J. Marks, *J. Am. Chem. Soc.*, **114**, 10338, (1992). J. O. Morley, P. Pavlides, and D. Pugh, *Int. J. Quantum. Chem.*, **43**, 7, (1992). A. Ulman, *J. Phys. Chem.*, **92**, 2385, (1988). D. R. Kanis, M. A. Ratner, T. J. Marks, and M. C. Zerner, *Chem. Mater.*, **3**, 19, (1991). M. Jain and J. Chandrasekhar, *J. Phys. Chem.*, **97**, 4044, (1993). S. Di Bella, M. A. Ratner, and T. J. Marks, *J. Am. Chem. Soc.*, **114**, 5842, (1992).

34. S. Ramasesha *J. Comput. Chem.*, **11**, 545, (1990).
35. J. J. P. Stewart *J. Comput. Aided Mol. Des.*, **4**, 1, (1990). *J. Comput. Chem.*, **5**, (1987).
36. N. Mataga and K. Nishimoto, *Z. Phys. Chem.*, **13**, 140, (1957).
37. K. Ohno, *Theor. Chim. Acta*, **2**, 219, (1964). G. Klopman, *J. Amer. Chem. Soc.*, **86**, 4550, (1964).
38. S. P. Karna, G. B. Talapatra, W. M. K. P. Wijekoon, and P. N. Prasad, *Phys. Rev. A*, **45**, 2763 (1992).
39. A. Willets, J. E. Rice, D. M. Burland, and D. P. Shelton, *J. Chem. Phys.*, **97**, 7590, (1992).
40. L-T. Cheng, W. Tam, S. H. Stevenson, G. R. Meridith, G. Rikken, and S. R. Marder, *J. Phys. Chem.*, **95**, 10631, (1991).
41. R. C. Eckardt, H. Masuda, Y. X. Fan and R. L. Byer, *IEEE J. Quantum Electron.*, **26**, 922 (1990).
42. S. P. Karna, P. N. Prasad, M. Dupuis, *J. Chem. Phys.*, **94**, 1171, (1991); H. Agren, O. Vahtras, H. Koch, P. Jorgensen, T. Helgaker, *J. Chem. Phys.*, **98**, 6417, (1993).
43 P. N. Prasad and S. P. Karna, *Int. J. Quantum Chem.*, **S28**, 395, (1994).
44. F. Sim, S. Chin, M. Dupuis, J. E. Rice, *J. Phys. Chem.*, **97**, 1158, (1993).

RECEIVED December 29, 1995

Chapter 7

Frequency-Dependent Polarizabilities and Hyperpolarizabilities of Polyenes

Prakashan Korambath[1] and Henry A. Kurtz

Department of Chemistry, University of Memphis, Memphis, TN 38152

Polarizabilities (α) and second hyperpolarizabilities (γ) as a function of frequency are calculated for a series of polyenes, $H(C_2H_2)_nH$ with n ranging from 2 to 20, by the TDHF method with AM1 parameterization. For the second hyperpolarizabilities, third harmonic generation, electric-field induced second harmonic generation, and intensity dependent refractive index quantities are calculated. The frequency dependencies are discussed and comparisons made amongst the different γ values. The saturation behavior of these quantities is also examined and limiting values for α and γ per subunit are computed.

Several computational studies have been done to explore the behavior of the nonlinear optical properties of polyenes as the length increases. Many early papers used a power-law expression (an^b) to fit the polarizability and hyperpolarizability *(1,2)*. Such a power-law behavior with a constant exponent is not adequate to describe the limiting behavior of polyenes. Both experimental *(3)* and theoretical *(4)* evidence have shown that the polarizability and second hyperpolarizability approach linearity with large numbers of subunits, i.e. the exponent approaches unity. The behavior is usually indicated by examining the value/subunit -- which approaches a constant at large n. Almost all previous theoretical work examined only the static values of the polarizability and second hyperpolarizability. The goal of this study is to examine the behavior of the frequency dependent quantities.

[1]Current address: Department of Chemistry, Ohio State University, Columbus, OH 43210

0097–6156/96/0628–0133$15.00/0

Method

Each $H(C_2H_2)_nH$ oliogmer, for n = 1 to 20, was first fully optimized using the AM1 *(5)* parameter set and then the properties calculated. The frequency dependent optical properties were calculated using the TDHF *(6,7)* method which has been implemented by us in both MOPAC *(8)* and GAMESS *(9)*. The properties of interest are the polarizability (α), first hyperpolarizability (β), and second hyperpolarizability (γ). These quantities can be defined from a series expansion of the response of a system to an external electric field (F) as

$$E(F) = E(0) - \sum_i \mu_i F_i - \frac{1}{2!}\sum_{i,j} \alpha_{ij} F_i F_j - \frac{1}{3!}\sum_{i,j,k} \beta_{ijk} F_i F_j F_k - \frac{1}{4!}\sum_{i,j,k,l} \gamma_{ijkl} F_i F_j F_k F_l - \cdots \quad (1)$$

where the indices i, j, k, and l run over the Cartesian components (x, y, and z). One or more of the external fields are usually frequency dependent (i.e. lasers) and each combination of different fields leads to slightly different values for the coefficients in Equation 1. This leads to a classification of the different optical properties of interest which correspond to different experimental situations. The quantities calculated by our programs are listed in Table I.

For a detailed definition of this quantities, see a text book on nonlinear optics such as the one by Boyd *(10)*.

Table I. Quantities Calculated by MOPAC and GAMESS

Polarizability:		
	$\alpha(-\omega;\omega)$	Frequency Dependent Polarizability
First Hyperpolarizabilities:		
	$\beta(-2\omega;\omega,\omega)$	Second Harmonic Generation (SHG)
	$\beta(-\omega;\omega,0)$	Electroptic Pockels Effect (EOPE)
	$\beta(0;\omega,-\omega)$	Optical Rectification (OR)
Second Hyperpolarizabilities:		
	$\gamma(-3\omega;\omega,\omega,\omega)$	Third Harmonic Generation (THG)
	$\gamma(-2\omega;\omega,\omega,0)$	Electric Field Induced Second Harmonic (EFISH)
	$\gamma(-\omega;\omega,\omega,-\omega)$	Intensity Dependent Refractive Index (IDRI) or Degenerate Four-Wave Mixing (DFWM)
	$\gamma(-\omega;\omega,0,0)$	Optical Kerr Effect (OKE)

The programs automatically provide all components of α, β, and most of γ in whatever molecular coordinate system the molecule was input. In order to provide a comparison with other work and to provide a unique set of data, we present our results as "averaged" values according to the following definitions.

$$\alpha = \frac{1}{3}\left(\alpha_{xx} + \alpha_{yy} + \alpha_{zz}\right) \quad (2)$$

$$\gamma^{THG} = \frac{1}{5}\left\{\gamma_{xxxx} + \gamma_{yyyy} + \gamma_{zzzz} + 2\left(\gamma_{xxyy} + \gamma_{yyzz} + \gamma_{yyzz}\right)\right\} \tag{3}$$

$$\begin{aligned}\gamma^{EFISH} = \frac{1}{15}\{&3(\gamma_{xxxx} + \gamma_{yyyy} + \gamma_{zzzz}) \\ &+ 2(\gamma_{xxyy} + \gamma_{xxzz} + \gamma_{yyzz} + \gamma_{yyxx} + \gamma_{zzxx} + \gamma_{zzyy}) \\ &+ (\gamma_{xyyx} + \gamma_{xzzx} + \gamma_{yxxy} + \gamma_{yzzy} + \gamma_{zxxz} + \gamma_{zyyz})\}\end{aligned} \tag{4}$$

$$\begin{aligned}\gamma^{IDRI} = \frac{1}{15}\{&3(\gamma_{xxxx} + \gamma_{yyyy} + \gamma_{zzzz}) \\ &+ 2(\gamma_{xyxy} + \gamma_{xzxz} + \gamma_{yzyz} + \gamma_{yxyx} + \gamma_{zxzx} + \gamma_{zyzy}) \\ &+ (\gamma_{xxyy} + \gamma_{xxzz} + \gamma_{yyzz} + \gamma_{zzxx} + \gamma_{zzxx} + \gamma_{zzyy})\}\end{aligned} \tag{5}$$

Values of the first hyperpolarizability (β) will not be reported in this work as they are all near or equal to zero. Molecules must have an asymmetry in their electronic distributions to exhibit β or a dipole moment (μ).

For large oligomers, the above averages are always dominated by a single component along the molecular axis, i.e. α_{XX} or γ_{XXXX}, and other works often only report these components. In the limit when all other components can be ignored,our values can be related to the others by $\alpha = 1/2\ \alpha_{XX}$ and $\gamma = 1/5\ \gamma_{XXXX}$.

Polarizability

A plot of the calculated $\alpha(-\omega;\omega)$ versus ω for each oligomer is given in Figure 1. These results show the following general trends -- 1) for a given oligomer α increases (slowly at first and then more rapidly) until a pole (infinity) is reached which corresponds to an excitation energy; 2) at a given frequency, as the oligomer length increases, α increases; and 3) as the oligomer length increases, the excitation energy moves lower. The lowering excitation energies approach a limiting value corresponding to the band gap of the ideal polyacetylene polymer. From AM1 optimized structures this value is approximately 2.0 eV and is in reasonable agreement with the experimental values *(11)*. The main reason for this agreement is the very good geometries provided by the AM1 parameterization for these systems. Bredas and co-workers *(12)* have shown that the optical properties of these conjugated systems are very sensitive to the bond alternation and the AM1 parameterization works very well for polyenes.

In order to demonstrate the saturation effect for α, it is more convenient to look at the value per subunit, shown in Figure 2. In our work, this quantity is defined as $\alpha/\text{sub}(n) = \alpha(n) - \alpha(n-1)$. It is often defined by other workers as $\alpha(n)/n$. These definitions are identical in the limit of very large n but the incremental definition used here shows more rapid convergence to the limiting value and is a better approximation to the appropriate numerical derivative. All the curves for frequencies below the lowest limiting excitation energy are clearly approaching a constant at large n. All the curves for frequencies above the excitation energy diverge as n increases.

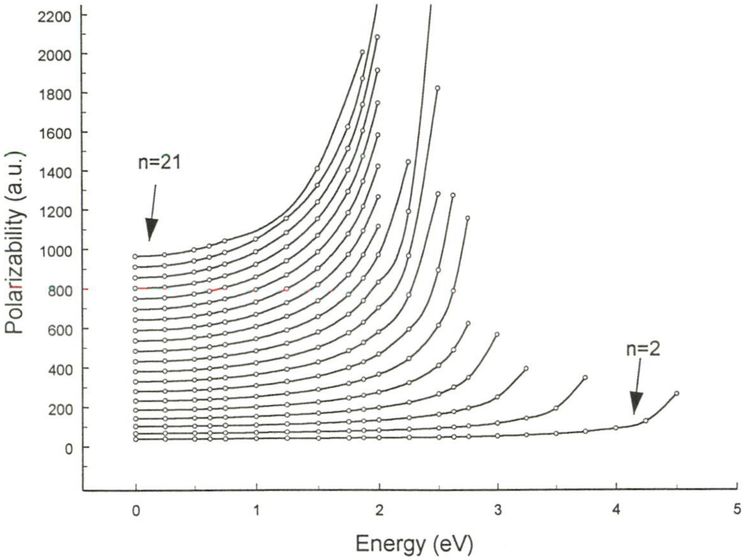

Figure 1. Calculated polarizabilities as a function of photon energy (frequency) for polyene oligomers $H(C_2H_2)_nH$ with n from 2 (lowest curve) to 21 (highest curve).

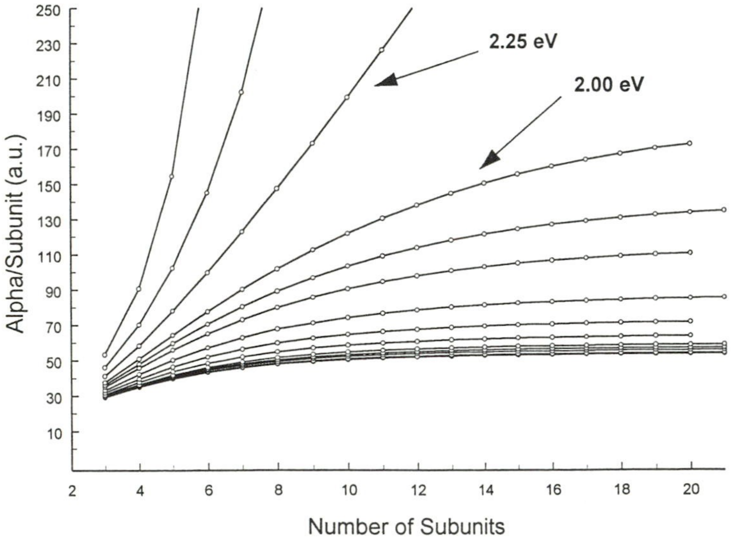

Figure 2. α/subunit as a function of the number of subunits, n, for different photon energies (frequencies) from 0.00 eV to 2.75 eV, in steps of 0.25 eV.

For the below excitation energy curves, the limiting α/sub values have been estimated by others by fitting log(α/sub), or α/sub, to a polynomial in 1/n *(13,14)*,

$$\log\left(\frac{\alpha}{\text{subunit}}\right) = a + \frac{b}{n} + \frac{c}{n^2} + \frac{d}{n^3} \,. \qquad (6)$$

The log(α/sub) is often used instead of α/sub to provide a better function for extrapolation which is a tenuous operation at best. The limiting value of α/sub is obtained from equation 6 as 10^a. With the data for our oligomers, this fitting procedure often lead to negative values for some of the parameters which results in a function that has maxima prior to infinite n. In order to use a function that increases monatonicaly, in this work, we have chosen the following function for extrapolation,

$$\log(\alpha / \text{sub}) = a - be^{-cn^d} \qquad (7)$$

The limiting value is given by the 10^a. This function works very well for extrapolation of the quantities of interest here. In our fits, the sum of the squares of the errors is always below 10^{-7}.

Our estimates of the limiting α/sub are given in Table II. This data was generated by fitting Equation 7 to the last 9 points (n=11 to 20). Nine points were chosen to be 2*(number of parameters) + 1. Fits were also done with more and less points and the results did not differ significantly from those from 9 points.

Table II: α/sub limiting values

E(eV)	α/sub limit	E(eV)	α/sub limit
0.00	54.2	1.25	72.7
0.25	55.1	1.50	87.4
0.50	57.3	1.75	114.5
0.625	57.8	1.875	139.9
0.75	59.5	2.00	183.7
1.00	64.5		

The static limiting value of 54.23 a.u. is in good agreement with the value of 60.8 a.u. estimated from *ab initio* oligomer calculations done by Hurst, Dupuis, and Clementi *(13)*.

Second Hyperpolarizability

The second hyperpolarizability results for each frequency (in eV) as a function of oligomer length n are given in Tables III, IV, and V, for THG, EFISH, and IDRI, respectively, and are illustrated in Figure 3. It is clear from these plots that the results follow the expected behavior with poles for THG at 1/3 the excitation energy obtained from the α calculation and EFISH and IDRI at 1/2 the excitation energy. By

Table III: Calculated γ^{THG} Values

E(eV):	0.00	0.25	0.50	0.625	0.75	0.875
2	3704	3852	4352	4795	5436	6380
3	26391	27844	33008	37897	45557	58254
4	94731	101259	125511	150059	191855	271397
5	239064	258515	333735	414822	565119	900499
6	483812	528450.0	707383	911752	1323993	2428958
7	842166	927784	1281988	1708294	2642510	5742108
8	1316219	1460530	2074768	2849361	4689295	12560437
9	1900402	2121978	3087187	4356889	7615949	
10	2583328	2899698	4306884	6228630	11541203	
11	3353187	3780737	5717298	8451881	16567715	
12	4196561	4749790	7295630	10999177	22758505	
13	5102918	5794772	9022930	13845804		
14	6060796	6902188	10876137	16956110		
15	7062249	8062643	12838582	20303083		
16	8099730	9267210	14893959	23858638		
17	9165937	10507022	17025068	27590078		
18	10256446	11776808	19221666	31477874		
19	11369500	13074378	21479006	35512237		
20	12498879	14392149	23781896	39661762		
21	13642596	15727718	26125031	43914425		
E(eV):	1.00 eV	1.25 eV				
2	7830	14757				
3	81753					
4	463192					
5	2115407					

Table IV: Calculated γ^{EFISH} Values

	Values Below Staturation Limit				
E(ev)	0.00	0.25	0.50	0.75	1.00
2	3704	3777	4008	4444	5183
3	26391	27102	29415	33959	422
4	94731	97910	108444	130040	172658
5	239064	248494	280262	347979	491882
6	483812	505367	579056	741680	1112060
7	842166	883368	1026023	1350587	2138337
8	1316219	1385506	1628060	2194740	3651934
9	1900398	2006436	2381175	3277113	5705218
10	2583322	2734392	3272610	4585419	8315876
11	3353197	3556971	4288018	6102561	11484165
12	4196552	4459798	5409912	7804249	15184366
13	5102918	5431659	6624320	9670176	19391636
14	6060796	6460100	7915322	11675344	24059662
15	7062249	7536499	9271726	13801598	29152576
16	8099730	8652671	10682812	16031099	34630267
17	9165937	9800580	12137798	18344920	40441425
18	10256446	10975392	13630266	20731761	46555586
19	11369500	12175175	15157445	23186355	52957275
20	12498879	13393052	16710080	25691930	59592845
	Values Above Staturation Limit				
E(ev)	1.25	1.50	1.75	2.00	2.25
2	6433	8686	13338	26179	124835
3	58217	94173	21530		
4	266913	555681			
5	861864	2640526			
6	2228011				
7	4942092				
8	9832760				
9	18113460				
10	31593781				

Table V: Calculated γ^{IDRI} Values

Values Below Staturation Limit					
E(ev)	0.00	0.25	0.50	0.75	1.00
2	3703	3752	3902	4174	4601
3	26388	26860	28343	31069	35502
4	94724	96826	103510	116051	137152
5	239055	245274	265236	303422	369712
6	483803	497994	543905	633225	792596
7	842162	869246	957472	1131620	1449768
8	1316219	1361709	1510752	1808582	2363776
9	1900398	1969949	2198952	2661370	3538353
10	2583322	2682328	3009690	3676596	4960114
11	3353187	3486640	3929464	4838476	6610126
12	4196552	4368846	4942318	6127196	8461616
13	5102918	5317969	6035613	7526710	10492119
14	6060796	6321886	7195155	9018426	12674050
15	7062249	73722134	8411042	10589163	14987399
16	8099730	8461005	9673850	12226269	17412566
17	9165937	9580452	10974185	13916801	19928755
18	10256446	10725899	12306475	15653110	22523673
19	11369500	11895467	13668349	17431784	25191516
20	12498879	13082482	15051787	19241705	27914031

Values Above Staturation Limit					
E(eV)	1.25	1.50	1.75	2.00	2.25
2	5247	6223	7732	10176	14445
3	42523	53833	72966		
4	172273	232991			
5	485268	698942			
6	1082073				
7	2048684				
8	3441604				
9	5286551				
10	7577414				

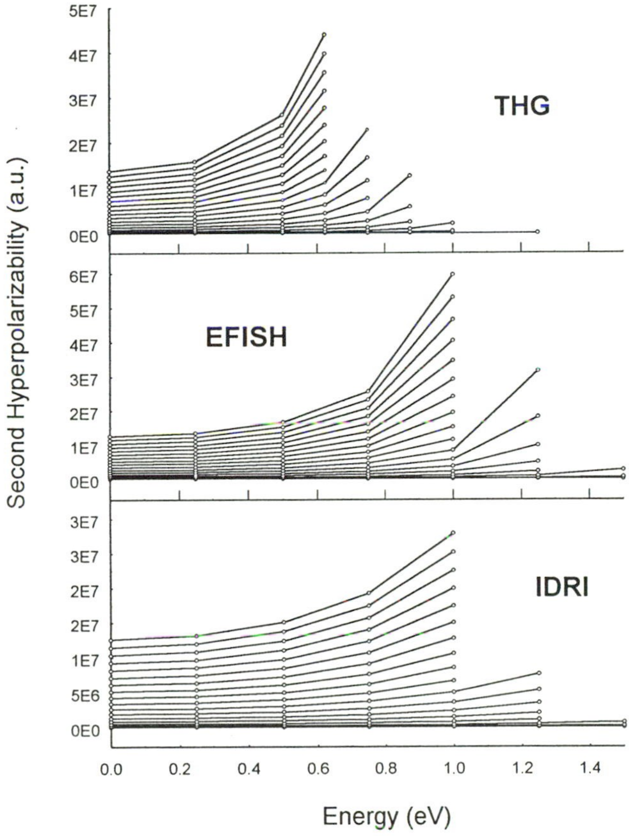

Figure 3. Calculated second hyperpolarizabilities (THG, EFISH, and IDRI) as a function of photon energy (frequency).

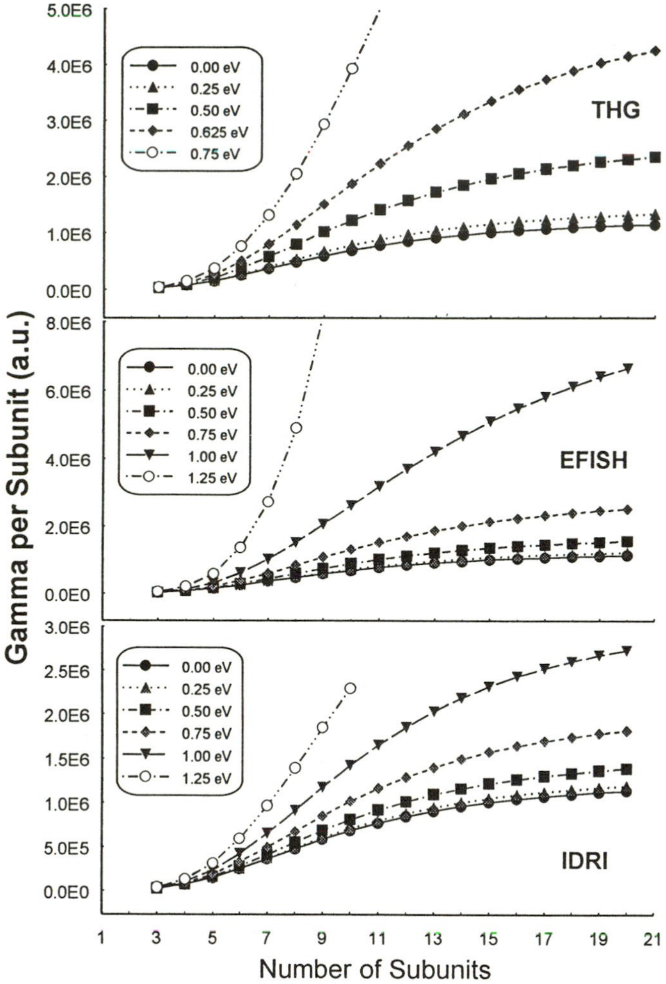

Figure 4. γ/sub as a function of the number of subunits, n, for different photon energies (frequencies). Lowest curve corresponds to the smallest oligomer and highest curve is the largest oligomer.

comparing these plots, it can also be seen the γ(THG) > γ(EFISH) > γ(IDRI) as noted by others *(6)*.

As with the polarizability, the saturation behavior is demonstrated by examining γ/sub (defined as γ(n) - γ(n-1)). Figure 4 shows the behavior of these values for the different types of γ. In this figure are all the curves that show convergence with increasing length plus one curve to illustrate divergent behavior. The curves that show convergence are those with energies less than the appropriate pole for the infinite (polymer) system, i.e. 1/3 the band gap for THG and 1/2 the band gap for EFISH and IDRI. At 20 subunits (40 carbons) the γ/subunit values do not show complete saturation. Those values at 0.0 eV are essentially converged n=20 but as the energy gets closer to the pole it is clear that the γ/subunit converges for much larger systems (n>20). This, like other theoretical works, is not in accord with the recent experimental estimate that the static (0.0 eV) g does not show saturation until near n=120 *(3)*.

The data shown in Figure 4 is used to find the limiting values of γ/subunit by fits to equation 7 in the same manner as the polarizability fits. The limiting values obtained from γ/sub values are much less reliable than α/sub as the data are much further from saturation. Data in Table VII gives the results of our extrapolation predictions, again using the last 9 points in each curve.

Table VI: γ/subunit limiting values

Energy	THG	EFISH	IDRI
0.0	1.211	1.213	1.212
0.25	1.411	1.304	1.274
0.50	2.525	1.674	1.479
0.625	4.880		
0.75		2.751	1.949
1.00		8.275	3.001

Conclusions

These semiempirical calculations clearly show the proper saturation behavior of α and γ for polyenes of increasing length. These AM1 results also agree quite well with the ab initio results (13) but further work needs to be done to understand the descrepancy with the experimental estimates.

Acknowledgments

Support for this work is gratefully acknowledged from the Air Force Office of Scientific Research (AFOSR-90-0010).

References

1. De Melo , C. P.; Silbey, R. *Chem. Phys. Lett.* **1987**, *140*, 537
2. Samuel, I. D. W.; Ledoux, I.; Dhenaut, C.; Zyss, J.; Fox, H. H.; Schrock, R. R.; Silbey, R. J. *Science* **1994**, *265*, 1070.
3. Etemad, S.; Heeger, A. J.; MacDiarmid, A. G. *Ann. Rev. Phys. Chem.* **1982**, *33*, 443.
4. H. A. Kurtz, *Int. J. Quantum Chemistry* Symp. **1990**, *24*, 791.
   B. Champagne, J. G. Fripiat, and J.-M. Andre, *J. Chem. Phys.* **1992**, *96*, 8330.
   E. F. Archibong and A. J. Thakkar, *J. Chem. Phys.* 1993, *98*, 8324.
5. Dewar, M. J. S.; Zoebisch, E. G.; Healy, E. F.; Stewart, J. J. P. *J. Am. Chem. Soc.* **1985**, *107*, 3902.
6. Sekino, H; Barlett, R. J. *J. Chem. Phys.* **1986**, ***85***, 976.
7. Karna, S. P.; Dupuis, M. *J. Comp. Chem.* **1991**, *12*, 487.
8. The TDHF capabilities have been released in J. J. P. Stewart's MOPAC 93 from Fujitsu Ltd and MOPAC 7 from QCPE. The work in this paper is based on our own modified version of MOPAC 6.
9. .Schmidt, M.W.; Baldridge, K. K.; Boatz, J. A.; Elbert, S. T.; Gordon, M. S.; Jensen, J. H.; Koseki, S.; Matsunaga, N.; Nguyen, K. A.; Su, S. J.; Windus, T.L.; Dupuis, M.; Montgomergy, J. A. *J.Comput.Chem.* **1993,** *14*, 1347
10. Boyd, R. W. Nonlinear Optics; Academic Press: San Diego, 1992.
11. Heflin, J. R.; Wong, K. Y.; Zamani-Khamiri, O.; Garito, A. F. *Phys. Rev. B.* **1988**, *38*, 1573.
12. Meyers, F.; Marder, S. R.; Pierce, B. M.; Brédas, J. L. *J. Am. Chem. Soc.* **1994**, *116*, 10703.
13. Hurst, Graham; Dupuis, M.; Clementi, E. *J. Chem. Phys.* **1988**, *89*, 385.
14. Kirtman, B. *Chem. Phys. Lett.* **1988**, *143*, 81; Kirtman. B. *Int. J. Quantum Chem.* **1992**, *43*, 147.

RECEIVED December 29, 1995

Chapter 8

Optical Properties from Density-Functional Theory

Mark E. Casida, Christine Jamorski, Fréderic Bohr[1], Jingang Guan, and Dennis R. Salahub

Département de chimie, Université de Montréal, C.P. 6128, Succursale centre-ville, Montréal, Québec H3C 3J7, Canada

Density-functional theory (DFT) is a promising method for the calculation of molecular optical properties, since it is less computationally demanding than other *ab initio* methods, yet typically yields results of a quality comparable to or better than those from the Hartree–Fock approximation. The calculation of static molecular response properties via DFT has now been studied for several years, whereas work on the corresponding dynamic properties is only just beginning, since none of the previously existing molecular DFT codes were capable of treating them. The present article gives a brief summary of some of our work in this area. This includes an illustration of the quality of results that can be expected from DFT for static molecular response properties (dipole moments, polarizabilities, and first hyperpolarizabilities), as well as illustrative early results (dynamic polarizabilities and excitation spectra) from our code deMon-DynaRho, the first molecular time-dependent density-functional response theory program.

The search for stable materials with enhanced nonlinear optical properties for use in telecommunications and computer information transmission and storage has spurred a renewed [1,2] interest by chemists in recent years [3–7] in the nonlinear optical properties of molecules. It is hoped that quantum chemical calculations will help in the design and preselection of candidate materials. However, several requirements will have to be met if the results of quantum chemical calculations are to find direct application to problems currently of interest in materials science. These requirements include the ability to handle some reasonably large molecules, the ability to treat the response to a dynamic field, and proper consideration of

[1]Current address: Laboratoire de Chimie Physique, Université de Reims, Faculté des Sciences, Moulin de la Housse, B.P. 347, 51062 Reims Cedex, France

0097–6156/96/0628–0145$15.00/0

solvent or matrix effects. To this list, we can add that good *ab initio* calculations require large basis sets and an adequate treatment of correlation, and that accurate comparisons with experiment can be complicated by the need to include vibrational and orientational contributions. Because of its scaling properties with respect to calculations on increasingly larger molecules and its ability to treat correlation in a simple way, density-functional theory (DFT) is a promising method for quantitative calculations of the optical properties of molecules in a size range of practical interest. This article is a report of where we stood in the Fall of 1994 in generating and calibrating essential DFT machinery for treating optical problems. Given that linear as well as nonlinear optical properties are a topic which has long been of interest to chemists and is likely to remain so for some time, we will not restrict the topic only to nonlinear properties but will also discuss the use of DFT for calculating simple polarizabilities and excitation spectra. Our current work on vibrational contributions [8] and solvent effects [9] will be discussed elsewhere.

Let us first situate DFT among the variety of quantum chemical methods available for calculating the optical properties of molecules. At one extreme, impressively quantitative *ab initio* methods, based upon, for example, Møller-Plesset perturbation theory [10], equation-of-motion [11], or coupled cluster [12] techniques, have been developed whose application tends to be limited to very small molecules. At the other extreme, semiempirical methods allow the consideration of much larger molecules, but the reliance on parameterizations limits the type of molecules and variety of properties to which any given semiempirical method can be applied with confidence. DFT offers the advantages of an *ab initio* method, yielding results for a variety of properties that are typically better than those obtained from the Hartree–Fock (HF) approximation, but with less computational effort. Although DFT is more computationally demanding than semi-empirical methods, it gives much more reliable results when a broad range of molecular types and properties is considered. Thus DFT represents a promising approach to the quantitative treatment of the optical properties of molecules for systems complex enough to be of interest to bench chemists and materials scientists.

At present, the potential of DFT remains largely untapped in this respect. Although there have been numerous applications of DFT to the calculation of electric response properties of atoms and solids (see Ref. [13] for a review), much less work has been done on molecular systems. Studies assessing DFT for calculation of molecular electric response properties have been for static properties, primarily dipole-polarizabilities and hyperpolarizabilities [9,14–19] of small molecules. Work on DFT calculations of dynamic molecular response properties has been limited to a few calculations using either spherically-averaged pseudopotentials [20,21] or single-center expansions [22,23], in order to make use of atomic-like algorithms, but which are not of any general utility for molecular calculations. After a brief look at how well DFT works for static molecular response properties, we focus on the first implementation of time-dependent density functional response theory using an algorithm appropriate for general molecular calculations, giving a summary of our method and results for N_2. The present results are at the level of the random phase approximation. Implementation of the fully-coupled time-dependent local density approximation is in progress. A more complete description of our methodology will be published elsewhere [24].

METHODOLOGY

Since excellent reviews of density-functional theory (DFT) are readily accessible [25,26], we will restrict our attention to what is needed for a discussion of the current status of DFT for the calculation of molecular optical properties.

Static Properties. With few exceptions, molecular applications of DFT are based upon the Kohn–Sham formalism, in which the exact ground state energy and charge density of a system of N electrons in an external local potential are obtained using the exact exchange-correlation functional. In practice this exchange-correlation functional must be approximated. The terms "local potential" and "ground state" are important. The former excludes a full, rigorous treatment of magnetic effects, though useful results can be obtained in practice [27–29]. The latter, together with the fact that the Kohn–Sham formalism is time-independent, excludes the treatment of dynamic response properties in the traditional theory. Extensions of the formalism to the time-dependent domain have been made, and dynamic response properties will be discussed in the next subsection. The standard Kohn–Sham formalism is, however, exact for static electronic electric response properties, in the limit of the exact exchange-correlation functional.

The charge density is obtained in Kohn–Sham theory as the sum of the charge densities of Kohn–Sham orbitals ψ_i^σ with occupation numbers f_i^σ. That is

$$\rho(\mathbf{r}) = \rho^{\uparrow}(\mathbf{r}) + \rho^{\downarrow}(\mathbf{r}), \tag{1}$$

where

$$\rho^\sigma(\mathbf{r}) = \sum_i f_i^\sigma |\psi_i^\sigma(\mathbf{r})|^2 . \tag{2}$$

(Hartree atomic units are used throughout.) The orbitals are found by solving the self-consistent Kohn–Sham equations

$$\left[-\frac{1}{2}\nabla^2 + v_{\text{eff}}^\sigma(\mathbf{r})\right] \psi_i^\sigma(\mathbf{r}) = \epsilon_i^\sigma \psi_i^\sigma(\mathbf{r}), \tag{3}$$

where the effective potential v_{eff}^σ is the sum of an external potential which, in molecular applications, is the sum of nuclear attraction terms and any applied potential v_{appl}^σ, and a self-consistent field (SCF) term,

$$v_{\text{SCF}}^\sigma(\mathbf{r}) = \int \frac{\rho(\mathbf{r}')}{|\mathbf{r}-\mathbf{r}'|}\, d\mathbf{r}' + v_{\text{xc}}^\sigma[\rho^{\uparrow}, \rho^{\downarrow}](\mathbf{r}), \tag{4}$$

which differs from the corresponding quantity in the Hartree–Fock approximation in that the Hartree–Fock exchange operator has been replaced with the density-functional exchange-correlation potential v_{xc}^σ. No practical exact form of the exchange-correlation potential is known, so it is approximated in practice. Popular approximations include the widespread local density approximation (LDA) and gradient-corrected functionals such as the B88x+P86c functional.

If the applied potential corresponds to a uniform electric field,

$$v^{\sigma}_{\text{appl}}(\mathbf{r}) = \mathbf{F} \cdot \mathbf{r}\,, \tag{5}$$

then the dipole moment is

$$\mu_i(\mathbf{F}) = -\int r_i \rho(\mathbf{r};\mathbf{F})\, d\mathbf{r} + \sum_{A}^{\text{nuclei}} \mathbf{r}_{Ai} Z_A\,, \tag{6}$$

the dipole-polarizability is

$$\alpha_{ij}(\mathbf{F}) = \frac{\partial \mu_i(\mathbf{F})}{\partial F_j}\,, \tag{7}$$

the first dipole-hyperpolarizability is

$$\beta_{ijk}(\mathbf{F}) = \frac{\partial \alpha_{ij}(\mathbf{F})}{\partial F_k}\,, \tag{8}$$

etc. Since the Hellmann-Feynman theorem holds in DFT [17], these are equivalent to energy derivatives. Of course, reported dipole moments, polarizabilities, and hyperpolarizabilities are evaluated at zero field strength. The mean polarizability ($\bar{\alpha}$), polarizability anisotropy ($\Delta\alpha$) and mean first hyperpolarizability ($\bar{\beta}$) are defined in the usual manner as

$$\bar{\alpha} = \frac{1}{3}\text{tr}\boldsymbol{\alpha}\,, \tag{9}$$

$$(\Delta\alpha)^2 = \frac{1}{2}\left[3\text{tr}\,(\boldsymbol{\alpha}^2) - (\text{tr}\,\boldsymbol{\alpha})^2\right]\,, \tag{10}$$

$$\bar{\beta} = \frac{9}{5}\frac{\partial\bar{\alpha}}{\partial F_z}\,, \tag{11}$$

where the permanent dipole moment of the molecule is directed in the $+z$ direction.

The easiest method to program for obtaining the static response properties is the finite field method in which the Kohn–Sham equations are solved for several values of the applied field and the derivatives are determined numerically, either by finite difference [14,15] or by least-squares fit to a polynomial expansion [16,17,19]. Unfortunately the need to optimize the choice of field strengths used and the method's susceptibility to numerical noise arising from e.g. grids used in DFT programs, greatly complicate routine calculations of higher-order polarizabilities, which however can still be extracted with some care [16,17,19].

Another approach, one which avoids these problems, is to calculate the derivatives (7) and (8) analytically by solving a set of coupled perturbed Kohn–Sham (CPKS) equations [18,30–33]. The method is formally analogous to the coupled perturbed Hartree-Fock (CPHF) method, but while the CPHF method is approximate, the CPKS method becomes exact in the limit of the exact exchange-correlation functional. The CPKS equations are normally derived by taking derivatives of the total energy or dipole moment in a finite basis set representation. In the next section, we show a different way to obtain the CPKS equations.

Dynamic Properties. Dynamical effects are essential to the description of the interaction of light and matter. Taking them into account requires the extension of DFT to the time-dependent domain. For a review of the formal foundations of time-dependent DFT, see Ref. [34]. The result is that the time-dependent charge density

$$\rho^\sigma(\mathbf{r},t) = \sum_{i\sigma} f_i^\sigma |\psi_i^\sigma(\mathbf{r},t)|^2 \tag{12}$$

is calculated from orbitals satisfying a time-dependent Kohn–Sham equation

$$\left[-\frac{1}{2}\nabla^2 + v_{\text{eff}}^\sigma(\mathbf{r},t)\right]\psi_i^\sigma(\mathbf{r},t) = i\frac{\partial}{\partial t}\psi_i^\sigma(\mathbf{r},t) \tag{13}$$

where the effective potential v_{eff}^σ is now the sum of the time-dependent external potential and a self-consistent field term

$$v_{\text{SCF}}^\sigma(\mathbf{r},t) = \int \frac{\rho(\mathbf{r}',t)}{|\mathbf{r}-\mathbf{r}'|}\,d\mathbf{r}' + v_{\text{xc}}^\sigma[\rho^\uparrow,\rho^\downarrow](\mathbf{r},t)\,, \tag{14}$$

which involves a new time-dependent exchange-correlation functional v_{xc}^σ. Naturally Eq. (13) reduces to Eq. (3) in the limit of the ground state, static problem.

Working in the energy ($= \hbar\times$ frequency) representation, the linear response of the charge density is related to the perturbation $\delta v_{\text{appl}}^\sigma$ through the generalized susceptibility $\chi^{\sigma,\tau}$,

$$\delta\rho^\sigma(\mathbf{r},\omega) = \sum_\tau \int \chi^{\sigma,\tau}(\mathbf{r},\mathbf{r}';\omega)\delta v_{\text{appl}}^\tau(\mathbf{r}',\omega)\,d\mathbf{r}'\,. \tag{15}$$

The dynamic polarizability is calculated as

$$\alpha_{ij}(\omega) = -\frac{1}{F(\omega)}\int \mathbf{r}_i\delta\rho(\mathbf{r},\omega)\,d\mathbf{r}\,, \tag{16}$$

when $\delta v_{\text{appl}}(\omega) = \mathbf{r}_j F(\omega)$. Since the Kohn–Sham equations have the form of one-particle, orbital equations, we can rewrite Eq. (15) as

$$\delta\rho^\sigma(\mathbf{r},\omega) = \sum_\tau \int \chi_{\text{KS}}^{\sigma,\tau}(\mathbf{r},\mathbf{r}';\omega)\delta v_{\text{eff}}^\tau(\mathbf{r}',\omega)\,d\mathbf{r}'\,, \tag{17}$$

where

$$\chi_{\text{KS}}^{\sigma,\tau}(\mathbf{r},\mathbf{r}';\omega) = \delta_{\sigma,\tau}\sum_{i,j}\frac{f_i^\sigma - f_j^\sigma}{\omega - (\epsilon_j^\sigma - \epsilon_i^\sigma)}\left[\psi_j^\sigma(\mathbf{r})\psi_i^{\sigma*}(\mathbf{r})\right]\left[\psi_j^\sigma(\mathbf{r}')\psi_i^{\sigma*}(\mathbf{r}')\right]^* \tag{18}$$

has the form of the generalized susceptibility for a system of independent particles, and the response of the effective potential $\delta v_{\text{eff}}^\sigma$ is the sum of the perturbation $\delta v_{\text{appl}}^\sigma$ and the response of the self-consistent field term,

$$\delta v_{\rm SCF}^{\sigma}(\mathbf{r},\omega) = \int \frac{\delta\rho^{\sigma}(\mathbf{r}',\omega)}{|\mathbf{r}-\mathbf{r}'|}\,d\mathbf{r}' + \sum_{\tau}\int f_{\rm xc}^{\sigma,\tau}(\mathbf{r},\mathbf{r}';\omega)\delta\rho^{\tau}(\mathbf{r}',\omega)\,d\mathbf{r}' . \tag{19}$$

The exchange-correlation kernel is given by

$$f_{\rm xc}^{\sigma,\tau}(\mathbf{r},\mathbf{r}';\omega) = \int e^{i\omega(t-t')}\frac{\delta v_{\rm xc}^{\sigma}(\mathbf{r},t)}{\delta\rho^{\tau}(\mathbf{r}',t')}\,d(t-t') . \tag{20}$$

It reduces to

$$f_{\rm xc}^{\sigma,\tau}(\mathbf{r},\mathbf{r}') = \frac{\delta v_{\rm xc}^{\sigma}(\mathbf{r})}{\delta\rho^{\tau}(\mathbf{r}')} \tag{21}$$

in the static limit ($\omega = 0$), in which case Eqs. (17)-(21) become the CPKS equations.

When $\omega \neq 0$, solving the dynamic coupled equations allows the dynamic polarizability $\bar{\alpha}(\omega)$ to be calculated. The method can also be extended to other dynamic properties, including higher-order polarizabilities and excitation spectra. In practice, we obtain excitation spectra by noting that the exact dynamic dipole-polarizability can be expanded in a sum-over-states representation as

$$\bar{\alpha}(\omega) = \sum_{I}^{\text{excited states}} \frac{f_I}{\omega_I^2 - \omega^2} \tag{22}$$

where the ω_I are vertical excitation energies and the f_I are the corresponding oscillator strengths. Since practical calculations use approximate exchange-correlation functionals, the calculated dynamic polarizability will also be approximate. Nevertheless, it still has the same analytic form as the exact dynamic polarizability, so the poles and residues of the calculated dynamic polarizability can be identified as (approximate) excitation energies and oscillator strengths. Note that the Thomas–Reiche–Kuhn (TRK) sum rule [35]

$$\sum_{I} f_I = N \tag{23}$$

should also be satisfied in the limit of the exact (time-dependent) exchange-correlation functional.

The problem of finding good time-dependent exchange-correlation functionals is still in its infancy. This problem does not arise at the level of the independent particle approximation (IPA), which consists of taking $\delta v_{\rm SCF} = 0$. The next level of approximation is the random phase approximation (RPA), where the response of the exchange-correlation potential (second term in Eq. (19)) is taken to be zero, which turns out to be a reasonably good approximation for some purposes (*vide infra*). Note that the RPA includes some exchange-correlation effects, namely those which enter through the orbitals and orbital energies of Eq. (18). A notation such as RPA/LDA gives a more complete description of the level of approximation (i.e. approximation used for the response / approximation used for the unperturbed orbitals and orbital energies.) A problem with the RPA is that it does not reduce to the CPKS equations in the static limit. This requirement is met by the adiabatic approximation (AA) in which the reaction of the exchange-correlation potential to changes in the charge density is assumed to be instantaneous,

$$\frac{\delta v_{\mathrm{xc}}^{\sigma}(\mathbf{r},t)}{\delta \rho^{\tau}(\mathbf{r}',t')} \approx \frac{\delta v_{\mathrm{xc}}^{\sigma}(\mathbf{r},t)}{\delta \rho^{\tau}(\mathbf{r}',t)} \delta(t-t') . \tag{24}$$

This assumption is rigorous in the static case and is at least reasonable in the low frequency limit. When the exchange-correlation functional is local, the AA is usually referred to as the time-dependent local density approximation (TDLDA). Since the orbitals and orbital energies used are also at the LDA level, the notation TDLDA/LDA gives a more complete description of this AA. An approximation which goes beyond the AA has also been suggested [34].

The dynamic results reported here were calculated at the RPA/LDA level. Implementation of the TDLDA is in progress.

COMPUTATIONAL DETAILS

The calculations reported here were carried out using two programs written at the University of Montreal. The first program, **deMon** (for "*densité de Montréal*") [36–38], is a general purpose density-functional program which uses the Gaussian-type orbital basis sets common in quantum chemistry. The second program, **DynaRho** (for "Dynamic Response of ρ"), is a post-**deMon** program which we are developing to calculate properties which depend on the dynamic response of the charge density. In **DynaRho**, the formal equations of the previous section are solved in a finite basis set representation. Although a full description of how this is accomplished is beyond the scope of the present paper, some insight into the operational aspects of **DynaRho** can be obtained by considering the particularly simple case of the H_2 molecule oriented along the z-axis and described using a minimal basis set. There are only two molecular orbitals in this case. The occupied σ-bonding combination will be denoted by the index i, while the unoccupied σ-antibonding combination will be denoted by a. In the molecular orbital representation, the response of the density matrix, δP_{ia}^{σ}, to an electric field $F_z \cos(\omega t)$ is found, within the RPA, by solving a matrix equation,

$$\left\{ \omega^2 \mathbf{1} - \begin{bmatrix} (\epsilon_a-\epsilon_i)^2 + 2(\epsilon_a-\epsilon_i)(ia;ia) & 2(\epsilon_a-\epsilon_i)(ia;ia) \\ 2(\epsilon_a-\epsilon_i)(ia;ia) & (\epsilon_a-\epsilon_i)^2 + 2(\epsilon_a-\epsilon_i)(ia;ia) \end{bmatrix} \right\} \begin{pmatrix} \delta P_{ia}^{\uparrow} \\ \delta P_{ia}^{\downarrow} \end{pmatrix}$$
$$= (\epsilon_a - \epsilon_i) \begin{pmatrix} z_{ia} \\ z_{ia} \end{pmatrix} \tag{25}$$

where

$$(rs;tu) = \int\int \psi_r(\mathbf{r})\psi_s(\mathbf{r}) \frac{1}{|\mathbf{r}-\mathbf{r}'|} \psi_t(\mathbf{r}')\psi_u(\mathbf{r}')\, d\mathbf{r} d\mathbf{r}' \tag{26}$$

and z_{ia} is a matrix element of the multiplicative operator z. Since 4-index integrals are not implemented in **deMon**, **DynaRho** uses auxiliary functions to construct the needed electron repulsion integrals from at most 3-center integrals [24]. Once Eq. (25) is solved, the dynamic polarizability is given by

$$\alpha_{zz}(\omega) = -z_{ia}\left(\delta P_{ia}^{\uparrow} + \delta P_{ia}^{\downarrow}\right) . \tag{27}$$

The matrix shown in square brackets in Eq. (25) resembles a small "configuration interaction (CI) matrix" in so far as its columns and rows, in the general case, correspond to the different single excitations possible out of the ground state configuration. The excitation energies are obtained by diagonalizing this matrix, to obtain the singlet-singlet excitation energy

$$\omega^S = \sqrt{(\epsilon_a - \epsilon_i)(\epsilon_a - \epsilon_i + 4(ia;ia))} , \tag{28}$$

and the singlet-triplet excitation energy

$$\omega^T = \epsilon_a - \epsilon_i . \tag{29}$$

Although the ground state of H_2 at its equilibrium geometry may be described, to a first approximation, by a single determinantal wavefunction, linear combinations of at least two determinants are needed to describe the singlet and triplet excited state wavefunctions. The difficulty of describing such states with conventional DFT, which is a fundamentally single determinantal theory, is sometimes known as the "multiplet problem." Notice how response theory leads to a natural solution involving a small "CI matrix." Notice also that the RPA singlet-triplet excitation energy expression is just the independent particle approximation (IPA) excitation energy. This is an example of the more general result that singly-excited triplet configurations which are degenerate at zero-order are not coupled at the RPA level. However this problem is resolved when the response of the exchange-correlation potential is included in the theory. A more detailed account of the algorithm used in DynaRho is given elsewhere [24].

Finite field calculations of static response properties were carried out with deMon. The least squares fitting procedure of Ref. [17] was used when hyperpolarizabilities were desired (i.e. for the results in Tables I and III), otherwise the simple finite difference method using deMon's default field step size of 0.0005 a.u. was used. Note that there are no significant differences between these two methods for calculating simple polarizabilities.

Dynamic properties, including spectra, were calculated with DynaRho at the level of the random phase approximation (RPA/LDA). Static properties were also calculated using DynaRho at both the RPA and independent particle approximation (IPA) levels.

Two functionals were considered, for the static properties. The parameterization of the local density approximation (LDA) used in deMon is that of Vosko, Wilk, and Nusair [39]. This was supplemented with Becke's 1988 gradient-correction for exchange [40] and Perdew's 1986 gradient-correction for correlation [41] to give the B88x+P86c functional. Since Becke's gradient correction is designed to give the correct asymptotic behavior of the exchange energy density, some improvement over the LDA is expected in properties such as polarizabilties and hyperpolarizabilities which are sensitive to the long range behavior of the charge density.

Unless otherwise specified, all calculations have been carried out at the experimental geometries (references are given in Ref. [17]) using deMon's extra-fine random grid, the hydrogen atom [3,1;3,1] auxiliary basis set and heavy atom [4,4;4,4] auxiliary basis sets from the deMon basis set library. The convergence parameters used in our calculations tend to be tighter than those normally used in deMon calculations, since experience indicates that such parameters are needed for well-converged polarizability calculations.

Several different orbital basis sets are used in the present work. The DZVP, DZVP2, and TZVP basis sets [42] are from the deMon basis set library. The STO-3G and Sadlej basis sets are from Refs. [43] and [44,45] respectively. The TZVP+ orbital basis set is the VTZP+ basis set of Ref. [17]. The Sadlej+ basis set consists of the Sadlej basis set supplemented with the field-induced polarization functions of Ref. [17]. Finally, the Diff basis set, used for N_2, is the TZVP+ augmented with two diffuse s-type Gaussian primatives with exponents $\alpha_s = 0.028$ and 0.0066 and with one set of diffuse p-type Gaussian primatives with exponent $\alpha_p = 0.025$.

RESULTS

Static results. In principle, DFT static electronic electrical response properties become exact in the limit of the exact exchange-correlation functional. This is in contrast to the Hartree–Fock approximation which neglects electron correlation effects and whose accuracy, for this reason, decreases markedly as the order of the response property increases (Table I). In practice, the use of approximate functionals (and finite basis sets) place restrictions on the quality of the results which can be obtained from DFT. Nevertheless, a number of studies using the finite field method to calculate molecular static electrical response properties [9,14–19,46], indicate that modern DFT is capable of producing accurate molecular response properties. The quality of results which can be expected and some of the concerns which enter into DFT calculations of dipole moments, polarizabilities, and hyperpolarizabilities are illustrated here by some example calculations on water and nitrogen.

Table I shows the dipole moment, mean polarizability, polarizability anisoptropy, and mean first hyperpolarizability of water, as a function of basis set. The minor differences shown for different deMon calculations with the same orbital basis set arise from small variations in the auxiliary functions, grid, and convergence criteria used in the calculations. The basis sets are arranged in order of decreasing energy *in the absence of an applied field.* To the extent that this measures the degree of convergence of first-order properties, one would expect the dipole moment to converge with decreasing energy. Additional field-induced polarization (FIP) functions are required in order to describe the change in the dipole moment due to an applied electric field needed for an accurate description of polarizabilities and hyperpolarizabilities. Results from two types of FIP basis sets are given in Table I. The Sadlej basis set [44,45] was designed as a medium sized basis set for *ab initio* polarizability calculations. Basis sets whose name includes a plus sign (+) have been augmented with the FIP functions of Ref. [17].

It is well-known that LDA dipole moments are excellent. This is illustrated in the case of water where the LDA value is 1 to 2% larger than the experimental value (depending upon the basis set used) and in much better agreement with experiment than the corresponding Hartree–Fock (HF) value which is about 7% too large.

TABLE I. Convergence of water static response properties with respect to basis set. All density-functional calculations in this table were performed at the LDA level, at the experimental geometry, and using the finite field method with least squares fitting. The energy, dipole moment, polarizability, and first hyperpolarizability are expressed in atomic units. Basis sets used in this paper are described in the text. Other basis sets are described in the references cited. Note that the VTZP+ basis of Ref. [17] and the TZP+ basis of Ref. [47] are identical with the present TZVP+ basis set; see text for other differences between these calculations.

Basis	Size[a]	Energy[b]	μ	$\bar{\alpha}$	$\Delta\alpha$	$\bar{\beta}$	Reference
			Previous deMon				
TZVP	29	-75.8996	0.853	6.20	3.39		[47]
TZVP	29		0.852	6.20	3.45	-47.8	[17]
TZVP+	42	-75.9031	0.7455	10.12	0.30		[47]
TZVP+	42		0.743	10.13	0.32	-21.8	[17]
ANO+	73	-75.9127	0.733	10.18	0.47		[47]
NHF+	104	-75.9132	0.737	10.47	0.17		[47]
			Present work				
Sadlej	44	-75.8951	0.732	10.56	0.27	-19.1	
Sadlej+	56	-75.8962	0.732	10.70	0.10	-23.4	
TZVP+	42	-75.9015	0.740	10.16	0.37	-21.7	
			Experiment				
			0.727[e]	9.921[f]	0.66[f]	-21.8(9)[g]	
			Conventional *ab initio* methods				
HF			0.780	8.51	1.08	-9.73	[17]
HF			0.7789	8.531		-10.86	[10]
SDQ-MP4			0.724	9.64		-17.96	[10]

[a]Number of contracted Gaussian orbitals.
[b]The so-called "fitted energy" in deMon.
[e]From Ref. [48].
[f]From Ref. [49].
[g]From Ref. [50].

TABLE II. Convergence with respect to basis set of the energy and mean polarisability of N_2 (in atomic units).

Basis set	Size	Energy[a]	Mean polarizability ($\bar{\alpha}$)		TRK sum
			LDA finite field	RPA/LDA	
STO-3G	10	-107.147182	3.879	3.516	3.362
DZVP	30	-108.657734	9.917	8.491	10.729
DZVP2	30	-108.657734	9.292	8.450	10.661
Sadlej	52	-108.664703	11.943	10.954	10.435
TZVP	38	-108.676105	9.653	8.759	11.851
TZVP+	38	-108.680713	11.794	10.780	11.480
Diff	62	-108.680790	11.795	10.868	11.488
			Experiment		
			11.74 [b]		14. [c]

[a]The so-called "numerical energy" in deMon.
[b]Ref. [51].
[c]Number of electrons.

Polarizabilities calculated at the LDA level are also quite good. In the case of water, the LDA mean polarizability is 2 to 7% larger than the experimental value while the HF value is about 14% too low. Figures 1 and 2 compare theoretical vs. experimental values of the mean polarizability and polarizability anisotropy for a few small molecules, at several levels of approximation. The polarizability anisotropy, being a small difference of larger quantities, is considerably more difficult to calculate accurately than is the mean polarizability. The independent particle approximation (IPA), which consists of neglecting δv_{SCF} entirely, is clearly inadequate for quantitative polarizability calculations. The random phase approximation (RPA) includes the coulomb part of the response δv_{SCF} but neglects the exchange-correlation part of δv_{SCF}, and gives results comparable to those obtained from the Hartree–Fock approximation. The fully coupled TDLDA includes both the coulomb and exchange-correlation contributions to δv_{SCF} and would be equivalent to the finite field LDA results shown here. The fact that the RPA results are far more similar to the finite field results than to the IPA indicates that, as would be expected on physical grounds, the response of the coulomb part of v_{SCF} to an applied electric field is an important part of the polarizability, whereas the response of the exchange-correlation potential is a relatively small contribution. Table II shows the convergence of the mean polarizability values for N_2 with respect to basis set. The discrepancy between the calculated value and theoretical limit of the TRK sum arises from the limitations of the basis sets used here, which are oriented towards a good description of the 10-electron valence space of the ground state molecule, but not necessarily of the core. These basis sets are expected to describe only the low lying excited states reasonably well.

Less data is available to judge the quality of DFT calculations of molecular hyperpolarizabilties, but indications to date [9,16–19] are that mean first hyperpolarizabilities are pretty good at the LDA level. The LDA value of $\bar{\beta}$ in Table I is in much better agreement with experiment than is the HF value. Neverthe-

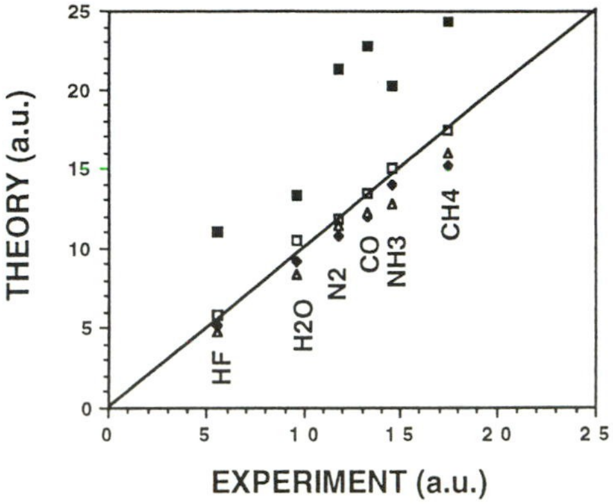

FIG. 1. Comparison of theoretical and experimental mean polarizabilities for N_2, CO, CH_4, H_2O, NH_3, and HF: independent particle approximation, solid squares; random phase approximation, solid diamonds; finite field, open squares; coupled Hartree-Fock, open triangles. The density-functional calculations used the local density approximation and the TZVP+ basis set. The coupled Hartree-Fock and experimental values are taken from Ref. [17]. See text for additional details.

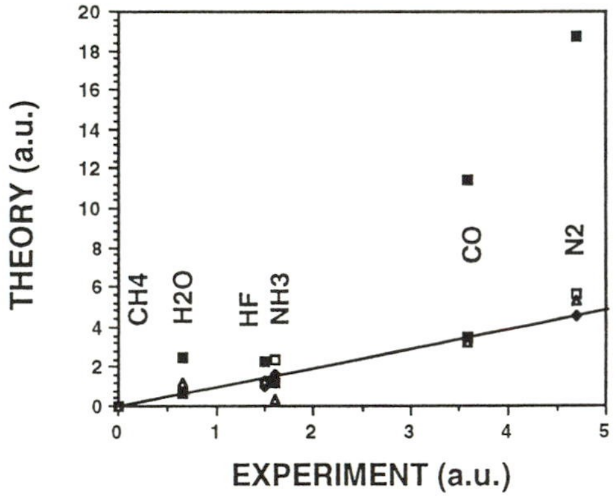

FIG. 2. Comparison of theoretical and experimental polarizability anisotropies for N_2, CO, CH_4, H_2O, NH_3, and HF: independent particle approximation, solid squares; random phase approximation, solid diamonds; LDA finite field, open squares; coupled Hartree-Fock, open triangles. The density-functional calculations used the local density approximation and the TZVP+ basis set. The coupled Hartree-Fock and experimental values are taken from Ref. [17]. See text for additional details.

TABLE III. Sensitivity of calculated dipole moment, mean polarizability, polarizability anisotropy, and mean hyperpolarizability of H_2O (in a.u.) to geometry and choice of functional. All calculations use the Sadlej basis set (see text).

Functional	Geometry	μ	$\bar{\alpha}$	$\Delta\alpha$	$\bar{\beta}$
LDA	Optimized	0.728	10.80	0.46	-20.0
B88x+P86c	Optimized	0.713	10.68	0.54	-18.4
LDA	Experimental	0.732	10.56	0.27	-19.1
B88x+P86c	Experimental	0.708	10.46	0.35	-17.4

less, it should be emphasized that a truly rigorous comparison with experiment would require the inclusion of finite frequency effects and vibrational contributions. Comparison with the singles, doubles, quadruples fourth-order Møller-Plesset perturbation theory results of Maroulis [10] (Table I) suggests that the LDA static electronic hyperpolarizability is too large.

Table III shows the sensitivity of our water results to the geometry used and choice of functional. Neither the mean polarizability nor the mean first hyperpolarizability is very sensitive to small changes in geometry. Roughly speaking, this is because the mean polarizability is a volume-like quantity and the mean hyperpolarizability is just its derivative. The polarizability anisotropy, being related to molecular shape, is much more sensitive to small changes in geometry. The B88x+P86c gradient-corrected functional is expected to yield improvements over the LDA for properties which depend upon the long range behavior of the charge density. However, although the improvements for water (and sodium clusters [46]) are in the right direction, they are not dramatic.

Dynamic results. Results are given here at the RPA/LDA level. A treatment including coupling of exchange-correlation effects will be reported in due course. We now have preliminary RPA/LDA results for a half dozen small molecules. For purposes of the present summary, we focus on N_2, an important benchmark molecule for calculation of excitation spectra [11,12,52], and one for which the experimental dynamic polarizability [51] and experimental excitation energies [53] are readily available.

Figure 3 shows our calculated dynamic mean polarizability in comparison with the experimental quantity. The frequency dependence is calculated at the RPA/LDA level, and is combined with the finite field LDA static value to give

$$\bar{\alpha}(\omega) = \left(\bar{\alpha}^{\mathrm{RPA}}(\omega) - \bar{\alpha}^{\mathrm{RPA}}(0)\right) + \bar{\alpha}^{\mathrm{FF}}(0)\,. \tag{30}$$

A similar procedure is sometimes adopted to graft the dynamic behavior from the time-dependent Hartree–Fock approximation (TDHFA) calculations onto better post-Hartree–Fock static calculations. The agreement with experiment is reasonably good.

Excitation spectra represent a considerably more challenging test of the RPA/LDA. We restrict our attention to the singlet-singlet transitions since, as was noted earlier, the singlet-triplet transitions are uncoupled at the RPA level. They are also "dark" states in the sense of having oscillator strengths which are

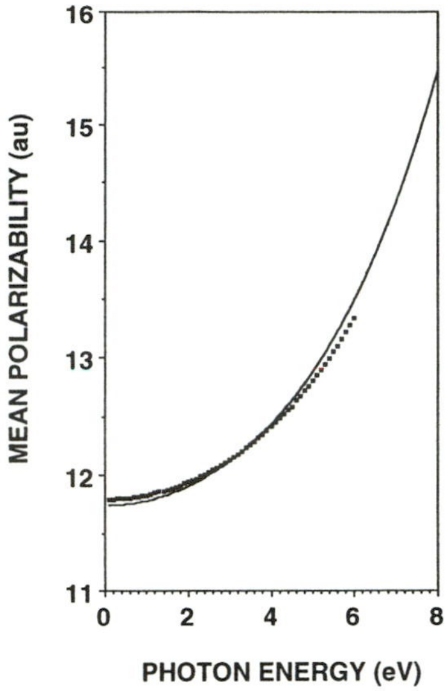

FIG. 3. Frequency dependence of the mean polarizability of N_2. The theoretical curve (dashed) is for the hybrid finite field–RPA/LDA calculation described in the text, with the TZVP+ basis set. The experimental curve (solid) is constructed from data taken from Ref. [52].

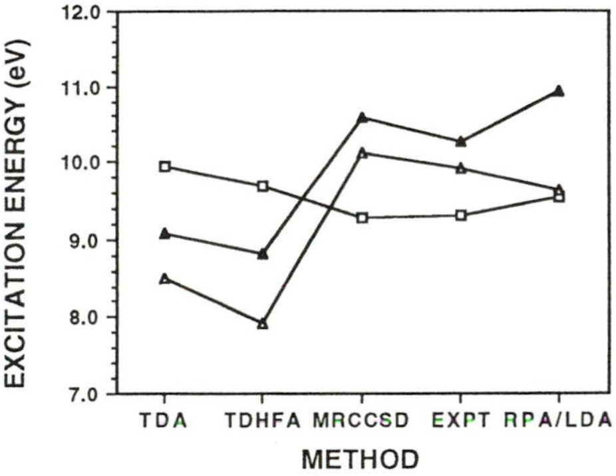

FIG. 4. Comparison of the first three singlet-singlet excitation energies of N_2 calculated by various methods with experiment. The Tamm–Dancoff approximation (TDA), time-dependent Hartree–Fock approximation (TDHFA) and singles and doubles multireference coupled cluster values are taken from Ref. [12]. The experimental values are taken from Ref. [52]. The RPA/LDA values were calculated using the Sadlej basis. The excited states and their dominant one-electron contributions are: a $^1\Pi_g$ ($3\sigma_u \rightarrow 1\pi_g$), open square; a $^1\Sigma_u^-$ ($1\pi_u \rightarrow 1\pi_g$), open triangle; and w $^1\Delta_u$ ($1\pi_u \rightarrow 1\pi_g$), solid triangle.

TABLE IV. Oscillator strengths for the first four vertical transitions of N_2 having nonzero oscillator strength. The RPA/LDA values are calculated with the Sadlej basis set and do *not* include a degeneracy factor of 2 for the $^1\Pi_u$ states. The time-dependent Hartree-Fock approximation (TDHFA) and second-order equations-of-motion (EOM2) oscillator strengths are taken from Ref. [11].

Excitation			States	TDHFA	EOM2	RPA/LDA
$3\sigma_g$	$\rightarrow$	$2\pi_u$	$c_3\ ^1\Pi_u$	0.091	0.12	0.02
$3\sigma_g$	$\rightarrow$	$3\sigma_u$	$c_4'\ ^1\Sigma_u^+$	0.65	0.094	0.11
$2\sigma_u$	$\rightarrow$	$1\pi_g$	$b\ ^1\Pi_u$	0.32	0.49	0.07
$1\pi_u$	$\rightarrow$	$1\pi_g$	$b'\ ^1\Sigma_u^+$	0.15	0.19	0.15

zero by symmetry. Figure 4 shows a comparison of the first 3 singlet-singlet vertical excitation energies for N_2, calculated by various methods, with the experimental values. Both the Tamm–Dancoff approximation (TDA), which is equivalent to a singles configuration interaction treatment of the excited states, and the TDHFA give the wrong ordering of these states. The RPA/LDA gives the correct ordering but, not surprisingly, does not do as well as multireference coupled cluster (MRCSD) calculations. These excitations are to spectroscopically "dark states". The excitation energies of the first four "bright states" calculated at the RPA/LDA level are compared in Figure 5 with excitation energies calculated using the TDHFA and using a second-order equation-of-motion (EOM2) method and with experimental transition energies. Calculated vertical transition energies for these states are strongly influenced by the presence of nearby avoided crossings of the excited state potential energy surfaces. Nevertheless, the RPA/LDA excitation energies are quite reasonable and all within about 1 eV of the experimental results. A comparison of oscillator strengths is given in Table IV. Experimental values are difficult to extract with precision and so have been omitted. Our RPA/LDA oscillator strengths do not seem to be fully converged with respect to basis set saturation, and should be viewed with caution.

The good quality of the results for N_2 are particularly noteworthy in view of the fact that conventional (time-independent) Kohn–Sham theory is a fundamentally single-determinantal theory. One of the important advantages of the present time-dependent density-functional response theory approach is that it provides a multi-determinantal treatment of the excitations. All of the excited states of N_2 treated here have an important multideterminantal character. This is especially true of the $^1\Sigma_u^-$ and $^1\Delta_u$ states each of which requires a minimum of four determinants simply to obtain a wavefunction of the correct symmetry. Our RPA/LDA calculation automatically includes not only those determinants required by symmetry, but also contributions from other determinants as well.

For the half dozen molecules studied so far, the singlet-singlet excitation energies obtained at the RPA/LDA level are generally within 1eV of the experimental values. It is interesting to note that the sum-over-states expression (22) implies a relationship between the quality of the excitation spectrum and the quality of the polarizability. Thus, for a molecule such as N_2, an absolute error of ≤ 1 eV in the excitation energies translates into a reasonably small error in the polarizability, yet for a molecule such as Na_2 with extraordinarily low excitation energies (first

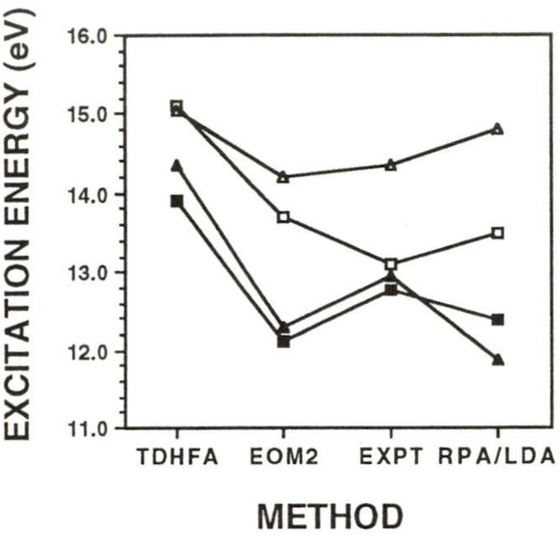

FIG. 5. Comparison of the first four excitation energies with nonzero oscillator strength of N_2 calculated by various methods with experiment. The time-dependent Hartree–Fock approximation (TDHFA) and second-order equation-of-motion (EOM2) values are taken from Ref. [11] and the experimental values were taken from Ref. [52]. The RPA/LDA values were calculated using the Sadlej basis set. The excited states and their dominant one-electron contributions are: b $^1\Pi_u$ $(2\sigma_u \rightarrow 1\pi_g)$, open square; b' $^1\Sigma_u^+$ $(1\pi_u \rightarrow 1\pi_g)$, open triangle; c_3 $^1\Pi_u$ $(3\sigma_g \rightarrow 2\pi_u)$, solid square; c_4' $^1\Sigma_u^+$ $(3\sigma_g \rightarrow 3\sigma_u)$, solid triangle.

bright state at about 1.8 eV [54]), the RPA/LDA polarizability is considerably worse.

CONCLUSION

This paper has given a summary of where we stand (as of fall 1994) in generating and calibrating essential DFT methodology for optical problems. In some ways the methods used here bear a close resemblence to Hartree–Fock-based techniques. However, whereas the Hartree–Fock method is an approximation, DFT electronic electrical response properties are formally exact in the limit of the exact exchange-correlation functional. This, together with efficiencies arising from the use of only local potentials in DFT, makes DFT a promising method for quantitative calculations of optical (and other) properties of molecules in a size range comparable to or greater than that now attainable with the Hartree–Fock method, provided, of course, that the approximate exchange-correlation functionals used in practical calculations are sufficiently accurate. That this is the case has been illustrated by the quality of static dipole moments, dipole polarizabilities, and first dipole polarizabilities of small molecules.

Since optical measurements are made with finite frequency electric fields, the extension to the time-dependent regime is important. Thus dynamic polarizabilities and excitation spectra, calculated at the RPA level, using the first general molecular implementation of time-dependent DFT (the DynaRho program), have been reported here for the first time. The results to date are quite encouraging, and a full treatment, including the response of the exchange-correlation potential, is already underway. This approach promises to become a powerful technique, applicable to a wide range of complex molecules and materials models.

ACKNOWLEDGMENTS

FB would like to thank the French Ministry of Foreign Affairs for financial support. Financial support from the Canadian Centre of Excellence in Molecular and Interfacial Dynamics (CEMAID), from the Natural Sciences and Engineering Research Council (NSERC) of Canada, and from the Fonds pour la formation de chercheurs et l'aide à la recherche (FCAR) of Quebec is gratefully acknowledged. We thank the Services informatiques de l'Université de Montréal for computing resources.

LITERATURE CITED

[1] Böttcher, C.J.F. *Theory of Electric Polarization. Volume I: Dielectrics in static fields*; Elsevier Scientific Publishing Company: Amsterdam, Holland, 1973.

[2] Böttcher, C.J.F.; Bordewijk, P. *Theory of Electric Polarization. Volume II: Dielectrics in time-dependent fields*; Elsevier Scientific Publishing Company: Amsterdam, Holland, 1978.

[3] *Nonlinear Optical Effects in Organic Polymers* Messier, J.; Kajzar, F.; Prasad, P.; Ulrich, D., Eds.; NATO Advanced Science Institutes Series E: Applied Sciences, Vol. 162; Kluwer Academic Publishers: Boston, Massachussets, 1989.

[4] *Organic Materials for Non-linear Optics: The Proceedings of a Conference Organised by the Applied Solid State Chemistry Group of the Dalton Division of the Royal Society of Chemistry* Hann, R.A.; Bloor, D., Eds.; Special Publication No. 69; Royal Society of Chemistry: London, England, 1989.

[5] *Materials for Nonlinear Optics, Chemical Perspectives* Marder S.R.; Sohn, J.E.; Stucky, G.D., Eds.; ACS Symposium Series 455; American Chemical Society: Washington, D.C., 1991.

[6] Special issue on nonlinear optics: *Int. J. Quant. Chem.* **1992**, *43*, 1.

[7] Special issue on optical nonlinearities in chemistry: *Chemical Reviews* **1994**, *94*, 1.

[8] Lefebvres, S.; Carrington, T.; Guan, J.; Casida, M.E.; Salahub, D.R. *work in progress.*

[9] Bohr, F.; Casida, M.E.; Guan, J.; Salahub, D.R. *in preparation.*

[10] Maroulis, G. *J. Chem. Phys.* **1991**, *94*, 1182.

[11] Oddershede, J.; Grüner, N.E.; Diercksen, G.H.F. *Chem. Phys.* **1985**, *97*, 303.

[12] Pal, S.; Rittby, M.; Bartlett, R.J.; Sinha, D.; Mukherjee, D. *J. Chem. Phys.* **1988**, *88*, 4357.

[13] Mahan, G.D.; Subbaswamy, K.R. *Local Density Theory of Polarizability*; Plenum Press: New York, New York, 1990.

[14] Jasien, P.G.; Fitzgerald, G. *J. Chem. Phys.* **1990**, *93*, 2554.

[15] Sim, F.; Salahub, D.R.; Chin, S. *Int. J. Quant. Chem.* **1992**, *43*, 463.

[16] Chong, D.P. *J. Chin. Chem. Soc.* **1992**, *39*, 375.

[17] Guan, J.; Duffy, P.; Carter, J.T.; Chong, D.P.; Casida, K.C.; Casida, M.E.; Wrinn, M., *J. Chem. Phys.* **1993**, *98*, 4753.

[18] Colwell, S.M.; Murray, C.W.; Handy, N.C.; Amos, R.D. *Chem. Phys. Lett.* **1993**, *210*, 261.

[19] Chong, D.P. *Chem. Phys. Lett.* **1994**, *217*, 539.

[20] Rubio, A.; Balbás, L.C.; Alonso, J.A. *Phys. Rev. B* **1992**, *45*, 13657.

[21] Balbás, L.C.; Rubio, A.; Torres, M.B. *Computational Materials Science* **1994**, *2*, 509.

[22] Levine, Z.H.; Soven, P. *Phys. Rev. Lett.* **1983**, *50*, 2074.

[23] Levine, Z.H.; Soven, P. *Phys. Rev. A* **1984**, *29*, 625.

[24] Casida, M.E. In *Recent Advances in Density Functional Methods*, Chong, D.P., Ed.; Recent Advances in Computational Chemistry; World Scientific, Singapore, *to appear*

[25] Dreizler, R.M.; Gross, E.K.U., *Density Functional Theory*; Springer-Verlag, New York, New York, 1990.

[26] Parr, R.G.; Yang, W. *Density-Functional Theory of Atoms and Molecules*; Oxford University Press, New York, New York, 1989.

[27] Malkin, V.G.; Malkina, O.L.; Salahub, D.R. *Chem. Phys. Lett.* **1993**, *204*, 80

[28] Malkin, V.G.; Malkina, O.L.; Salahub, D.R. *Chem. Phys. Lett.* **1993**, *204*, 87

[29] Malkin, V.G.; Malkina, O.L.; Casida, M.E.; Salahub, D.R. *J. Am. Chem. Soc.* **1994**, *116*, 5898

[30] Fournier, R. *J. Chem. Phys.* **1990**, *92*, 5422.
[31] Dunlap, B.I.; Andzelm, J. *Phys. Rev. A* **1991**, *45*, 81.
[32] Komornicki, A.; Fitzgerald, G. *J. Chem. Phys.* **1993**, *98*, 1398.
[33] Colwell, S.M.; Handy, N.C. *Chem. Phys. Lett.* **1994**, *217*, 271.
[34] Gross, E.K.U.; Kohn, W. *Adv. Quant. Chem.* **1990**, *21*, 255.
[35] Berkowitz, J. *Photoabsorption, Photoionization, and Photoelectron Spectroscopy*; Academic Press: New York, 1979; pp. 64-68.
[36] St-Amant, A.; Salahub, D.R. *Chem. Phys. Lett.* **1990**, *169*, 387.
[37] A. St-Amant, Ph.D. Thesis, Université de Montréal, 1992.
[38] Salahub, D.R.; Castro, M.; Proynov, E.I. In Malli, G.L. *Relativistic and Electron Correlation Effects in Molecules and Solids*; Nato Advanced Study Institute, Series B: Physics; Plenum: New York, New York, 1994, Vol. 318.
[39] Vosko, S.H.; Wilk, L.; Nusair, M. *Can. J. Phys.* **1980**, *58*, 1200.
[40] Becke, A.D. *Phys. Rev. A* **1988** *38*, 3098.
[41] Perdew, J.P. *Phys. Rev. B* **1986** *33*, 8822.
[42] Godbout, N.; Salahub, D.R.; Andzelm, J.; Wimmer, E. *Can. J. Chem.* **1992**, *70*, 560.
[43] Hehre, W.J.; Stewart, R.F.; Pople, J.A. *J. Chem. Phys.* **1969**, *51*, 2657.
[44] Sadlej, A.J. *Coll. Czech. Chem. Comm.* **1988**, *53*, 1995.
[45] Sadlej, A.J. *Theor. Chim. Acta* **1991**, *79*, 123.
[46] Guan, J.G.; Casida, M.; Köster, A.; Salahub, D.R., *Phys. Rev. B*, *submitted.*
[47] Duffy, P.; Chong, D.P.; Casida, M.E.; Salahub, D.R., *Phys. Rev. A* **1994**, *50*, 4707.
[48] Clough, S.A.; Beers, Y.; Klein, G.P.; Rothman, L.S. *J. Chem. Phys.* **1993**, *59*, 490.
[49] Murphy, W.F. *J. Chem. Phys.* **1977**, *67*, 5877.
[50] Ward, J.F.; Miller, C.K. *Phys. Rev. A* **1979**, *19*, 826.
[51] Zeiss, G.D.; Meath, W.J. *Mol. Phys.* **1977**, *33*, 1155.
[52] Kaldor, U.; Ben-Shlomo, S.B. *J. Chem. Phys.* **1990**, *92*, 3680.
[53] Lofthus, A.; Krupenie, P.H. *J. Phys. Chem. Ref. Data*, **1977**, *6*, 113.
[54] Huber, K.P.; Herzberg, G. *Molecular Spectra and Molecular Structure. IV. Constants of Diatomic Molecules*; Van Nostrand Reinhold Company: New York, New York, 1979.

RECEIVED August 14, 1995

Chapter 9

A Combined Hartree–Fock and Local-Density-Functional Method To Calculate Linear and Nonlinear Optical Properties of Molecules

Brett I. Dunlap[1] and Shashi P. Karna[2,3]

[1]Naval Research Laboratory, Code 6179, Washington, DC 20375–5342
[2]Photonics Research Laboratory, Department of Chemistry, State University of New York, Buffalo, NY 14214

The differences that are relevant to the calculation of optical properties by perturbation theory between the density-functional and Hartree-Fock one-electron equations are discussed. Most importantly, local density-functionals and Hartree-Fock underestimate and overestimate, respectively, the HOMO-LUMO gap defined as the difference between the ionization potential and the electron affinity of a molecule. This suggests averaging the density-functional and Hartree-Fock eigenvalues for use in a time-dependent Hartree-Fock calculation. The method is used to compute the linear and nonlinear static polarizabilities of HF, H_2O, and CO. The results are compared with the experimental as well as other theoretical data.

Accurate prediction of molecular linear and nonlinear optical properties by first-principles quantum chemical methods presents a major challenge to computational theoretical chemistry. The two most important factors influencing the accuracy of the predicted linear and nonlinear polarizabilities are the (i) atomic basis set and (ii) the amount of electron correlation included in the calculation. While the basis set problem in the accurate *ab initio* prediction of molecular polarizabilities has been somewhat alleviated by the development of the so called "direct" methods (*1*) that allow the use of a considerably extended space of one-electron functions and by the availability of appropriately optimized semidiffuse and diffuse polarization functions, adequate treatment of electron correlation (EC) remains a challenging problem. Several recent studies have indicated that the first-principles methods based on the Hartree-Fock (HF) theory generally underestimate molecular polarizability and hyperpolarizabilities compared to the gas phase experimental data (*2-4*). These studies have also shown that including EC may change the HF results by as much as 50%. Unfortunately, the post-HF methods based on perturbative

[3]Current address: Space Electronics Division, U.S. Air Force Phillips Laboratory, 3550 Aberdeen Avenue, Southeast, Kirtland Air Force Base, NM 87117–5776

or variational treatment of EC are too expensive to be used to investigate the nonlinear optical (NLO) properties of the larger molecular systems of practical interest.

The practical difficulties in the use of conventional first-principles quantum chemical methods to investigate optical nonlinearities of medium to large molecular systems has created a need for alternative, chemically accurate computational tools and theoretical models. One such theoretical model that offers a clear computational advantage over the conventional quantum chemical methods, while retaining high predictive chemical accuracy, is density functional theory (DFT). In recent years, there has been an increasing effort to extend the capabilities of DFT-based methods to predict molecular linear and NLO properties. Static linear and nonlinear polarizabilities of atoms and molecules have been predicted with reasonable accuracy both by the finite-field (FF) (*5-9)* and analytical derivative methods (*10)* within the DFT formalism. Although these calculations have been successful in demonstrating the capability of the DFT-based methods to predict NLO properties, the real challenge of predicting the experimentally observable quantity, which involves a non-zero optical frequency, has yet to be met.

Subbaswamy and coworkers (*11)* have used a perturbative scheme within the local-density approximation of DFT to predict the linear and nonlinear response of closed-shell atoms and ions. Their formalism uses a perturbation expansion of the Kohn-Sham equations (*12)* in the presence of a static electric field and self-consistent solution of the perturbed Kohn-Sham equations. In their formulation, the fourth-order energy, which gives the second-hyperpolarizability, γ, is obtained by solving the perturbed equations only up to the second-order. Although, a similar method including the perturbation due to an optical field has not yet been reported, the work of Subbaswamy and coworkers could form the basis of a computationally viable method within DFT to predict accurately NLO properties of systems of practical interest.

While there have been a number of promising efforts to apply and extend DFT to time-dependent phenomena (*12-14)*, a less dramatic departure from current NLO technology would be to correct the highly-developed time-dependent-Hartree-Fock (TDHF) method for known deficiencies. As will be shown below, the only zeroth-order quantities appearing in the coupled perturbed TDHF equations (*15)* are the self-consistent field (SCF) eigenvector coefficient matrix, $\mathbf{C}^{(0)}$, and the eigenvalues, $\epsilon^{(0)}$. The eigenvectors, and the wavefunctions that they are used to construct, are highly constrained by electronic orthogonality; they change relatively little with each order of perturbation theory. On the other hand, the rate of convergence of the perturbation expansion can be significantly improved by including the first and even higher-order corrections (as in the Feenberg formula (*16)*) to the zeroth order energies in the denominators of the various formulae. This suggests that a promising approach might be to correct in some practical manner the fact that the HF eigenvalues have too large a gap between the highest occupied molecular orbital (HOMO) and lowest occupied molecular orbital (LUMO).

In this chapter, we present a mixed HF-DFT method to compute molecular linear and NLO properties using Gaussian-type atomic basis functions. The motivation for this work has been to devise a method which combines the conceptual and mathematical framework of the time-dependent Hartree-Fock theory with the EC inherent in zeroth-order time-independent Kohn-Sham DFT. Becke has shown that averaging HF and DFT total energies leads to a more accurate description of a representative class of small molecules than either method alone (*17*). One might expect, therefore, that averaged zeroth-order eigenvalues could also lead to improved higher order quantities needed in the calculation of (hyper)polarizabilities.

The difference between the HF and density-functional eigenvalues is discussed in Section 2. The working equations to solve the first-order TDHF equations are described in section 3. Section 4 describes the molecules and the corresponding atomic Gaussian basis sets used in both the HF and the local-density-functional (LDF) SCF calculations. In section 5, the results of the application of the present method to the selected molecules are presented and compared with experimental and other theoretical results. Section 6 contains the conclusions of this preliminary investigation.

1. One-Electron Equations

Quantum-chemical methods begin with an expression for the total energy,

$$E = E_1 + E_C + E_{XC}, \tag{1}$$

which can be divided into three parts. The term, E_1 is the one-electron contributions to the energy, which includes the kinetic energy, the nuclear attraction energy, and the energy caused by external fields. The Coulombic energy of all the electrons E_C is naturally expressed in density-functional form,

$$E_C = \int \rho(\underline{\mathbf{r}}_1)\rho(\underline{\mathbf{r}}_2)\, d^3r_1 d^3r_2/(2r_{12}) = [\rho|\rho]. \tag{2}$$

The remainder E_{XC} is called the exchange-correlation energy.

In both HF and DFT, the density is the sum of the magnitudes squared of the one-electron orbitals,

$$\rho(\underline{\mathbf{r}}) = \sum_a n_a\, \phi_a^*(\underline{\mathbf{r}})\phi_a(\underline{\mathbf{r}}). \tag{3}$$

where n_a is the occupation number (0 or 2 herein) of the ath orbital $\phi_a(\underline{\mathbf{r}})$. These one-electron orbitals may be approximated as a linear-combination-of-Gaussian-type-orbital (LCGTO) expansion on each atom,

$$\phi_a(\underline{\mathbf{r}}) = \sum_i C_{ai} O_i(\underline{\mathbf{r}}). \tag{4}$$

Thus the density can be expressed as the product of the LCGTO density matrix and the set of orbital basis-set pairs,

$$\rho(\underline{\mathbf{r}}) = \sum_{ij} D_{ij} O_i^*(\underline{\mathbf{r}}) O_j(\underline{\mathbf{r}}), \tag{5}$$

where

$$D_{ij} = \sum_a n_a\, C_{ai}^* C_{aj}. \tag{6}$$

The optimal LCGTO coefficients are found by varying the total energy expression with respect to each coefficient in each occupied orbital. This variation leads to the one-electron equations,

$$[h_1 + V_C(\underline{r}) + V_{XC\,a}(\underline{r})] \sum_i C_{ai} O_i(\underline{r}) = \epsilon_{ab} \sum_j C_{bj} O_j(\underline{r}), \tag{7}$$

where h_1 contains the kinetic-energy operator, the nuclear potential, and the external electromagnetic potential. The variation leading to the Coulomb potential V_C can be rewritten as a density-functional variation,

$$V_C(\underline{r}) = \frac{\delta E_C}{\delta \rho(\underline{r})}. \tag{8}$$

Similarly in density-functional theory,

$$V_{XC}(\underline{r}) = \frac{\delta E_{XC}}{\delta \rho(\underline{r})}, \tag{9}$$

as a consequence of the Kohn-Sham theorem *(18)*. HF is not a DFT, and thus the exchange potential is nonlocal. The possibility that the XC potential is nonlocal is indicated by an orbital-dependent subscript in equation 7. One consequence of this difference is that the eigenvalue matrix ϵ_{ab} can always be diagonalized in DFT, but must contain off-diagonal terms in HF for certain open-shell systems *(19)*.

In DFT all electrons see the same local one-electron potential. In HF the occupied orbitals experience individual potentials due to the $N-1$ other electrons while the virtual orbitals experience a potential due to all N electrons. This causes the eigenvalue difference between the HOMO and LUMO to be underestimated in DFT and overestimated in HF.

The (diagonal) eigenvalues are very important because they are used in the denominators of perturbation theory. The eigenvalues approximate only to zeroth order the energetics of electron rearrangements in both theories. In DFT the eigenvalues are the derivatives of the total energy with respect to occupation number *(20)*. In HF the eigenvalues approximate these rearrangements via Koopman's theorem *(21)*. In both theories a more accurate calculation of electronic rearrangement energies requires proper treatment of relaxation, *i.e.*, an SCF calculation on both the initial and final states (ΔSCF).

For closed-shell molecules the lowest energy electronic excitation involves transfer of an electron from the HOMO to the LUMO. All other things being equal, this excitation would be expected to be the dominant process in perturbation theory due to its smallest energy denominator. Neglecting relaxation, this excitation can be divided into an ionization from the HOMO followed by electron capture into the LUMO, *i.e.* the difference between the ionization potential (IP) and the electron affinity (EA) of the molecule. DFT underestimates this "band gap" *(22,23)*, while HF overestimates it. This can be seen by comparing the relevant quantities for C_{60}. The experimental IP is 7.61 eV and the experimental EA is 2.65 eV *(24)*. The Perdew-Zunger local-density functional (LDF) *(25)* C_{60} HOMO and LUMO eigenvalues are -5.94 and -4.26 eV, respectively *(26)*. The HF C_{60} HOMO and LUMO eigenvalues are -7.97 and -0.65 eV, respectively (*27*). Thus the LDF, experimental, and HF "band gaps" of molecular C_{60} are 1.68, 4.96, and 7.32 eV, respectively.

Higher order terms in perturbation theory can be combined into lower-order terms to have the effect of moving the eigenvalues into closer agreement with the corresponding experimental energies (*28*). Thus one can often improve low-order perturbation theory results by replacing the HF eigenvalues with experimental excitation energies. The bracketing nature of DFT and HF eigenvalue differences between occupied and virtual orbitals suggests that replacing the HF eigenvalues by the the average of HF and DFT eigenvalues might similarly improve the results of low-order perturbation theory, without resorting to experimental quantities.

2. Perturbation Theory

The elements of the linear polarizability tensor α are obtained as

$$\alpha_{ab}(-\omega_a;\quad \omega_b) = -Tr[h^{(1)}_a D^{(1)}(\omega_b)];\quad a,b = x,y,z \tag{10}$$

Where, the perturbation $h^{(1)}_a(\equiv \mu_a)$ is the ath component of the dipole moment matrix and

$$D^{(1)}(\omega_a) = C^{(1)}(\omega_a)nC^{(0)\dagger} + C^{(0)}nC^{(1)\dagger}(-\omega_a) \tag{11}$$

is the first-order density matrix. The perturbed eigenvector matrix $C^{(1)}$ is obtained from the iterative solution of the first-order TDCPHF equation

$$F^{(1)}(\omega_a)C^{(0)} + F^{(0)}C^{(1)}(\omega_a) + \omega S^{(0)}C^{(1)}(\omega_a) = S^{(0)}C^{(0)}\epsilon^{(1)}(\omega_a) + S^{(0)}C^{(1)}(\omega_a)\epsilon^{(0)} \tag{12}$$

subject to the orthonormalization condition

$$C^{(1)\dagger}(\omega_a)nC^{(0)} + C^{(0)\dagger}nC^{(1)}(\omega_a) = 0 \tag{13}$$

The perturbed eigenvector matrix $C^{(1)}(\omega_a)$ can be written as

$$C^{(1)}(\omega_a) = C^{(0)}R^{(1)}(\omega_a) \tag{14}$$

where the transformation matrix $R^{(1)}(\omega_a)$ is defined as

$$\begin{aligned} R^{(1)}_{ij}(\omega_a) &= \frac{G^{(1)}_{ij}(\omega_a)}{\epsilon^0_j - \epsilon^0_i - \hbar\omega};\quad i \in occ;\quad j \in virt, \\ R^a_{ji}(\omega) &= -R^a_{ij}(\omega) \end{aligned} \tag{15}$$

and the MO basis Fock matrix, $G^{(1)}(\omega_a)$ is defined as

$$G^{(1)}(\omega_a) = C^{(0)\dagger}F^{(1)}(\omega_a)C^{(0)} \tag{16}$$

The first order F matrix has the structure

$$F^{(1)}(\omega_a) = h^{(1)}_a + D^{(1)}(\omega_a)\ [2J - K] \tag{17}$$

Where, J anad K are the two-electron Coulomb and exchange integrals. In the present formulation, it is assumed that the two-electron integrals (as well as the atomic overlap integrals, S) are not affected by the external perturbation.

The flow of the first-order TDHF calculation can be written as: $F^{(1)}(\omega_a)$ ($\equiv h^{(1)}$ first iteration)$\longrightarrow G^{(1)} \longrightarrow R^{(1)} \longrightarrow C^{(1)} \longrightarrow D^{(1)}(\omega_a)$ until the convergence of $D^{(1)}(\omega_a)$ and therefore of $R^{(1)}(\omega_a)$.

It is clear that apart from the atomic integrals ($h^{(1)}$, J and K), the only quantities needed to obtain the first-order density matrix are the zeroth-order eigenvalues, ϵ^0, (in equation 15) and the eigenvectors, $C^{(0)}$ (in equations 11-16) which, in the TDCPHF formulation are obtained from the usual SCF HF calculation. In the present HF-DFT formulation, we replace the HF ϵ^0 of equation 15 with the average of the HF and LDF eigenvalues.

Once the first-order transformation matrix, $R^{(1)}$, MO basis Fock matrix $G^{(1)}$ and the eigenvalue matrix $\epsilon^{(1)}$ (diagonal blocks of $G^{(1)}$ matrix) have been obtained self-consistently, the elements of the first-hyperpolarizability tensor, β are calculated as

$$\begin{aligned}
&\beta_{abc}(-\omega_a;\omega_b,\omega_c) \\
&= -Tr[n\{R(-\omega_a)G(\omega_b)R(\omega_c) + R(-\omega_a)G(\omega_c)R(\omega_b) \\
&\quad + R(\omega_b)G(\omega_c)R(-\omega_a) + R(\omega_b)G(-\omega_a)R(\omega_c) \\
&\quad + R(\omega_c)G(-\omega_a)R(\omega_b) + R(\omega_c)G(\omega_b)R(-\omega_a)\}] \\
&\quad - Tr[n\{R(-\omega_a)R(\omega_b)\epsilon(\omega_c) + R(-\omega_a)R(\omega_c)\epsilon(\omega_b) \\
&\quad + R(\omega_b)R(\omega_c)\epsilon(-\omega_a) + R(\omega_b)R(-\omega_a)\epsilon(\omega_c) \\
&\quad + R(\omega_c)R(-\omega_a)\epsilon(\omega_b) + R(\omega_c)R(\omega_b)\epsilon(-\omega_a)\}]
\end{aligned} \tag{18}$$

where,

$$\omega_a = \omega_b + \omega_c \tag{19}$$

The full derivation of equation 9 has been given previously *(29)*. In it the superscript indicating the order of perturbation (first in this case) has been droped for the sake of simplicity.

The experimentally useful quantities reported in the tables are:

$$< \alpha >= \frac{1}{3}\sum_i \alpha_{ii}; \quad i = x, y, z \tag{20}$$

$$\beta_{\parallel} = \frac{\mu \cdot \beta}{|\mu|} \tag{21}$$

In the above equation,

$$\beta_i = \frac{1}{5}\sum_j (\beta_{ijj} + \beta_{jij} + \beta_{jji}); \quad i, j = x, y, z \tag{22}$$

3. SCF Calculations

The experimental geometry of the three molecules as listed in Ref. 4 was used in the present calculations. The Hartree-Fock calculations were performed by the PHOTON system of electronic structure program *(30)* runnning on IBM RS/6000-550 and Silicon Graphics INDIGO machines. The LDF calculations were performed by the LCGTO-DF code *(31)*. In the present implementation of the above described strategy, the zeroth-order eigenvalues obtained from each of the HF and

LDF calculations are read from separate files and averaged. HF and LDF one-electron orbitals are largely determined by the orthogonality constraints and differ very little (*32*). Thus, consistent with our desire to make minimal changes to existent technology, the zeroth-order HF eigenvectors and average of the zeroth-order HF and LDF eigenvalues were input to the higher-order calculations.

Almost identical Gaussian primitive orbital basis sets were used in both the HF and LDF calculations. The only difference was that a sixth d_{r^2}-function in each *d* primitive-basis-set shell was not used in the LDF calculation. The LDF calculations require two additional basis sets to fit the charge density and exchange-correlation energy density (*33*). The primitive basis sets for H, C, O, and F were the 6s, 11s/7p, 12s/7p, and 12s/7p bases, respectively of Ref. 34. Each set of orbital exponents were divided into two groups, a core and valence set. The core set was contracted according to an atomic LDF calculation. The valence set was left uncontracted. The contraction schemes for H, C, O, and F were 1,4; 2,3/1,2; 2,3/1,2; and 1,5/1,4; respectively, where the slash separates angular momenta, and for each angular momentum the first number gives the number of core contractions and the second give the number of uncontracted diffuse functions. The polarization function exponents for H, C, O, and F were chosen as 1.0, 0.6, 1.2, and 1.3 atomic units (a.u., in this case a_o^2), respectively.

The fitting basis sets—required only for the LDF calculations—were for the most part scaled from the orbital basis set. All *s* orbital exponents were scaled by 2 and 2/3 for the charge density and exchange-correlation basis sets, respectively, and used without contraction. Three r^2 fitting functions with exponents that were double the third, fifth, and sixth most diffuse *p* exponents were used in the the charge density fitting basis of O. Similarly, three r^2 fitting functions with exponents that were double the second, fourth, and sixth most diffuse *p* exponents were used in the the charge density fitting fitting basis of F. In both fitting basis sets for H, a *p* exponent of 1.0 a.u. was used. In both fitting basis sets for C, five *p* and *d* exponents of 0.25, 0.37, 0.7, 2.0, and 5.0 a.u. were used. In both fitting basis sets for O and F, five *p* and *d* exponents of 0.1, 0.3, 0.6, 2.0, and 5.0 a.u. were used.

4. Application to H_2O, HF, and CO

The results of the present calculations are listed in Tables 1 – Tables 3. Also listed in the tables are the singles and doubles coupled cluster (triples) [CCSD(T)] (*4*) and gradient-corrected DFT using finite fields DFT-FF (*7*) theoretical results along with the available experimental data for comparison.

There is good agreement on the linear polarizabilities, α, of these three molecules among all methods. Consistent with our analysis that DFT underestimates the HOMO-LUMO gap, the DFT-FF values are consistently slightly higher than both CCSD and experiment. For the three molecules studied, the present method shows excellent agreement with the experiment and the CCSD(T) results for H_2O and CO. For the HF molecule, the present method underestimates the spherically averaged polarizability value, $< \alpha >$, by about 20%. In fact, the calculated α_{zz} value of HF are in excellent agreement with the experiment and also with other theoretical calculations. However, the perpendicular components, α_{xx} and α_{yy}, are heavily underestimated, by almost a factor of 2. The reason for this apparent discrepancy

Table 1. Comparison with literature values of the static linear and nonlinear optical properties for H_2O.

Molecule			H_2O	
Method	CCSD(T)	DFT(FF)	PW	EXP
α_{xx} (a.u.)	9.6362	10.748	7.55	9.55
α_{yy} (a.u.)	10.02	10.532	10.28	10.31
α_{zz} (a.u.)	9.79	10.370	8.53	9.90
$<\alpha>$ (a.u.)	9.79	10.550	8.79	9.81
β_{zzz} (a.u.)	-13.7		-11.91	
β_{zxx} (a.u.)	-6.2		2.90	
β_{zyy} (a.u.)	-10.2		-22.90	
$\beta_{\|\|}$ (a.u.)	-18.0		-19.56	-22±6

CCSD(T): Coupled Cluster (Triple excitations) (*4*).
DFT(FF): Density Functional Theory using Finite-Fields (*7*).
PW: Present work—average HF and LDF eigenvalues.
EXP: Experimental work cited in Refs. 4 and 7.

Table 2. Comparison with literature values of the static linear and nonlinear optical properties for HF.

Molecule			HF	
Method	CCSD(T)	DFT(FF)	PW	EXP
α_{xx} (a.u.)	5.3398	6.251	2.92	5.08
α_{yy} (a.u.)	5.3398	6.251	2.92	5.08
α_{zz} (a.u.)	6.4378	6.764	6.18	6.40
$<\alpha>$ (a.u.)	5.71	6.422	4.01	5.52
β_{zzz} (a.u.)	- 9.62		-15.06	
β_{zxx} (a.u.)	- 1.27		-1.50	
$\beta_{\|\|}$ (a.u.)	- 7.30		-10.84	-10.9±0.95

in the values of α_{xx} and α_{yy} for HF is not clear at this moment, but it should be noted that our orbital basis sets have been selected only for their ability to accurately reproduce the total uncontracted LDF energy of the isolated atoms.

There is much less data with which to compare our first hyperpolarizabilities, β. The calculated $\beta_{||}$ values for HF and H_2O show excellent agreement with experiment. The $\beta_{||}$ values calculated by the present method also show good accord with their CCSD(T) counterparts, although, there seems to be substantial difference in the individual components of β obtained by the two methods.

In these studies we have modified our coupled-perturbed TDHF computer code to read in LDF eigenvalues, which are ordered according to increasing energy, and then to average these with the HF eigenvalues, which are ordered similarly. This average corrects known deficiencies of both methods for eigenvalues about the

Table 3. Comparison with literature values of the static linear optical properties for CO.

Molecule	CO		
Method	CCSD(T)	PW	EXP
α_{xx} (a.u.)	11.7332	14.28	
α_{yy} (a.u.)	11.7332	14.28	
α_{zz} (a.u.)	15.6522	19.23	
$< \alpha >$ (a.u.)	13.04	15.94	13.08

Fermi level. This simplest procedure is not optimal for the highly excited virtual levels which have considerable d_{r^2} character, which is missing in our present LDF treatment. In fact the number of molecular orbitals in both methods do not agree. However, our calculated quantities are unchanged when the highest lying orbitals included in the HF calculation and missing from the LDF calculation are effectively removed from the perturbation-theory calculations by tripling their eigenvalues.

5. Conclusions

Considering the simplicity of the theoretical model presented in this work, the calculated results are very encouraging. The results of the three molecules investigated suggest that a combination of DFT and coupled perturbed HF methods does in fact yield accurate linear and nonlinear optical properties of molecules. The attractive feature of the approach presented here is that the computational cost of the perturbation calculations are much less than EC methods, such as CCSD(T) *(4)* or MP/2 *(4)*. The drawback of the method is that it needs two separate zeroth-order calculations. Although, for small to medium size molecules, this does not present any major bottleneck. As the size of the molecules increases, however, the cost of the calculations scale as does HF, instead of the significantly less demanding scaling of DFT calculations that use variational fitting *(33,35)*. Thus, it would be desirable to devise a full DFT-based method, perhaps similar to those of Subbaswamy et al, so that one can avoid the time-consuming two-electron integral evaluation of the Hartree-Fock method. The work in that direction is in progress in our laboratories and will be communicated in a forthcoming paper.

Acknowledgement

This work was supported by the Office of Naval Research through the Naval Research Laboratory and the Air Force Office of Scientific Research.

Literature Cited

1. Karna, S. P.; *Chem. Phys. Lett.* **1993**, *214*, 201.
2. Perrin, E.; Prasad, P. N.; Mougenot, P.; Dupuis, M.; *J. Chem. Phys.* **1989**, *91*, 4728.

3. Rice, J. E.; Handy, N. C. *Int. J. Quantum Chem.* **1992**, *43*, 91.
4. Sekino, H.; Bartlett, R. J. *J. Chem. Phys.* **1993**, *98*, 3022.
5. Moullet, I.; Martins, J. L. *J. Chem. Phys.* **1989**, *92*, 527.
6. Jasien, P. G.; Fitzgerald, G. *J. Chem. Phys.* **1990**, *93*, 2554.
7. Sim, F.; Salahub D. R.; Chin, S. *Int. J. Quantum Chem.* **1992**, *43*, 463.
8. Pederson, M.; Quong, A. A. *Phys. Rev.* B **1992**, *46*, 13548.
9. Guan, J.; Duffy, P.; Carter, J. T.; Chong, D. P.; Casida, K. C.; Casida, M. E.; Wrinn, M. *J. Chem. Phys.* **1993**, *98*, 4753.
10. Colwell, S. M.; Murray, C. W.; Handy N. C.; Amos, R. D.; *Chem. Phys. Lett.* **1993**, *210*, 261.
11. Subbaswamy, K. R.; Mahan, G. D. *J. Chem. Phys.* **1986**, *84*, 15; Senatore G.; Subbaswamy, K. R. *Phys. Rev.* A **1986**, *34*, 3619.
12. Stott, M. J. Zaremba, E. *Phys. Rev.* A **1980**, *21*, 12.
13. Levine, Z. H. Soven, P. *Phys. Rev.* A **1984**, *29*, 625.
14. Gross, E. K. U.; Kohn, W. *Phys. Rev. Lett.* **1985**, *55*, 2850.
15. Sekino, H.; Bartlett, R. J. *J. Chem. Phys.* **1986**, *85*, 976.
16. Morse, P. M.; Feshbach, H. *Methods of Theoretical Physics*; McGraw-Hill; New York, NY, 1953; Part 2, pp 1010-1033.
17. Becke, A. D. *J. Chem. Phys.***1993**, *98*, 1272.
18. Kohn, W.; Sham, L. J.; *Phys. Rev.* **1965**, *140*, A1133.
19. Fischer, C. F. *The Hartree-Fock Method for Atoms*; Wiley: New York, NY, 1977.
20. Janak, J. F. *Phys. Rev.* B **1978**, *18*, 7165.
21. Szabo, S.; Ostlund, N. S. *Modern Quantum Chemistry*; McMillan: New York, NY, 1982.
22. Perdew, J. P.; Levy, M. *Phys. Rev. Lett.* **1983**, *51*, 1884.
23. Sham L. J.; Schlüter, M. *Phys. Rev. Lett.* **1983**, *51*, 1888.
24. Diogo, H. P.; Minas da Piedade, M. E.; Dennis, T. J. S.; Hare, J. P.; Kroto, H. W.; Taylor, R.; Walton, D. R. *J. Chem. Soc. Faraday Trans.* **1993**. *89*, 3541.
25. Perdew, J. P.; Zunger, A. *Phys. Rev.* B **1981**, *23*, 5048.
26. Dunlap B. I.; Brenner D. W.; Mintmire J. W.; Mowrey R. C.; White C. T. *J. Phys. Chem.* **1991**, *95*, 8735.
27. Scuseria, G. E. *Chem. Phys. Lett.* **1991**, *176*, 423.
28. Kelly, H. P. *Physica Scripta* **1987**, *T17*, 109.
29. Karna, S. P.; Dupuis, M., *J. Comp. Chem.* **1991**, *12*, 487.
30. PHOTON system of ab initio electronic structure code has been written by Shashi Karna and is based on the initial works of Professor P. Chandra and his group at Banaras Hindu University, Varanasi, India, 1983. See, Chandra, P.; Buenker, R.J., *J. Chem. Phys.* **1983**, *79*, 358, 366.
31. Dunlap B. I.; Rösch, N. *Adv. Quantum Chem.* **1990**, *21*, 317.
32. Bursten, B. E.; Fenske, R. F. *J. Chem. Phys.* **1977**, *67*, 3138.
33. Dunlap, B. I.; Connolly, J. W. D.; Sabin, J. R. *J. Chem. Phys.* **1979**, *71*, 3396; 4993.
34. van Duijneveldt, F. B. *IBM Research Report RJ945* **1971**.
35. Dunlap, B. I.; Andzelm, J.; Mintmire, J. W. *Phys. Rev.* A **1990**, *42*, 6354.

RECEIVED October 2, 1995

Chapter 10

Theory of Nonlinear Optical Properties of Quasi-1D Periodic Polymers

Janos J. Ladik

Institute for Theoretical Chemistry, Friedrich-Alexander University Erlangen-Nuremberg, Egerland Strasse 3, D–91058 Erlangen, Germany

A theory is described for the effect of static and dynamic electric fields on the electronic structure of a quasi-1D polymer. For the static field a simple perturbational method and for the dynamic one the coupled Hartree-Fock equations were formulated. After introducing a basis set the resulting hypermatrix equations have been simplified by using the periodic symmetry of the polymer. Thus, all matrices have only the rank of the number of basis functions per unit cell. After solving the problem one obtains the crystal orbitals in the presence of both electric fields. Applying them the total energy per cell can be calculated also with correlation. The derivatives of the energy according to the field components give the static and dynamic (hyper)polarizability tensor elements. Preliminary calculations for the static case resulted in much larger polarizabilities than using different extrapolation methods from olygomers. Finally, the extension of the method for the interaction of a polymer with a laser pulse is shown.

Non-linear optics has a great practical importance in electrooptics, in optical switches and modulators (1,2,3). The discovery of laser has provided the ideal tool to study non-linear optical phenomena in molecules and polymers experimentally.

The theoretical treatment of a molecule or a polymer in the presence of an electric field generally and of a laser beam presents a formidable problem. In this paper we shall remain within the framework of the Born-Oppenheimer approximation (we shall not consider the change of the phonons in the presence of a laser pulse because we shall work in a fixed nuclear framework). Further, we shall not take into account the effect of the interaction between linear polymers on their polarizabilities and hyperpolarizabilities, though both effects are non-neglible (4,5,6).

We shall not review here the different semiempirical (not very successful)

0097–6156/96/0628–0174$15.00/0

calculations on non-linear optical properties of molecules (for a partial review for them see the Introduction of (7)). In the ab initio case there are some rather successful calculations for static and dynamic polarizabilities and hyperpolarizibilities of smaller molecules (8,9,10). It is questionable, however, how well would work the perturbational method used by the authors for larger molecules interacting with laser light.

With polymers there is the additional problem that the potential of an electric field $\underline{E}$, $\underline{E}\,\underline{r}$ is unbounded and this destroys the translational symmetry of a periodic polymer. Because of this difficulty in a larger number of calculations various authors have applied different extrapolation methods for the (hyper)polarizibilities starting from oligomers with increasing number of units. Only in a few cases has been attempted to treat infinite polymers in the tight binding and ab initio Hartree-Fock level. The latter calculations use, however, a formalism which is so complicated that its application to polymers with larger unit cells seems to be prohibitive (for a review see the Introduction of (11)).

The purpose of the present paper is to present a full theory for static and dynamic (hyper)polarizabilities of periodic quasi 1D polymers at an ab initio Hartree-Fock + correlation level. The theory will be described at two different levels: 1). interaction of an electric field $\underline{E}$ with a periodic polymer, 2). interaction of a laser pulse (taking into account both electric and magnetic field strengths) with a quasi 1D periodic polymer.

To be able to formulate the theory one has to treat first of all the problem of the unbounded operator $\underline{E}\,\underline{r}$. If $\underline{E}$ is homogeneous (which is fulfilled in a good approximation within a laser pulse) we can apply following Mott and Jones (12) (see also: (11)) the Nabla-operator ∇_k to a Bloch function

$$\varphi_n(\underline{k},\underline{r}) = e^{i\underline{k}\underline{r}} u_n(\underline{k},\underline{r}) \tag{1}$$

(here, as it is well-known $u_n(\underline{k},\underline{r})$ is lattice periodic):

$$\begin{aligned}\nabla_k \varphi_n(\underline{k},\underline{r}) &= i\underline{r}\varphi_n(\underline{k},\underline{r}) + e^{i\underline{k}\underline{r}}\, \nabla_k u_n(\underline{k},\underline{r}) = \\ &= i\underline{r}\varphi_n(\underline{k},\underline{r}) + e^{i\underline{k}\underline{r}}\, \nabla_k e^{-i\underline{k}\underline{r}} \varphi_n(\underline{k},\underline{r})\end{aligned} \tag{2}$$

Multiplying both sides of equation (2) by $-ie\underline{E}$ one obtains after reordering the terms

$$-e\underline{E}\,\underline{r}\varphi_n(\underline{k},\underline{r}) = -ie\underline{E}e^{i\underline{k}\underline{r}}\, \nabla_k e^{-i\underline{k}\underline{r}} \varphi_n(\underline{k},\underline{r}) + ie\underline{E}\, \nabla_k \varphi_n(\underline{k},\underline{r}) \tag{3}$$

If we multiply (3) by a Bloch function belonging to band m with a value $\underline{k}'$, we find for the matrix elements of the first term of (3) on the r.h.s.

$$\begin{aligned}-ie\left\langle \varphi_m(\underline{k}',\underline{r}) \right| \underline{E}e^{i\underline{k}\underline{r}}\, \nabla_k e^{-i\underline{k}\underline{r}} \left| \varphi_n(\underline{k},\underline{r}) \right\rangle &= \\ = -ie\underline{E}\int d\underline{r}\, e^{i(\underline{k}-\underline{k}')\underline{r}} u_m(\underline{k}',\underline{r})\, \nabla_k u_n(\underline{k},\underline{r})\end{aligned} \tag{4}$$

vanishes unless $\underline{k}'=\underline{k}$. In the latter case the remaining integrand and with it the integral is lattice periodic. Since it allows interband mixing (generally $m \neq n$), it describes the polarization of the system in the presence of $\underline{E}$. On the other hand, the second matrix element originating from the r.h.s. of (3)

$$\left\langle \varphi_m(\underline{k}',\underline{r}) \mid \nabla_k \varphi_n(\underline{k},\underline{r}) \right\rangle \tag{5}$$

is not lattice periodic. Since, however, this term allows the change of the quasi momentum vektor, $\underline{k}' \neq \underline{k}$, $\Delta\underline{k}=\underline{k}'-\underline{k}$, this term corresponds to a polarization current. Therefore, it has not to be taken into account if we want to treat theoretically the non-linear optical properties of a periodic system (13) which depend on the (hyper) polarizabilities (charge redistribution) in a molecule or a chain in the presence of an electric field, but not on the movement of charges (current).

Interaction of Static and Time-dependent Electric Fields with Quasi 1D Polymers

a) Derivation of the Coupled Hartree-Fock Equations. Let us assume that we have a homogeneous electric field $\underline{E}$,

$$\begin{aligned}\underline{E} &= \underline{E}_{st} + \underline{E}_{\omega} = \underline{E}_{st} + \sum_{m=1}^{M} \underline{E}_0 (e^{im\omega t} + e^{-im\omega t}) = \\ &= \underline{E}_{st} + \sum_{m=1}^{M} \underline{E}_0 2\cos(m\omega t)\end{aligned} \tag{6}$$

Here $\underline{E}_{st}$ is the static field and in the case of the time-dependent field $\underline{E}_{\omega}$ with frequency ω and amplitude $\underline{E}_0$ we have taken into account also the overtones. At the same time we assume that in this case that no magnetic field $\underline{H}$ is present, that is $\underline{A}=\underline{O}$ ($\underline{H}=\mathrm{curl}\underline{A}$).

In this case the total Hamiltonian of an n-electron system can be written as

$$\hat{H} = \hat{H}_0 + \hat{H}' \tag{7}$$

where $\hat{H}_0$ is the unperturbed Hamiltonian of the n-electron system. The field-dependent part of $\hat{H}$, $\hat{H}'$ can be expressed, if we take into account equation (3) (and the text after it) as well (6), as

$$\begin{aligned}&\hat{H}'(\underline{r}_1,\ldots,\underline{r}_n,\underline{E}_{st},\underline{E}_0,\omega,t) = \\ &= \sum_{i=1}^{n} [\underline{E}_{st} e^{ikr_i} \nabla_k e^{-ikr_i} + \underline{E}_0 e^{ikr_i} \nabla_k e^{-ikr_i} \sum 2\cos(m\omega t)]\end{aligned} \tag{8}$$

One can substitute (7) with (8) into Frenkel's variational principle (14) which provides the condition for the existence of a stationary state (see also (15))

$$J = \frac{\left\langle \Phi \middle| \hat{H} - i\dfrac{\partial}{\partial t} \middle| \Phi \right\rangle}{\langle \Phi | \Phi \rangle} \; ; \; \delta J = 0 \tag{9}$$

We apply for the field and time-dependent n-electron wave function the Ansatz

$$\Phi(\underline{r}_1,\ldots,\underline{r}_n,\underline{E}_{st},\underline{E}_0,\omega,t) = e^{-iW_0 t} \hat{A} \prod_{i=1}^{n} \tilde{\varphi}_i(\underline{r}_i,\underline{E}_{st},\underline{E}_0,\omega,t) \tag{10}$$

Here W_0 is the total energy of the n-electron system in the field-free case (the eigenvalue of $\hat{H}_0$), $\hat{A}$ is the antisymmetrizer and the one electron orbitals in the presence of the fields are

$$\tilde{\varphi}_i(\underline{r}_i,\underline{E}_{st},\underline{E}_0,\omega,t) = \varphi_i(\underline{r}_i) + \varphi_i^{st}(\underline{r}_i,\underline{E}_{st}) + \\ + \sum [\Delta\varphi^+_{i,m}(\underline{r}_i,\underline{E}_0)\, e^{im\omega t} + \Delta\varphi^-_{i,m}(\underline{r}_i,\underline{E}_0)\, e^{-im\omega t}] \tag{11}$$

One should mention that the effect of the static field $\underline{E}_{st}$ one can calculate with the help of a modified Fock operator ($\hat{F}_0$ is the Fock operator belonging to $\hat{H}_0$)

$$\hat{F} = \hat{F}_0 + \hat{F}_{st}\ , \quad where\ \hat{F}_{st} = -i|e|\underline{E}_{st} e^{i\underline{k}\underline{r}}\, \nabla_{\underline{k}} e^{-i\underline{k}\underline{r}} \tag{12}$$

with the help of simple first-order perturbation theory. In this way one obtains instead of

$$\hat{F}_0\varphi_i = \epsilon_i^{(0)}\varphi_i \tag{13}$$

a shift $\Delta\epsilon_i^{st}$ with respect to $\epsilon_i^{(0)}$ and the correction $\varphi_i^{st}(\underline{r},\underline{E}_{st})$ with respect to $\varphi_i(\underline{r})$.

To determine the effect of the time-dependent-field $\underline{E}_\omega$ one has to substitute equation (10) with (11) into equation (9), and perform the variation of J with respect to the unknown functions $\Delta\varphi^+_{i,m}$ and $\Delta\varphi^-_{i,m}$, respectively. One obtains in this way the coupled Hartree-Fock (RPA) equations for a closed shell system

$$[\hat{F} - (\epsilon_i^{(0)} + \Delta\epsilon_i^{st}(\underline{E}_{st})) \pm m\omega]\,|\Delta\varphi^{\pm}_{i,m}(\underline{r}_1,\underline{E}_0)\rangle + \hat{h}|\varphi'_i(\underline{r}_1,\underline{E}_{st})\rangle + \\ + \sum_j^{\frac{n}{2}} [(\langle\varphi'_j(\underline{r}_2,\underline{E}_{st}|\frac{2-\hat{P}_{1\leftrightarrow 2}}{r_{12}}|\Delta\varphi^{\pm}_{j,m}(\underline{r}_2,\underline{E}_0)\rangle_2 + h.c.)]\,|\varphi'_i(\underline{r}_1,\underline{E}_{st})\rangle = 0 \\ (i = 1,2,\ldots,n\ ;\ m = 1,2,\ldots,M) \tag{14}$$

$$Here \quad \varphi'_i(\underline{r},\underline{E}_{st}) = \varphi_i(\underline{r}) + \varphi_i^{st}(\underline{r},\underline{E}_{st}) \\ and \quad \hat{h} = -i|e|\underline{E}_0 e^{i\underline{k}\underline{r}_1}\, \nabla_{\underline{k}}\, e^{-i\underline{k}\underline{r}_1} \tag{15}$$

Further in the case of a quasi 1D periodic polymer

$$\hat{F}_0 = \sum_{j=1}^{\frac{n}{2}} \langle\varphi_j(\underline{r}_2)|\frac{2-\hat{P}_{1\leftrightarrow 2}}{r_{12}}|\varphi_j(\underline{r}_2)\rangle_2 + \hat{H}^N \tag{16a}$$

$$\hat{H}^N = -\frac{1}{2}\Delta_1 - \sum_{l=1}^{2N+1}\sum_{\alpha=1}^{N_\alpha} \frac{Z_\alpha}{|\underline{r}_1 - \underline{R}_\alpha^\ell|} \tag{16b}$$

where (2N+1) is the number of unit cells in the polymer, N_α the number of nuclei in the unit cell, Z_α the charge of the α-th nucleus and, finally, $\underline{R}_\alpha^\ell$ is the position vector of the α-th nucleus in the l-th cell.

Inspecting the system of equations (14) one should observe that because of the occurrence of the unknown functions $\Delta\varphi_{i,m}^{\pm}$ in the integrals (last two terms of the l.h.s. of (14)) these coupled Hartree-Fock (HF) equations (as the simple HF ones) are non-linear and, therefore, have to be solved in an iterative way.

b) LCAO Approximation for the one-electron wave functions. Introducing a basis set $\{\chi_s^q(\underline{r})\}$ for the whole polymer chain, where χ_s^q is the s-th basis function in the q-th cell, one can write

$$\varphi_i'(\underline{r},\underline{E}_{st}) = \sum_{q=-N}^{+N}\sum_{s=1}^{\tilde{m}} C_{i,st,s}^{q}(\underline{E}_{st})\,\chi_s^q(\underline{r}) \tag{17a}$$

$$\text{and}\quad \Delta\varphi_{i,m}^{\pm} = \sum_{q=-N}^{N}\sum_{s=1}^{\tilde{m}} C_{i,m;s}^{q\pm}(\underline{E}_0)\,\chi_s^q(\underline{r}) \tag{17b}$$

($\tilde{m}$ is the number of basis functions per unit cell).

The coefficients $C_{i,st,s}^{q}(\underline{E}_{st})$ can be obtained by solving the generalized matrix equation

$$\begin{aligned}&[\boldsymbol{F}_0 + \boldsymbol{F}_{st}]\,\underline{C}_i'(\underline{E}_{st}) =\\ &= \boldsymbol{F}\,\underline{C}_i'(\underline{E}_{st}) = (\epsilon_i^{(0)} + \Delta\epsilon_i^{st}(\underline{E}_{st}))\;\boldsymbol{S}\,\underline{C}_i'(\underline{E}_{st})\end{aligned} \tag{18}$$

in the same way as one does in the case of periodic systems in the absence of $\underline{E}_{st}$ (16,17,18). [Since $\mathbf{F}_0$ and $\mathbf{F}_{st}$ are, if we use periodic boundary conditions, both cyclic hypermatrices, they can be blockdiagonalized and the problem of a long finite or infinite chain can be reduced to the problem of $\tilde{m}$x$\tilde{m}$ matrices. For further details see: (17)]. Substituting the expansion (17b) into the coupled HF equations (14) one arrives at the hypermatrix equation

$$\begin{pmatrix}\boldsymbol{A}_{i,m}^{+} & \boldsymbol{B}_{i,m}^{+}\\ \boldsymbol{B}_{i,m}^{-} & \boldsymbol{A}_{i,m}^{-}\end{pmatrix}\begin{pmatrix}\underline{C}_{i,m}^{+}\\ \underline{C}_{i,m}^{-}\end{pmatrix} = \begin{pmatrix}\underline{D}_i\\ \underline{D}_i\end{pmatrix} \tag{19}$$

Here the matrices **A** and **B** have the elements

$$A_{i,m;r,s}^{0,q\pm} = \langle\chi_r^0|\hat{F} - (\epsilon_i^{(0)} + \Delta\epsilon_{i,st}(\underline{E}_{st}))\pm m\omega|\chi_s^q\rangle + B_{i,m;r,s}^{0,q\pm} \tag{20}$$

$$\begin{aligned}&B_{i,m;r,s}^{0,q\pm} =\\ &= 2\sum_{j=1}^{\frac{n}{2}}\sum_{u,v}\sum_{q_1,q_2} C_{j,m;u}^{q_1\pm} C_{i,st,s}^{q}(\underline{E}_{st})\, C_{j,st,v}^{q_2}\times\\ &\times\langle\chi_r^0(\underline{r}_1)\,\chi_u^{q_1}(\underline{r}_2)\,|\frac{2-\hat{P}_{1\leftrightarrow 2}}{r_{12}}|\chi_s^q(\underline{r}_1)\,\chi_v^{q_2}(\underline{r}_2)\rangle\end{aligned} \tag{21}$$

$$\text{and}\quad D_{i;r}^{q} = \sum_{s=1}^{m} C_{i,st,s}^{q}(\underline{E}_{st})\,\langle\chi_r^0|\hat{h}|\chi_s^q\rangle \tag{22}$$

Equation (19) can be rewritten as

$$\mathbf{A}^{+}_{i,m}\underline{C}^{+}_{i,m} + \mathbf{B}^{+}_{i,m}\underline{C}^{-}_{i,m} = \underline{D}_{i} \tag{23a}$$

$$\mathbf{B}^{-}_{i,m}\underline{C}^{+}_{i,m} + \mathbf{A}^{-}_{i,m}\underline{C}^{-}_{i,m} = \underline{D}_{i} \tag{23b}$$

Since all the matrices $\mathbf{A}_{i,m}{}^{\pm}$ and $\mathbf{B}_{i,m}{}^{\pm}$ occurring in equ-s (23) are in the case of a linear chain with periodic boundary conditions, cyclic hypermatrices, they can be block-diagonalized with the help of the unitary matrix **U** [the p,q-th block of **U** is $U_{p,q} = 1/(2N+1) \exp [i2\pi pq]\mathbf{1}$ (17)]:

$$\mathbf{U}^{+}\mathbf{A}^{+}_{i,m}\mathbf{U}\mathbf{U}^{+}\underline{C}^{+}_{i,m} + \mathbf{U}^{+}\mathbf{B}^{+}_{i,m}\mathbf{U}\mathbf{U}^{+}\underline{C}^{-}_{i,m} = \mathbf{U}^{+}\underline{D}_{i} \tag{24a}$$

$$\mathbf{U}^{+}\mathbf{B}^{-}_{i,m}\mathbf{U}\mathbf{U}^{+}\underline{C}^{+}_{i,m} + \mathbf{U}^{+}\mathbf{A}^{-}_{i,m}\mathbf{U}\mathbf{U}^{+}\underline{C}^{-}_{i,m} = \mathbf{U}^{+}\underline{D}_{i} \tag{24b}$$

Using the notations $\mathbf{A}^{+}{}_{i,m}{}^{B.D.} = \mathbf{U}^{+}\mathbf{A}^{+}{}_{i,m}\mathbf{U}$ etc, $\underline{F}^{\pm}{}_{i,m} = \mathbf{U}^{+}\underline{C}^{\pm}{}_{i,m}$ and $\underline{G}_{i} = \mathbf{U}^{+}\underline{D}_{i}$ we obtain

$$\mathbf{A}^{+\,B.D.}_{i,m}\underline{F}^{+}_{i,m} + \mathbf{B}^{+\,B.D.}_{i,m}\underline{F}^{-}_{i,m} = \underline{G}_{i} \tag{25a}$$

$$\mathbf{B}^{-\,B.D.}_{i,m}\underline{F}^{+}_{i,m} + \mathbf{A}^{-\,B.D.}_{i,m}\underline{F}^{-}_{i,m} = \underline{G}_{i} \tag{25b}$$

This system of equations can be reduced easily in the usual way to such matrix equations in which each matrix has only the order $\tilde{m}\mathrm{x}\tilde{m}$. For $N \rightarrow \infty$ one can introduce the continuous variable $k = (2\pi p)/(a(2N+1))$ $(-\pi/a \leq k \leq \pi/a)$ and can rewrite equations (25) as

$$\mathbf{A}^{+}_{i,m}(k)\underline{F}^{+}_{i,m}(k) + \mathbf{B}^{+}_{i,m}(k)\underline{F}^{-}_{i,m}(k) = \underline{G}_{i}(k) \tag{26a}$$

$$\mathbf{B}^{-}_{i,m}(k)\underline{F}^{+}_{i,m}(k) + \mathbf{A}^{-}_{i,m}(k)\underline{F}^{-}_{i,m}(k) = \underline{G}_{i}(k) \tag{26b}$$

Putting back the two equations in the hypermatrix form one obtains

$$\begin{pmatrix} \mathbf{A}^{+}_{i,m}(k) & \mathbf{B}^{+}_{i,m}(k) \\ \mathbf{B}^{-}_{i,m}(k) & \mathbf{A}^{-}_{i,m}(k) \end{pmatrix} \begin{pmatrix} \underline{F}^{+}_{i,m}(k) \\ \underline{F}^{-}_{i,m}(k) \end{pmatrix} = \begin{pmatrix} \underline{G}_{i}(k) \\ \underline{G}_{i}(k) \end{pmatrix} \tag{27}$$

which can be solved at each point k and value $\underline{E}_0$ to obtain the vectors $\underline{C}_{i,m}{}^{q\pm}(k,\underline{E}_0)$. For this we need the back transformation

$$\underline{C}^{\pm}_{i,m}(k,\underline{E}_0) = \tilde{\mathbf{U}}\underline{F}^{\pm}_{i,m}(k,\underline{E}_0)$$

[Actually $\tilde{\mathbf{U}}$ and $\tilde{\mathbf{U}}^{+}$ are those blocks of **U** and $\mathbf{U}^{+}$, respectively, that are necessary for the transformations of these vectors].

Furthermore from the unitary transformations performed in equations (24) it follows (16-18) that

$$\mathbf{A}^{\pm}_{i,m}(k) = \sum_{q=-N}^{N} e^{ikqa}\mathbf{A}^{\pm}_{i,m}(q) \tag{28a}$$

$$\mathbf{B}^{\pm}_{i,m}(k) = \sum_{q=-N}^{N} e^{ikqa}\mathbf{B}^{\pm}_{i,m}(q) \tag{28b}$$

$$\underline{G}_{i}(k) = \sum_{q=-N}^{N} e^{ikqa}\underline{G}_{i}(q) \tag{28c}$$

Finally, the LCAO crystal orbitals can be written in the presence of the electric fields $\underline{E}_{st} + \underline{E}_{\omega}$ as

$$\tilde{\varphi}_i(k,\underline{E}_{st},\underline{E}_0,\omega,t,\underline{r}) =$$

$$= \frac{1}{2N+1}\sum_{q=-N}^{N}\sum_{s=1}^{\tilde{m}} e^{ikqa}[C_{i,st;s}(k,\underline{E}_{st}) + \tag{29}$$

$$+ \sum_{m=1}^{M}\left(C^{+}_{i,m;s}(k,\underline{E}_0)\,e^{im\omega t} + C^{-}_{i,m;s}(k,\underline{E}_0)\,e^{-im\omega t}\right)]\,\chi^{q}_{s}(\underline{r})$$

c) Moeller-Plesset Perturbation Theory. After having computed the quasi one-electron orbitals $\tilde{\varphi}_i$ and quasi one electron energies $\tilde{\epsilon}_i$, one can apply (following Rice and Handy (8,9)) the MP/2 expression of the second order correlation correction of the quasi total energy for given k, ω, t, $\underline{E}_{st}$, and $\underline{E}_0$. In this case the Moeller-Plesset perturbation operator will be

$$\hat{\tilde{H}}_{MP} = \sum_{i<j}^{m}\frac{1}{r_{ij}} - \sum_{occ}\left(\hat{\tilde{F}}(\underline{r},t,\omega,k,\underline{E}_{st},\underline{E}_0)\right) - ni\frac{\partial}{\partial t} = \sum_{i<j}^{m}\frac{1}{r_{ij}} - n(\hat{\tilde{F}} - i\frac{\partial}{\partial t}) \tag{30}$$

($n = \tilde{n}(2N+1)$) where the quasi Fock operators $\hat{\tilde{F}}$ are defined as

$$\hat{\tilde{F}}(\underline{E}_{st},\underline{E}_0,\omega,t,k) = \tilde{H}^{N}(\underline{E}_{st},\underline{E}_0,\omega,t,\underline{r}_1) + \sum_{j=1}^{\frac{n}{2}}(2\tilde{J}_j - \tilde{K}_j) \tag{31}$$

The quasi Coulomb operators $\tilde{J}_j$ and quasi exchange operators $\tilde{K}_j$ are defined in the usual way but not with the help of the unperturbed crystal orbitals φ_j, but with the perturbed ones $\tilde{\varphi}_j$. Furthermore,

$$\tilde{H}^{N} = \hat{H}^{N} + \sum_{m=1}^{M}[\underline{E}_{st} + \underline{E}_0(e^{im\omega t} + e^{-im\omega t})]\,e^{ik\underline{r}}\,\nabla_{\underline{k}}e^{-ik\underline{r}} \tag{32}$$

($\hat{H}^N$ is the one-electron operator of the unperturbed polymer).
With the definition (30), one can derive in the standard way (19) the second-order correction to the quasi total energy (for given t, ω, $\underline{E}_{st}$ and $\underline{E}_0$)

$$\tilde{E}_{MP}^{(2)}(t,\omega,\underline{E}_{st},\underline{E}_0) =$$

$$= \sum_{I,J,A,B} \frac{\left|\langle\tilde{\varphi}_I(\underline{r}_1,\ldots)\tilde{\varphi}_J(\underline{r}_2,\ldots)\left|\frac{1}{r_{12}}(2-\hat{P}_{1\to 2})\right|\tilde{\varphi}_A(\underline{r}_1,\ldots)\tilde{\varphi}_B(\underline{r}_2,\ldots)\rangle\right|^2}{\tilde{\varepsilon}_I(t)+\tilde{\varepsilon}_J(t)-\tilde{\varepsilon}_A(t)-\tilde{\varepsilon}_B(t)} \tag{33}$$

Here $\tilde{E}_{MP}^{(2)} = \tilde{E}_{MP}^{(2)}(t, \omega, \underline{E}_{st}, \underline{E}_0)$, because the perturbed crystal orbitals $\tilde{\varphi}_I(\underline{r}, \ldots)$ and quasi one-electron energies

$$\tilde{\varepsilon}_I(t) = \langle\tilde{\varphi}_I|\hat{F}|\tilde{\varphi}_I\rangle \tag{34}$$

depend on the same variables (this is meant in the argument of the $\tilde{\varphi}_I$, etc. by the three points). The k- and m-dependence is expressed by the combined indices $I = i,k_i,m_i$ etc. Therefore, the fourfold summation includes the threefold integration over k (because of the conservation of momenta $k_i + k_j = k_a + k_b$); for the details, how one calculates the matrix elements $\langle\tilde{\varphi}_I\,\tilde{\varphi}_J \mid \tilde{\varphi}_A\,\tilde{\varphi}_B\rangle$ occurring in (33) see again (19)).

After having calculated $\tilde{E}_{MP}^{(2)}$, we can write for the total energy with second-order correlation corrections

$$\tilde{E}(\omega,\underline{E}_{st},\underline{E}_0) =$$

$$= \frac{1}{T}\int_0^T \left[\frac{a}{2\pi}\int_{-\pi/a}^{\pi/a} \tilde{E}^{HF}(k,t,\omega,\underline{E}_{st},\underline{E}_0)\,dk + \tilde{E}_{MP}^{(2)}(t,\omega,\underline{E}_{st},\underline{E}_0)\right]dt \tag{35}$$

Here $\tilde{E}^{HF}(k, t)$ is the quasi Hartree-Fock total energy calculated in the presence of $\underline{E}_{st}$ and $\underline{E}_\omega$ at a certain value of k and at time t.

The development of the program package for the rather complicated formalism developed here and in the previous subsection is in progress in our Laboratory.

d) The Calculation of the Polarizabilities and Hyperpolarizabilities. Having the total energy (35) (following in this point also Rice and Handy (8,9)), we can formally express it in the presence of the electric field

$$\underline{E} = \underline{E}_{st} + \underline{E}_\omega = \underline{E}_{st} + \sum_{m=1}^{M} \underline{E}_0\left(e^{im\omega t} + e^{-im\omega t}\right) = \underline{E}_{st} + \sum_{m=1}^{M} \underline{E}_0 \cos(m\omega t) \tag{36}$$

as

$$\tilde{E}(\omega,\underline{E}_{st},\underline{E}_0) = E_{HF}^{(0)} + E_{MP}^{(2)} + \underline{\mu}_0^+\underline{E}_{st} + \underline{E}^+\alpha(\omega)\underline{E} + \frac{1}{2}\underline{E}^+\beta E\underline{E} + \frac{1}{3!}\underline{E}^+\gamma E^2\underline{E} + \ldots \tag{37}$$

where $E_{HF}^{(0)}$ and $E_{MP}^{(2)}$ are the Hartree-Fock and MP/2 energies of the polymer per unit cell in the absence of the field and

$$\mu_0 = \alpha_{st}\underline{E}_{st} + \frac{1}{2}\underline{E}^{+}_{st}\beta_{st}\underline{E}_{st} + \frac{1}{3!}\underline{E}^{+}_{st}\gamma_{st}E_{st}\underline{E}_{st} + \ldots \tag{38}$$

is the part of the induced dipole moment due to the static field. It can be easily calculated with the aid of the expression

$$\mu_0 = 2\frac{a}{2\pi}\sum_{i=1}^{n^*}\int_{-\frac{\pi}{a}}^{\frac{\pi}{a}}\Big\langle[\varphi_i(\underline{r}_i,k) + \varphi_i^{st}(\underline{r}_i,\underline{E}_{st},k)]\,\Big| \tag{39}$$
$$\times\sum_{i=1}^{n}\hat{\mu}_i\,|\varphi_i(\underline{r}_i,k) + \varphi_i^{st}(\underline{r}_i,\underline{E}_{st},k)\Big\rangle dk + corr.\ corr.$$

where the dipole moment operator $\hat{\mu}_i = |e|\underline{r}_i$ and corr. corr. stands for terms coming from taking into account correlation corrections in the wave function (9).

Using expression (35) one can apply the time-independent Hellmann-Feynmann theorem

$$\frac{\partial\tilde{E}}{\partial\underline{E}_{st}} = \Big\langle\tilde{\Phi}\,\Big|\,\frac{\partial\hat{H}}{\partial\underline{E}_{st}}\,\Big|\,\tilde{\Phi}\Big\rangle \tag{40}$$

(for the conditions of its validity see (9,16)). Here $\tilde{E}$ is the total energy (see equation (35)), $\tilde{\Phi}$ the many electron wave function [see equation (10)] and $\hat{H}$ the full perturbed Hamiltonian [see equations (7,8)]. Following Rice and Handy (8,9) one can expand the quasi total energy (together with the $(i\cdot\partial/\partial t)$-term occuring in Frenkel's variational principle) as a function of $\underline{E}_{st}$ and $\underline{E}_{\omega}$. One obtains in this way

$$\mu_0^{\lambda} = -\frac{\partial\tilde{E}}{\partial E_{st,\lambda}}\bigg|_{\underline{E}=\underline{0}} \tag{41a}$$

$$\alpha^{\lambda\mu}(0;0) = -\frac{\partial^2\tilde{E}}{\partial E_{st,\lambda}\partial E_{st,\mu}}\bigg|_{\underline{E}=\underline{0}} \tag{41b}$$

$$\beta^{\lambda\mu\nu}(0;0,0) = -\frac{\partial^3\tilde{E}}{\partial E_{st,\lambda}\partial E_{st,\mu}\partial E_{st,\nu}}\bigg|_{\underline{E}=\underline{0}} \tag{41c}$$

$$\alpha^{\lambda\mu}(-\omega;\omega)\cos(\omega t) = -\frac{\partial^2\tilde{E}}{\partial E_{st,\lambda}\partial E_{\omega,\mu}}\bigg|_{\underline{E}=\underline{0}} \tag{42a}$$

$$\frac{1}{2}\beta^{\lambda\mu\nu}(0;-\omega,\omega) + \frac{1}{2}\beta^{\lambda\mu\nu}(-2\omega;\omega,\omega)\cos(2\omega t) = \\ = -\left.\frac{\partial^3 \tilde{E}}{\partial E_{st,\lambda}\partial E_{\omega,\mu}\partial E_{\omega,\nu}}\right|_{\underline{E}=\underline{0}} \tag{42b}$$

and

$$\beta^{\lambda\mu\nu}(-\omega;0,\omega)\cos(\omega t) = -\left.\frac{\partial^3 \tilde{E}}{\partial E_{st,\mu}\partial E_{\omega,\lambda}\partial E_{\omega,\nu}}\right|_{\underline{E}=\underline{0}} \tag{42c}$$

In these equations $\underline{E}_{st,\mu}$ and $\underline{E}_{\omega,\lambda}$, respectively, are the μ-th component of $\underline{E}_{st}$ and the λ-th one of $\underline{E}_{\omega}$, respectively. The series expansion indicated above provides also the symmetry relation (9)

$$\beta^{\lambda\mu\nu}(0;\omega,-\omega) = \beta^{\nu\lambda\mu}(\omega;0,\omega) \tag{43}$$

There are, of course, more combinations of ω in β possible, but they are not measurable at the present time (9).

Formally one can write also for the elements of γ similar expressions as fourth derivatives of $\tilde{E}$. For instance, it can be proven in a similiar way as described above (8,9)

$$\gamma^{\lambda\mu\nu\kappa}(-\omega;0,0,\omega)\cos(\omega t) = -\left.\frac{\partial^4 \tilde{E}}{\partial E_{st,\mu}\partial E_{st,\nu}\partial E_{\omega,\lambda}\partial E_{\omega,\kappa}}\right|_{\underline{E}=\underline{0}} \tag{44}$$

For the numerical implementation of these formulae for polymers one has to use again the procedure recommended by Rice and Handy (8,9) to achieve computational simplifications. The most efficient computational algorithm for polymers can be found, however, only after some numerical tests have been performed.

Interaction of a Quasi-1D Polymer with a Laser Pulse

In a previous paper (7) this problem was already treated, but the paper contains a larger number of typographical errors and several inconsistencies. Further, the lattice periodic part of the operator $\underline{E}\,\underline{r}$ was not introduced. Since in the previous section (starting from equation (6)) a rather general formalism was described, it is easy to extend to the case of a laser pulse.

Let us assume to have a laser pulse of Gaussian shape,

$$\underline{E} = \underline{E}_0 e^{-\alpha^2 t'^2} \quad where \quad \alpha^2 = \frac{1}{\left(\frac{T}{2}\right)^2},$$

that is in time T/2 the value of $\underline{E}$(t) decreases to $\underline{E}_0$/e (see Figure 1). Further, in this case $\underline{E}_{st} = \underline{0}$ and $\underline{A} \neq \underline{0}$.

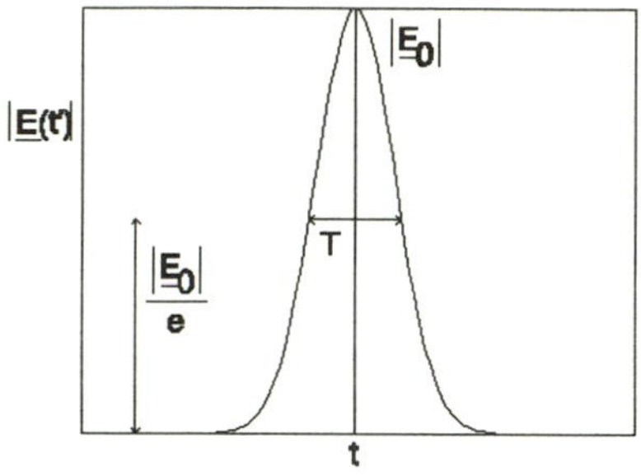

Fig.1 The Gaussian shape of a laser pulse with a peak electric field strength of $\underline{E}_0$.(Adapted from ref (7))

Further we can write

$$\underline{A}(t,\omega) = \sum_{m=1}^{M} \underline{A}_m(t,\omega) = \sum_{m=1}^{M} [\underline{A}_0^* e^{im\omega t} + \underline{A}_0 e^{-im\omega t}] \tag{45}$$

Using the expression

$$\underline{E} = -grad\varphi - \frac{1}{c}\frac{\partial \underline{A}(t,\omega)}{\partial t} \quad ; \quad V = -\varphi \tag{46}$$

(Coulomb gauge (20)) one obtains for V (taking into account equation (3).

$$V = \sum_{m=1}^{M} V_m = -|e|\underline{E}_0 e^{ik\underline{r}} \nabla_k e^{-ik\underline{r}} + \sum_{m=1}^{M} \frac{2m\omega}{c} Im\underline{A}^* e^{im\omega t} \tag{47}$$

If one introduces the generalized momentum $\underline{P}$-(e/c)$\underline{A}$(t,ω) to take into account the magnetic field $\underline{H}$(t,ω) one obtains for Ĥ' instead of equation (8)

$$\hat{H}'(\underline{r}_1,\ldots,\underline{r}_n,\underline{E}_0,\underline{A}_0,\omega,t) = e^{-\alpha^2 t'^2} \times$$

$$\times \sum_{m=1}^{M} \left\{ \sum_{i=1}^{n} \left[[-|e|\underline{E}_0 e^{ik\underline{r}_i} \nabla_k e^{-ik\underline{r}_i} + \frac{m\omega}{c} Im(\underline{A}_0^*) e^{im\omega t} + \frac{i}{c}\underline{A}_0^* \nabla_i] e^{im\omega t} + \right.\right.$$

$$\left. + [-|e|\underline{E}_0 e^{ik\underline{r}_i} \nabla_k e^{-ik\underline{r}_i} + \frac{m\omega}{c} Im(\underline{A}_0) e^{-im\omega t} + \frac{i}{c}\underline{A}_0 \underline{\nabla}_i] e^{-im\omega t} \right] +$$

$$\left. + \frac{1}{2c^2}[2|\underline{A}_0|^2 + \underline{A}_0^{*2} e^{2im\omega t} + \underline{A}_0^2 e^{-2im\omega t}] \right\} \tag{48}$$

The n-electron wave function will be in this case

$$\Phi(\underline{r}_1,\ldots,\underline{r}_n,t,\omega) = e^{-iW_0 t} \prod_{i=1}^{n} \tilde{\varphi}_i(\underline{r},t,\omega) \tag{49}$$

as before, but the definition of the one-electron orbitals in the presence of the electromagnetic field changes to

$$\tilde{\varphi}_i(\underline{r}_1, t, \omega, t') = \varphi_i(\underline{r}_i) + e^{-\alpha^2 t'^2} \sum_{m=1}^{M} \times$$

$$\times \left[\Delta\varphi^+_{i,m}(\underline{r}_i, \underline{E}_0, \underline{A}_0)\, e^{im\omega t}\, \Theta(t\in t') + \Delta\varphi^-_{i,m}(\underline{r}_i, \underline{E}_0, \underline{A}_0)\, e^{-im\omega t}\, \Theta(t\in t')\right] \tag{50}$$

Here W_0 is again the eigenvalue of $\hat{H}_0$ and

$$\Theta(t\in t') = \begin{cases} 1 & \text{if} \quad -100\left(\frac{T}{2}\right) \le t' \le 100\left(\frac{T}{2}\right) \\ 0 & \text{otherwise} \end{cases} \tag{51}$$

Substituting again $\hat{H}=\hat{H}_0+\hat{H}'$ with the definition (48) of $\hat{H}'$ and (49) with the definition (50) of the $\tilde{\varphi}_i$-s into equation (9) (Frenkel's variational principle) one obtains instead of (14) the generalized coupled HF equations

$$\Big[(\hat{F}_0 - \epsilon_i^{(0)} \pm m\omega)\; |\Delta\varphi^{\pm}_{i,m}(\underline{r}, \underline{E}_0, \underline{A}_0)\rangle + \hat{h}^{\pm} e^{-\alpha^2 t^2} |\,\varphi_i\rangle +$$

$$+ \sum_j \left(\langle\varphi_j(\underline{r}_2)\,|\, \frac{2-\hat{P}_{1\leftrightarrow 2}}{r_{12}}\, |\Delta\varphi^{\pm}_{j,m}(\underline{r}_2, \underline{E}_0, \underline{A}_0)\rangle_2 + h.c. \right) |\varphi_i(\underline{r}_1)\rangle +$$

$$+ 4\, Re\, \hat{Q}_2\; |Re\, \varphi_i\rangle + \begin{Bmatrix} 2\, Re[\hat{Q}_1 + \hat{Q}_3]\, |\Delta\varphi^-_{i,m}\rangle \\ 2\, Re[\hat{Q}_1 + \hat{Q}_3]\, |\Delta\varphi^+_{i,m}\rangle \end{Bmatrix} \Bigg]\; e^{-\alpha^2 t'^2} = 0 \tag{52}$$

Here $\hat{F}_0$ is the field free Fock operator of the polymer (equation 16a) and the new operators $\hat{Q}_1$, $\hat{Q}_2$, $\hat{Q}_3$ and $\hat{h}^{\pm}$ are defined as follows

$$\hat{Q}_1 \equiv \frac{1}{2c^2}\left[2\,|\underline{A}_0|^2 + \underline{A}_0^2\, \frac{e^{-2i} - 1}{-2i} + (\underline{A}_0^*)^2\, \frac{e^{2i} - 1}{2i}\right] \tag{53a}$$

$$\hat{Q}_2 \equiv \frac{1}{2c^2}\left[2\,|\underline{A}_0|^2\, \frac{e^{-i} - 1}{-i} + \underline{A}_0^2\, \frac{e^{-3i} - 1}{-3i} + (\underline{A}_0^*)^2\, \frac{e^{i} - 1}{i}\right] \tag{53b}$$

$$\hat{Q}_3 \equiv \frac{1}{2c^2}\left[2\,|\underline{A}_0|^2\, \frac{e^{2i} - 1}{2i} + \underline{A}_0^2 + (\underline{A}_0^*)^2\, \frac{e^{4i} - 1}{4i}\right] \tag{53c}$$

$$\hat{h}^+(\underline{r}, \underline{E}_0, \underline{A}_0) = \frac{i}{c}\underline{A}_0\, \nabla - |e|\underline{E}_0 e^{ik\underline{r}}\, \nabla_{k^-}\, e^{-ik\underline{r}} + \frac{2m\omega}{c}\, Im(\underline{A}_0^*) \tag{53d}$$

and $\hat{h}^-$ has the same definition only in the first and third terms on the r.h.s. $\underline{A}_0$ and $\underline{A}_0^*$ are exchanged.

If we use again an LCAO expansion for the $\Delta\varphi_{i,m}^{\pm}$ we obtain again the hypermatrix equation (19) (only the r.h.s. of it will stand $(\underline{D}_i^+\ \underline{D}_i^-)^t$ [t means transposed]) with the somewhat changed definition

$$A^{0,q\pm}_{i,m;r,s} = \langle\chi^0_r|\hat{F}_0 - \epsilon^{(0)}_i \pm m\omega|\chi^q_s\rangle + \\ + B^{0,q\pm}_{i,m;r,s} + e^{-\alpha^2 t'^2}\langle\chi^0_r|2Re(\hat{Q}_1 + \hat{Q}_3)|\chi^q_r\rangle \tag{54}$$

The matrix elements of the matrix $\mathbf{B}^{0\pm}_{i,m}$ are the same as in equation (21) only instead of $C^q_{i,st,s}(\underline{E}_{st})$-s etc. (because now $\underline{E}_{st}=\underline{0}$) the $C^{q(0)}_{i,s}$-s should stand. Finally, the components of the vectors $\underline{D}_i^{\pm}$ are now defined as

$$D^{q\pm}_{i;r} = \sum_{s=1}^{m} C^{q(0)}_{i,s}\langle\chi^0_r|\hat{h}^{\pm}|\chi^q_s\rangle \tag{55}$$

(The coefficients $C^{q(0)}_{i,s}$ are the eigenvector components of the Fock operator of the unperturbed ($\underline{E}_{st}=\underline{E}_{\omega}=\underline{0}$) polymer). The new matrices $\mathbf{A}_{i,m}^{\pm}$ are also cyclic hypermatrices and can be block diagonalized in the same way as before. Therefore, one obtains finally again equation (27), only on the r.h.s. stands now

$$\begin{pmatrix} \underline{G}^{+}_i(k) \\ \underline{G}^{-}_i(k) \end{pmatrix}$$

Equation (29) which defines the one-electron orbitals $\tilde{\varphi}_i$ is nearly the same as before only instead of $C_{i,st;s}(k,\underline{E}_{st})$ $C^{(0)}_{i,s}$ occurs and the coefficients $C_{i,m,s}{}^{q\pm}$ depend now also on $\underline{A}_0$. The expressions in point b) and c) remain the same only instead of $\underline{E}_{st}$ the quantities depend on $\underline{A}_0$. To be able to calculate the (hyper)polarizibilities (point d) of the previous section) one has to introduce again in all equations of this section a fictious $\underline{E}_{st}$ again.

Concluding Remarks

As has been mentioned before the programing of the rather complicated formalism, described in the previous two sections, is in progress. The calculation of $\alpha_{zz_{st}}$ for simple infinite polymers as poly(H_2), poly(H_2O), poly(LiH) have given values which are much larger than those obtained by different extrapolation procedures [for details see (11)]. Further calculations for $(\alpha_{zz_{st}})/(2N+1)$, the polarizability per unit cell of poly (HN), poly (O) and poly (CH_2) have shown similar results [see (21)].

This indicates that to find polymers with high polarizibilities and hyperpolarizabilities (both static and dynamic) one really has to apply solid state physical methods.

After finishing the programing of the new methods and performing the corresponding calculations one expects to be able to predict new polymers with advantageous non-linear optical properties. On the other hand, certainly one would obtain a much deeper insight into the interaction of a periodic polymer with an electromagnetic field (laser light), than it was possible until now.

On the theoretical side we plan to extend the methods for the case of interacting chains. If the arrangement of the units is such that there is a periodicity also in the second direction, one can use the same formalism as before only instead of k one has to introduce a two-component crystal momentum vector $\underline{k}$ and the crystal orbitals will describe a quasi-2D system. Otherwise, one has to take into account the interactions between chains with the help of other methods (22).

To take into account the effect of phonons on the (hyper)polarizibilities one can extend the methods of Bishop and Kirtman (5,6) to quasi-1D- and quasi-2D periodic systems.

Finally, one should mention that the question can be raised whether for the proper treatment of the interaction of an electromagnetic field with a molecule or polymer is it not necessary to quantize also the field and do not treat it classically. Since, however, in the case of laser pulses the field strengths are large, the second quantization most probably would not change considerably the results.

Acknowledgements

The author would like to express his gratitude to Professors P. Otto and B. Kirtman, Dr. D. Dudis and Dr. G. Das for the very useful discussions and to the Referee for his important remarks. Further, the author is indebted to Dipl.-Chem. R. Knab for his very large help in the preparation of the manuscript. Financial support of the US Air Force (Grant No.: F 49620-92-J-0253) and of the "Fond der Chemischen Industrie" is gratefully acknowledged.

References

(1) Heeger, A.J.; Moses, D.; Sinclair, M. *Synth. Metals* 1986, *15*, 95.
(2) Heeger, A.J.; Moses, P.; Sinclair, M. *Synth. Metals* 1987, *77*, 343.
(3) Kajzar, F.; Etemad, G.; Boker, G.L.; Merrier, J.; *Synth. Metals* 1987, *17*, 563.
(4) Kirtman, B., lecture at "Theoretical and Computational Modeling of NLO and Electronic Materials" Symp. in the framework of the XIV. ACS Meeting, Washington D.C., 1994.
(5) Bishop, D.M.; Kirtman, B. *J. Chem. Phys.* 1991, *95*, 2466.
(6) Bishop, D.M.; Kirtman, B. *J. Chem. Phys.* 1992, *97*, 5255.
(7) Ladik, J.J.; Dalton, L.R. *J. Mol. Structure* (Theochem), 1991, *231*, 77.
(8) Rice, J.; Handy, N.C. *J. Chem. Phys.* 1991, *94*, 4959.
(9) Rice, J.; Handy, N.C. *Int. J. Quant. Chem.* 1992, *43*, 91.
(10) Sekino, H; Bartlett, J. J. Chem. Phys. 1986, *84*, 2726.
(11) Otto, P. *Phys. Rev. B.* 1992-I, *45*, 10876.
(12) Mott, N.F.; Jones, H. *The theory of the properties of metals and alloys*; Oxford University Press: Oxford, 1936.
(13) Kittel, W. *Quantentheorie der Festkörper* (in German) 1st ed; Oldenburg Verlag: München-Wien, 1970.
(14) Frenkel, J. *Wave Mechanics, Advanced General Theory*, Dover Press: New York, 1950.
(15) Löwdin, P.O.; Mukherjee, P.K. *Chem. Phys. Lett.* 1972, *14*, 1.
(16) Löwdin, P.O. *Adv. Phys.* 1956, *5*, 1.
(17) Del Re, G.; Ladik, J; Biczo, G. *Phys. Rev.* 1967, *155*, 967.

(18) André, J.-M.; Goverheur, L.; Leroy, G. *Int. J. Quant. Chem.* 1967, *1*, 427, 451.
(19) Ladik, J. *Quantum Theory of Polymers as Solids, Plenum Press*: New York-London, 1988, Sec. 5.2.
(20) Haken, H. *Quantum Field Theory of Solids*, North Holland: Amsterdam, 1976, pp. 81.
(21) Otto, P. *Int. J. Quant. Chem.*, 1994, *52*, 353.
(22) See Chapter 6 of ref. (19).

RECEIVED June 14, 1995

Chapter 11

Model Hamiltonians for Nonlinear Optical Properties of Conjugated Polymers

Z. G. Soos[1], D. Mukhopadhyay[1], and S. Ramasesha[2]

[1]Department of Chemistry, Princeton University, Princeton, NJ 08544
[2]Solid State and Structural Chemistry Unit, Indian Institute of Science, Bangalore 650012, India

Quantum cell models for delocalized electrons provide a unified approach to the large NLO responses of conjugated polymers and π-π* spectra of conjugated molecules. We discuss exact NLO coefficients of infinite chains with noninteracting π-electrons and finite chains with molecular Coulomb interactions V(R) in order to compare exact and self-consistent-field results, to follow the evolution from molecular to polymeric responses, and to model vibronic contributions in third-harmonic-generation spectra. We relate polymer fluorescence to the alternation δ of transfer integrals $t(1 \pm \delta)$ along the chain and discuss correlated excited states and energy thresholds of conjugated polymers.

I. Introduction

Nonlinear optical (NLO) coefficients describe the response of an electronic system to radiation. Second and third-order coefficients $\chi^{(2)}$ and $\chi^{(3)}$ are quadratic and cubic, respectively, in the applied electric field $E(\omega,t)$. Formal developments in the dipole approximation are widely available(1,2) as sum over states (SOS) expressions and tend to represent different perspectives. Physicists and engineers focus on diverse possibilities afforded by different polarizations or frequencies of applied fields or on identifying higher-order responses. Chemists and materials scientists often concentrate on the electronic properties of the system, either to evaluate or to optimize NLO responses, while spectroscopists have devised ingenious new NLO techniques for extracting microscopic information. We will discuss NLO responses of extended systems such as conjugated polymers(3,4) in terms of π-electron and related quantum cell models(5-8). Several other contributions to this book are devoted to molecular responses and the challenges of quantitative NLO calculations. Hückel theory for conjugated hydrocarbons and tight-binding descriptions of metals illustrate simple models that combine physical insight and computational ease. Models remain a powerful

0097–6156/96/0628–0189$15.50/0

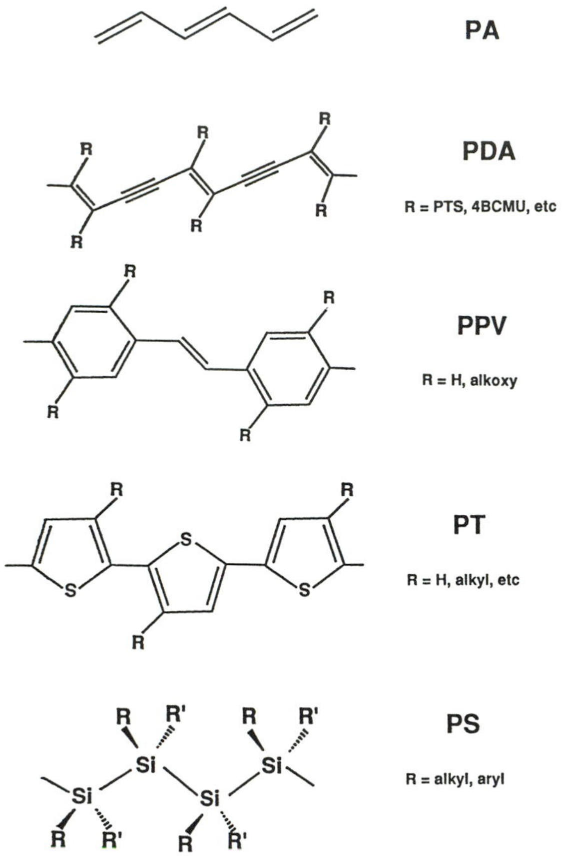

Figure 1. Idealized planar backbones of conjugated polymers.

approach to novel NLO phenomena, as discussed below, because they unify, clarify, and quantify behavior in extended systems that are otherwise quite different chemically.

Physical considerations are the primary motivation for models, although the possibilities of exact results and computational convenience are also important. The quantum cell models in Section II include π-electron models for conjugated molecules or polymers; Hubbard and extended Hubbard models for high-T_C materials, for inorganic complexes, and for organic ion-radical and charge-transfer salts; and exchange-coupled magnetic insulators containing spin-1/2 sites. Models pose well-defined mathematical problems that are central to collective phenomena. The statistical physics of magnetism, for example, is closely related to exchange-coupled networks of spins with dipolar and other interactions whose realizations in inorganic salts identify materials of special interest. Mathematical techniques are freely transferable between different fields and extended models have evolved into separate areas of research.

The connection of successful models to more fundamental descriptions is important theoretically. Such derivations are more suitable for molecules than for extended systems. The advent of powerful computers and efficient algorithms several decades ago enabled quantum chemists to include all electrons, or at least all valence electrons, in increasingly large molecules. Semiempirical and *ab initio* self-consistent-field (SCF) methods proved to be well suited for molecular geometries or the ground-state potential surface. Band structures are the corresponding applications to solids. These all-electron studies tend to complement models, which focus on optical, electric, or magnetic excitations and start with the observed geometry.

The Pariser-Parr-Pople (PPP) model for conjugated hydrocarbons(9,10), for example, was developed for π-π* spectra(11) and extends Hückel theory to include Coulomb interactions V(R). Similar π-π* spectra and NLO responses(12) of conjugated polymers are experimental observations that imply prominent π-electron contributions quite independently of theoretical justifications for σ-π separability. All-electron and model calculations of NLO properties pose different challenges associated with excited states. NLO coefficients are sums over excited states and transition moments, quantities that are not readily measured directly or evaluated at the SCF level. The need for configuration interaction (CI) is widely recognized for correlations, but the proper balance between rigor and truncation in *ab initio* calculations of NLO coefficient remains to be worked out. An adequate basis is clearly a prerequisite for real molecules. The finite basis of models allows exact solutions for oligomers, as discussed in Section III. The evolution of NLO responses from oligomers to bands for noninteracting models is a current topic summarized in Section IV.

Our discussion of NLO responses of model Hamiltonians illustrates a variety of applications to conjugated polymers, including exact results, symmetry arguments, and analysis of SOS expressions as well as coefficient calculations. The generality of half-filled quantum cell models is summarized in Section II. The PPP model provides a unified description of the conjugated polymers in Figure 1, which are among the most extensively studied. Exact NLO coefficients of interacting models are compared in Section III with approximate results. Their size dependencies are found in Section IV for noninteracting models and in excitonic models based on molecular crystals with reduced delocalization. We relate in Section V the photophysics of the polymers in Figure 1 to an excited-state crossover involving the lowest singlet excitation, S_1, and discuss in Section VI how to model vibronic or conformational contributions.

II. Hückel, Hubbard, and PPP Models

Quantum cell models are widely applied to low-dimensional extended systems, where their physical faithfulness is far less demonstrable than in molecules. We comment briefly of their justification and parametrization in conjugated molecules after introducing models whose NLO responses are our principal topic. We consider electron transfer integrals $t(R_{pp'})$ between adjacent or bonded sites p,p' in Figure 1, vanishing t's between more distant neighbors, and use the zero-differential-overlap (ZDO) approximation(11) to reduce the potential V(p,p') to one and two-center integrals. We obtain(8) $H = H_t + H_U + H_V + H_s$ for any conjugated molecule or polymer. H_t is the usual Hückel model for noninteracting electrons, H_U is the on-site or Hubbard interaction, H_V describes intersite interactions, and H_s describes systems with different kinds of sites. Their second-quantized expressions for an N-site linear system are

$$\begin{aligned} H_t &= \sum_{p=1,\sigma}^{N-1} t(R_{pp+1})[a_{p\sigma}^{+}a_{p+1\sigma} + a_{p+1\sigma}^{+}a_{p\sigma}] \\ H_U &= \sum_{p} U_p n_p (n_p - 1)/2 \\ H_V &= \sum_{pp'}{}' V(p,p')(z_p - n_p)(z_{p'} - n_{p'}) \\ H_s &= \sum_{p} \varepsilon_p n_p \quad . \end{aligned} \tag{1}$$

The fermion operators $a_{p\sigma}{}^{+}$, $a_{p\sigma}$ create and annihilate an electron with spin σ at site p, n_p is the number operator, z_p is the charge at p when $n_p = 0$, and the primed sum excludes p = p'. Intersite interactions V(p,p') typically depend on the distance $R_{pp'}$ and can be either short or long-ranged. The total spin S is conserved, as expected for nonrelativistic Hamiltonians. The models above include noninteracting Hückel or tight-binding systems as well as interacting Hubbard, extended Hubbard, and PPP systems whose U >> t limit corresponds to Heisenberg spin chains.

Conjugated hydrocarbons or polymers are often half-filled systems, with one electron per site, so that solutions of Eq.(1) with $N_e = N$ electrons are sought. They ususally have common on-site U = V(0) and site energy $\varepsilon = 0$ at all p, which amounts to taking energies relative to Hückel's α integral. The resulting half-filled chains are bipartite, with alternancy symmetry in Hückel models or electron-hole (e-h) symmetry (13,14) in Hubbard or PPP models. The Ohno potential(15) for PPP models is

$$V(p,p') = V(R_{pp'}) = e^2/(\rho^2 + R_{pp'}^2)^{1/2}. \tag{2}$$

V(0) = 11.26 eV is taken from the ionization potential and electron affinity of carbon and leads to ρ = 1.28 Å for the size of the quantum cell, while an unshielded Coulomb potential is found at large R. The systematics of π-π^* spectra motivated the PPP model with the Ohno potential and $t(R_o)$ = -2.40 eV taken from benzene.

Transferable parameters, a major goal of PPP theory, has been achieved for hydrocarbons(16,17) and partly so for heteroatoms(18,19), although several related parameter sets are used. The Ohno potential and distance dependence of $t(R_{pp'})$ in Eq.(1) completely fix the PPP Hamiltonian for the conjugated polymers in Fig. 1; atomic Si leads(20) to larger radius ρ. The PPP model is a special case of extended Hubbard models with fixed potential, Eq.(2), rather than arbitrary U, V_1, V_2, ... parameters in Eq.(1) taken from experiment. We note that the distance dependencies of t(R) or V(R) couple electronic and vibrational degrees of freedom, with the latter restricted to backbone motions in conjugated systems. The Su-Schrieffer-Heeger (SSH) model(21), for example, is an adiabatic approximation for H_t with linear t(R) and a harmonic lattice. It involves only C-C stretches, while CCC bends(22) also appear in PPP models through V(p,p').

In addition to the excitation spectrum, NLO coefficients require transition dipoles. The ZDO approximation for the dipole operator is

$$\vec{\mu} = \sum_p e(n_p - z_p)\vec{r}_p \quad , \qquad (3)$$

where r_p is the position of site p, n_p-z_p is the charge operator, and the origin is arbitrary for a neutral system. Matrix elements $\langle X|\mu|Y\rangle$ over the eigenstates of Eq.(1) completely fix the largest (electric dipole) response. Large transition dipoles in delocalized systems such as organic dyes dominate NLO responses. The contributions of the more numerous core electrons are roughly additive(23) in N, while the exaltation of π-electrons increases(24) far more rapidly. NLO applications are particularly well suited for PPP models developed for π-π^* spectra of conjugated molecules, whose parameters(12) also hold in conjugated polymers.

The scope of quantum cell models(5-8) has several advantages. Exact results are known for special cases such as Hubbard chains with uniform transfer integrals t, for Heisenberg chains with uniform exchange $J \sim 2t^2/U$, and for other linear spin or donor-acceptor models. These provide guidance for the alternating backbones of the polymers in Figure 1. Moreover, different physical realizations often sample different parameter regimes. Conjugated molecules or polymers have comparable V(0) and band widths 4t ~ 10 eV. Ion-radical organic salts are more correlated, with U ~ 1.5 eV and band widths 4t ~ 0.5-0.8 eV; in addition to N_e = N, they illustrate N_e = N/2, 2N/3, or other fillings. The band width is negligible in magnetic insulators, which for N_e = N have a localized spin at each site. The strong-correlation limit of Eq.(1) reduces for N_e = N to covalent valence bond (VB) diagrams(25), with n_p = 1 at every p. Ionic VB diagrams are also needed for the complete basis and these many-electron functions explicitly conserve the total spin S. Slater determinants, with fixed S_z, of molecular orbitals also provide a complete basis for Eq.(1) that is best suited for band or weakly-correlated systems. Continuum limits(26,27) of quantum cell models have also been extensively studied, but we will not discuss field-theoretical results here.

We return now to derivations of Eq.(1) for conjugated molecules, a major concern of early work(11) on the ZDO approximation in PPP or Hubbard models, and to accurate evaluation of parameters such as t(R). Comparisons to all-electron calculations(28) are inherently restricted to small molecules such as butadiene, or perhaps hexatriene, whose π^* and σ^* excitations are similar. Larger systems have lower-energy π-π^* excitations around 2-3 eV, less than half the σ-σ^* threshold. Greater delocalization and different excitation energies clearly improve σ-π separability, although no quantitative assessment has been achieved. Dipole-allowed excitations provide qualitative support: the 1B excited state is not planar for butadiene or hexatriene, but planar for octatetraene and longer polyenes(29). Transfer integrals, or Hückel β's, of 2-3 eV can be estimated directly, in contrast to the smaller parameters of Eq.(1) in ion-radical or magnetic systems whose microscopic constants are taken from experiment. As shown by Hubbard(30) for d-electron metals and by Soos and Klein(8) for ion-radical and CT organic salts, the convenient orbital interpretation of Eq.(1) is not mandatory. We may rigorously consider instead many-electron site functions associated with different charge distributions. The fundamental approximation is the restriction to four states per site. Core-electron relaxation is automatically included, for example, and poses challenges for molecular treatments not limited to rigid cores. Solid-state quantities such as effective masses or transfer integrals in fact incorporate many subtle electronic interactions.

III. Exact Dynamic NLO Coefficients

The quantum cell models in Eq.(1) have a large but finite basis: each site can only be empty, doubly occupied, or singly occupied with spin up or down. Subspaces with a fixed number N_e of electrons, total spin S, and spatial or electron-hole symmetries reduce the 4^N possible many-electron states for N sites. The dimensions P(R) of the Rth exact subspace of Eq.(1) thus depends on the system and increases exponentially with N. Although direct diagonalization of P(R)xP(R) matrices becomes impractical for large N, low-lying eigenstates are accessible through sparse-matrix methods(25) and all NLO responses can be found exactly(31) when the exact ground state $|G\rangle$ is known.

There are several reasons for choosing a basis of normalized VB diagrams, $|k\rangle$, in addition to their easy visualization and widespread use in organic chemistry. Second-quantized expressions for $|k\rangle$ yield N_e-electron functions that automatically conserve S, diagonalize Eq.(1) except for the Hückel term H_t, and can be symmetry adapted(32). Their representation in terms of 2N-digit binary integers facilitates the construction and evaluation of eigenstates, while resolving difficulties encountered in first-quantization with the nonorthogonality of VB diagrams. The desirable sparseness of VB representations is due to having ~N bonds for N sites and generating at most a few new diagrams per electron transfer. Quite generally, the eigenstates of Eq.(1) are linear combinations of VB diagrams, as originally proposed by Pauling for conjugated molecules. We rigorously have

$$|G\rangle = \sum_k c_k|k\rangle \tag{4}$$

and a similar expansion for any other eigenstate |X>.

NLO coefficients(1,2) are responses of |G> to applied fields $E_i\cos\omega t$ with polarization i and frequency ω. Perturbation theory in the applied fields generates SOS expressions. An n-th order response contains n+1 transition moments and polarizations, n energy denominators, and n sums over unperturbed energies. There is an equivalent formal development(33) in terms of corrections $|\phi_i(\omega)\rangle$ for a field with polarization i and time dependence exp(iωt). The inhomogeneous equation for the first-order correction is

$$(H - E_G + \hbar\omega)|\phi_i(\omega)\rangle = -\mu_i|G\rangle, \tag{5}$$

where E_G is the exact ground-state energy and μ_i is given by Eq.(3) in systems with an inversion center and by the dipole displacement operator(31) μ_i - $\langle G|\mu_i|G\rangle$ in polar systems. The SOS expression for the polarizability tensors $\alpha_{ij}(\omega)$ is simply(34)

$$\alpha_{ij}(\omega) = -[\langle G|\mu_i|\varphi_i(\omega)\rangle + \langle G|\mu_i|\varphi_j(\omega)\rangle]. \tag{6}$$

Just as the finite dimensionality of Eq.(1) makes possible an exact |G>, it allows(31) direct solution of $|\phi_i(\omega)\rangle$ in Eq.(5) by expanding in the VB basis and solving P(R) linear equations. The ground state of *trans*-polyenes, for example, is a covalent A_g singlet and the correction $|\phi_i(\omega)\rangle$ is a linear combination of ionic B_u singlets whose dimensionality fixes the number of linear equations in Eq.(5). In the absence of spatial symmetry, $|\phi_i(\omega)\rangle$ and |G> are in the same subspace and their orthogonality is ensured by the dipole-displacement operator.

Additional correction functions are needed for NLO coefficients and are equivalent to SOS expressions to any order. The second-order correction $|\phi_{ij}(\omega_2,\omega_1)\rangle$, with two polarizations and frequencies, is given by(31)

$$(H - E_G + \hbar\omega_2)|\phi_{ij}(\omega_2,\omega_1)\rangle = -\mu_j|\phi_i(\omega_1)\rangle, \tag{7}$$

again with the dipole-displacement operator for polar systems. Direct solution for $|\phi_{ij}(\omega_2,\omega_1)\rangle$ is again practical in the VB basis for finite dimensional models and leads to simple expressions for all $\chi^{(2)}$ and $\chi^{(3)}$ coefficients. Third-harmonic-generation (THG) for fields along the polymer axis x in Figure 1 is

$$\begin{aligned}\gamma_{xxxx}(-3\omega;\omega,\omega,\omega) = {} & [\langle\phi_x(-3\omega)|\mu_x|\phi_{xx}(-2\omega,-\omega)\rangle + \\ & \langle\phi_x(-\omega)|\mu_x|\phi_{xx}(2\omega,\omega)\rangle + \omega\rightarrow-\omega]/8\end{aligned}, \tag{8}$$

where $\omega \rightarrow -\omega$ indicates the same matrix elements with new arguments. Similarly, the two-photon transition moment M(Y) for state |Y> with excitation energy E_Y is

$$M(Y) = \langle Y|\mu_x|\phi_x(-\omega_Y)\rangle = \sum_R \langle Y|\mu_x|R\rangle\langle R|\mu_x|G\rangle / (E_R - \hbar\omega_Y) \quad (9)$$

for monochromatic radiation with $\omega_Y = E_Y/2\hbar$ and polarization x. The sum over intermediate states |R> with excitation energies E_R is explicitly shown.

Matrix elements over correction functions yield exact dynamic NLO coefficients for any quantum cell model in Eq.(1) whose |G> is known in terms of the VB expansion, Eq.(4). The finite dimensionality of the basis is crucial. We emphasize that all dipole transitions between excited states are implicitly contained in the correction functions, as can be verified(31) when excitations |R> have also been found exactly. Specific transition moments <R|μ|G> or M(Y) in Eq.(9) still require knowledge of the final state, but virtual or intermediate states can always be avoided. In practice, exact VB results for the lowest 10-15 states(35) in the relevant symmetry subspaces of Eq.(1) suffice for THG or two-photon spectra of linear systems. Exact transition moments among these states and SOS expressions are more convenient for simulating a spectrum than single-frequency results such as Eq.(8) and excited-state lifetimes $i\Gamma$ are easily introduced in the energy denominator of Eq.(9). The finite dimensionality of quantum cell models is also important in SOS expressions based on exact excitations and transition moments(35).

We illustrate in Table I several features of the second hyperpolarizability of N-site polyenes. The first column compares the xxxx component of Eq.(8) along the backbone for *trans* and *cis* geometries at $\hbar\omega$ = 0.65 eV and molecular PPP parameters. The other columns compare the transverse xxyy and yyyy components at $\hbar\omega$ = 0.30 eV of *trans* polyenes using PPP and Hückel models with equal band widths 4t = 9.6 eV and t(1 ± δ) with δ = 0.07. The different signs of the transverse components

Table I. Second hyperpolarizability, $\gamma(-3\omega,\omega,\omega,\omega)$ in Eq.(8), of PPP and Hückel models for N-site polyenes at $\hbar\omega$ = 0.65 or 0.30 eV. The polymers in Figure 1 define the xy plane, with x along and y perpendicular to the backbone; $\gamma(-3\omega,\omega,\omega,\omega) \times 10^3$ is in atomic units and Hückel results are in parenthesis; from ref. 31.
(Reproduced with permission from reference 31. Copyright 1989 American Institute of Physics.)

N	xxxx, PPP (0.65 eV)		trans, PPP (Hückel)(0.30eV)	
	trans	cis	xxyy	yyyy
6	3.85	2.71	0.126(-4.46)	0.019(-0.19)
8	12.72	8.74	0.265(-13.98)	0.035(-0.33)
10	31.26	21.61	0.456(-68.01)	0.055(-0.51)
12	63.25	44.42		

is a correlation effect related to alternancy or electron-hole symmetry. The far larger magnitude of the Hückel response is due to smaller excitation energies and larger transition dipoles. We summarize some implications of these results, starting with the signs of transverse components.

Alternant hydrocarbons have transfer integrals t in Eq.(1) that exclusively connect sites in opposite (starred and unstarred) sets. The resulting Hückel energies are symmetric about $\alpha = 0$, as sketched in Figure 2, with index $r > 0$ for antibonding molecular orbitals and $r < 0$ for bonding MOs. The MO expansion coefficients c_{pr} and $c_{p\text{-}r}$ at site p are equal with the same or opposite signs depending on which set contains p. The cation and anion of alternant hydrocarbons indeed have similar charge and spin distributions. Alternancy symmetry is compatible with V(p,p') in Eq.(1), but requires equal site energies and U = V(0). An SCF approximation for Eq.(1) thus preserves the symmetric distribution in Figure 2, even in extended systems. An applied electric field breaks alternancy symmetry, since the dipole operator, Eq.(4), generates site energies er_{pE} on projecting r_p on E. As seen in Figure 1, a transverse field on *trans* -PA gives equal but opposite site energies at even and odd carbons. Transition dipoles are then rigorously(34) between MO's with $\pm$ r. The system behaves as a collection of two-level models whose $\gamma < 0$ is simply understood in terms of saturation. Selection rules imply negative γ for *trans* polyenes in SCF treatment of Eq.(1) for two or more transverse polarizations. The xxyy, xyyy, and yyyy components are indeed negative in a variational-perturbation analysis(36) of the PPP model, while the same parameters and exact results in Table I give positive responses.

The linear spectrum is typically fit to quantum cell models. In conjugated polymers or linear polyenes, the intense π-π* absorption to 1^1B_u is polarized along the centrosymmetric backbone and provides a convenient reference E(1B). Thus Hückel and PPP parameters for a polymer do not coincide in general and the direct comparison of NLO coefficients in Table I may not be the most instructive. Reduced coefficients(37) for the same E(1B) and bond lengths provide clearer assessments for different models and polymers. An $E(1B)^3$ scaling of $\chi^{(3)}$ responses sharply reduces the difference between PPP and Hückel magnitudes in Table I, even for the dominant xxxx component. Differences among reduced coefficients are associated with transition dipoles, which are generally smaller in PPP calculations due to electron-hole attraction in excited states. Oscillator strengths of conjugated molecules are overestimated in Hückel or SCF theory. The Ohno potential, Eq.(2), yields correct oscillator strengths as well as molecular excitations(38,5) and leads to a singlet exciton for 1^1B_u in conjugated polymers. Although large NLO responses continue to motivate research, there are many experimental and theoretical difficulties with extracting accurate magnitudes. NLO spectra and other relative quantities afford(12) more useful comparisons.

IV. From Molecular to Polymeric Responses

The striking connection between delocalization and NLO responses has been amply documented. THG efficiency increases(24) roughly as N^5 with the number of

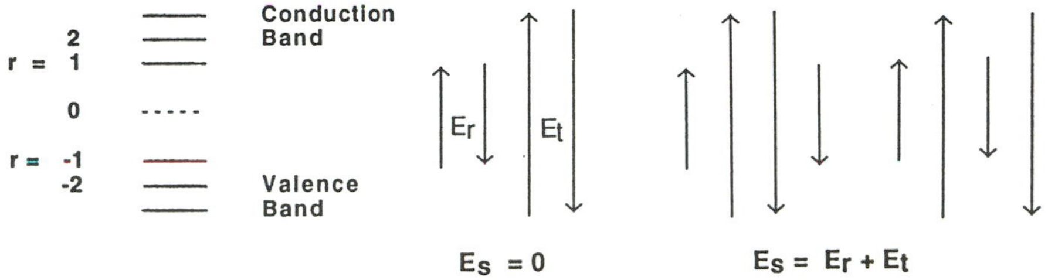

Figure 2. Symmetric valence and conduction band energies of alternant hydrocarbons. The virtual electron-hole excitations E_r and E_t are dipole allowed; the even-parity virtual state is $E_s = O$ for successive generation and $E_r + E_t$ for consecutive excitation of two pairs.

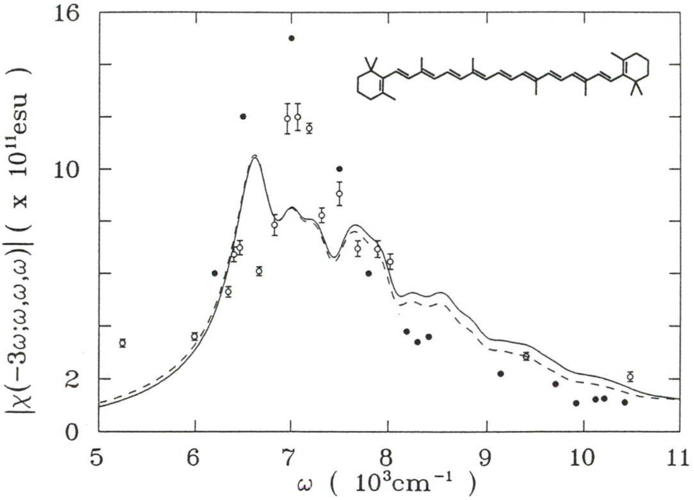

Figure 3. THG spectrum of β-carotene. Closed and open circles are data from refs. 41-43 and 44, respectively, while the theoretical fit discussed in ref. 40 is based on harmonic potentials for overlapping 1B and 2A resonances related to linear and polyene spectra.

conjugated carbons. Accurate length dependencies are difficult to measure, however, since even the most homologous series leads to conformational and solubility differences with increasing N and requires resonance as well as local-field corrections. Moreover, physical arguments require the per-site response to become independent of length in polymers. Their large THG responses indicate an N between 50 and 100 carbons for comparable molecular responses(12), consistent with recent THG data(39) on long conjugated molecules. The β-carotene spectrum(40) in Figure 3 is representative of both polyenes and polymers: the dominant feature is a three-photon resonance at E(1B)/3 that is within an order of magnitude of PDA responses. The data of Van Beek *et al.* (41-43) are shown as closed circles, while two separate data sets of Aramaki *et al.* (44) are open circles with estimated uncertainties. Both groups measured the phase as well as the magnitude of the THG response. THG spectra of polymers also show weaker two and three-photon resonances; vibronic contributions to the β-carotene spectrum are summarized in Section VI.

The evolution from localized to extended states also appears in atomic clusters, disordered conductors, and percolation problems. Idealized models are invoked to explore such questions and to identify central issues. Hückel models are the natural starting point. The molecular perspective of early studies(45) is apparent from the choice of *uniform* transfer integrals, or alternation $\delta = 0$ on taking $t(1 \pm \delta)$ for partial double and single bonds. Large finite-size splittings in short polyenes are not sensitive to small δ, but regular and alternating Hückel chains differ fundamentally at large N. Uniform t's lead to a one-dimensional metal. The perturbation expansion in applied fields then diverges and NLO coefficients become unphysical. Conversely, *alternating* chains with $\delta > 0$ are semiconductor for $N_e = N$ and the Hückel gap $4t\delta$ ensures well-defined NLO responses. The N and δ dependencies are linked. Molecular or finite-size effects are important for $N\delta << 1$, band results hold for $N\delta >> 1$, and crossover behavior is expected around $N\delta \sim 1$. Such arguments apply in general to quantum cell models and indeed to the full electronic structure, but detailed studies are restricted to noninteracting systems with fixed sites. The Peierls and other instabilities of one-dimensional metals are separate issues. Alternating site energies $\pm \varepsilon$ in Hückel chains model donor-acceptor properties(46), generate a gap in half-filled regular chains, and thus lead to similar considerations with increasing N.

NLO coefficients of Hückel or other noninteracting models are obtained by solving NxN secular determinants for N sites. As sketched in Figure 2, the ground state $|G_0\rangle$ has N_e electrons in the lowest MOs and there are $N - N_e/2$ empty orbitals at higher energy. SOS expressions for NLO coefficients are conveniently analyzed in terms of virtual excitations(37). We know from linked-cluster expansions that contributions in N^2 or higher powers must cancel exactly. In Hückel models, the polarizability, Eq.(6), involves one electron-hole (e-h) pair and scales as N on converting the sum to an integral. The four transitions for any $\chi^{(3)}$ response generate at most two e-h pairs. In centrosymmetric systems, the pathway through virtual states is $G \rightarrow rB \rightarrow sA \rightarrow tB \rightarrow G$. Odd-parity B states have one e-h pair in noninteracting models, while the even-parity state sA has 0, 1, or 2 e-h pairs. The first possibility corresponds to successive e-h pairs, with sA = G as sketched in Figure 2, and scales as N(N - 1). The N^2 part cancels in general(37) against sA states with the same two e-h pairs, but created and annihilated consecutively as sketched in Figure 2. The unlinked contributions to $\chi^{(3)}$ contain sA terms with one e-h pair, with scattering of either the electron or hole in rB, as well as one and two e-h pairs that scale as N. NLO coefficients for Hückel chains of N ~ 500 have been reported(47) using e-h pairs as virtual excitations.

Translational symmetry is normally used to simplify the analysis of extended systems. The evolution of molecular to polymeric states in alternating Hückel chains can be followed analytically for N = 4n+2, which gives nondegenerate $|G_0\rangle$ at $\delta = 0$. Care must then be taken with the applied field, either by using the velocity operator(48) or by explicitly correcting(49) for the angle between the field and the backbone. The static second polarizability per site, $\gamma(0;\delta,N)/N$, is shown(50) in Figure 4 in units of the dimer response, $\gamma_d = -e^2a^2/2^8t^3$, which is negative for a two-level system. Uniform t's lead to the simple analytical result(50)

$$\gamma(0;0,N) \quad = \quad \frac{2|\gamma_d|(\cos^{-2}\pi/N \quad -1)}{3\sin^7\pi/N} \quad , \tag{10}$$

which increases as N^5 for $N >> \pi$ and regains the size dependence found numerically. The response diverges in the unphysical limit of a half-filled metallic band. For $\delta > 0$, the curves in Figure 4 become independent of N. The analytical expression for $\gamma(0;\delta,N)/N$ goes as δ^{-6} and agrees(50) precisely with the band-theoretical result(51) based on the velocity operator. The marked enhancement of $\gamma(0;\delta,N)/N$ in the crossover region $\delta N \sim 1$ can be traced to cancellation between positive intraband processes in which the virtual state sA contains one e-h pair and negative interband processes. As seen in Eq.(10), the terms individually diverge as N^7, faster than their sum, and the cancellation is sensitive to the band-edge states around $\delta N \sim \pi$. To describe $\gamma(0;\delta,N)$ as N^b, the exponent becomes $b(\delta,N)$ and is simply the logarithmic derivative. We have b ~ 5 for short chains with $\delta N << 1$, b = 1 for infinite chains with any $\delta > 0$, and large(50) b ~ 8-9 in the crossover region for $\delta = 10^{-3}$. The interesting behavior of $\gamma(0;\delta,N)/N$ is associated with Hückel models and implications for actual molecules are indirect.

Size extensivity also holds for interacting models. NLO responses presuppose a semiconductor or insulator. An energy gap E(1B) in the linear spectrum implies a localized spectrum and greater localization with increasing alternation δ. Virtual excitations sufficiently far apart are then additive and the cancellation in Fig. 2 between successive and consecutive virtual excitations again follows. While first-order CI is size extensive, truncation at doubles, triples, or higher is not. Static $\gamma(0;\delta,N)$ of PPP models with the Ohno potential have been obtained by different methods, most often by direct-field SCF calculations. Exact results give and exponent b ~ 3.8 for γ_{xxxx} between N = 6 and 12; variational-perturbation(36) theory gives b ~ 4.25 up to N = 20, still well below the noninteracting value; CI including all double excitations gives(52) b ~ 5.4 up to N = 16, close to the noninteracting value. DCI is almost quantitative for N = 4, but becomes more approximate with increasing N. The evolution to band states shown in Fig. 4 for Hückel chains cannot presently be followed in PPP or other interacting models with $\delta \sim 0.1$.

The limited scope of exact solutions and the rapidly increasing basis of Eq.(1) clearly indicate the need for size-extensive approximations for interacting models. Coupled(53), coupled-perturbed(54), or time-dependent Hartree-Fock (TDHF)(55) methods illustrate current techniques for correlated approximations. An anharmonic oscillator model based on TDHF theory has recently been applied(56) to interacting models for NLO responses of conjugated polymers. These approaches all use a single-determinantal approximation, but multireference treatments(57) of quantum cell models have also been tried. Size extensivity is a major challenge for CI beyond first-order required for strong correlations. Coupled-cluster methods(58) maintain size extensivity while systematically including correlations in single as well as multireference approaches. Size-extensive corrections incorporate linked-diagram expansions and have primarily been applied(59,60) to small molecules.

We have recently considered(61) a different approach for correlated states of extended systems. We start with analytical excitations of dimers, construct crystal states using molecular exciton theory, and treat interdimer hopping or interactions in chains with $\delta < 1$ as perturbations. Molecular exciton analysis based on the dimer basis accounts for the number, position, and intensities of linear and two-photon spectra found by exact solution of oligomers. As shown in Table II, threshold energies for triplet, one-photon, and two-photon excitations converge rapidly with N at $\delta = 0.6$, where $t_+ = 4t_-$, in either Hubbard or PPP models. We used the Ohno potential, Eq.(2), the PA geometry in Fig. 1, and reduced the bandwidth 4t to 5.0 eV in Table II. This preserves the ordering E(2A) < E(1B) found(35) in polyenes, PA, and PDA's. The crossover of 2A and 1B with increasing correlations occurs at intermediate correlations $U \sim 2t$ where neither band nor spin-wave approximations hold. The suitability of oligomer calculations for strongly alternating inifnite chains is based on their short coherence length, which goes as δ^{-6} Hückel chains(51). The smaller alternation realized in conjugated polymers implies considerably more delocalization, but preserves the symmetry of quantum cell models, Eq.(1).

Table II. Energy thresholds, in eV, of N-site chains with $t_+ = -2.0$ eV, alternation $\delta = 0.60$, Ohno potential V(p,p') in Eq. (2), and PA geometry in Fig. 1; E_T is the lowest triplet, E(2A) and E(1B) the lowest even and odd-parity singlets, and $-E_g/N$ is the ground-state energy per site; from ref. 61.
(Reproduced with permission from reference 61.)

N	E_T	E(2A)	E(1B)	$-E_g/N$
8	2.5165	4.9187	5.4472	1.3213
10	2.5099	4.9097	5.3717	1.3224
12	2.5060	4.9054	5.3203	1.3231
14	2.5035	4.9031	5.2835	1.3237
polymer	2.50	4.90	5.20	1.335

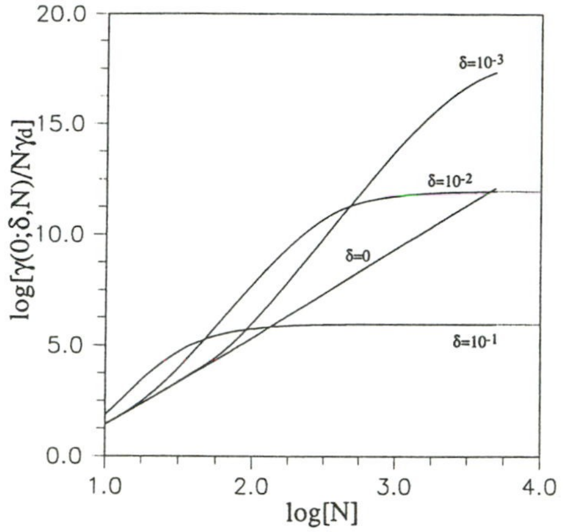

Figure 4. Static second hyperpolarizability per site, $\gamma(0;\delta,N)/N$, of $N = 4n+2$ site Hückel rings with alternating transfer integrals $t(1\pm\delta)$; from ref. 50. (Reproduced with permission from reference 50. Copyright 1993 American Institute of Physics.)

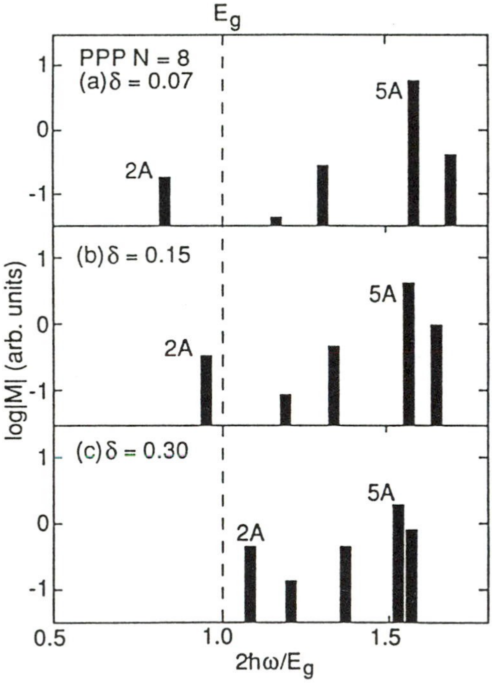

Figure 5. Exact two-photon transition moments, M in Eq.(9), of ^1A_g states of 8-site PPP chains with alternations δ appropriate for PA, PDA, and PS polymers in Fig. 1 and otherwise identical polyene parameters and geometry; from ref. 12.
(Reproduced with permission from reference 12. Copyright 1991 John Wiley & Sons.)

V. Even-Parity States and Excitons

The excitation thresholds E_T, E(2A), and E(1B) in Table II resemble molecular solids(62) rather than wide band semiconductors, whose excitations all coincide at the band gap. Conjugated polymers also have a threshold E_b > E(1B) associated with the generation of charge carriers, as found(63) most accurately in crystalline PDA's in Figure 1. The intense absorption at E(1B) is consequently associated with a singlet exciton. The triplet, singlet, two-photon, and charge-carrier thresholds have been measured separately for several PS polymers(62) in Figure 1, and only the triplet excitation remains to be found in PDA's. These observations indicate that SCF treatments of extended systems must minimally include correlations between an excited electron-hole pair to describe different energy thresholds. First-order CI for quantum-cell models with long-rage Coulomb interactions similar to Eq.(2) have been applied(64,65) to PS and PPV polymers, with adjustable parameters. First-order CI leads to E(2A) > E(1B) and is limited to less correlated polymers with larger effective alternation(66).

The occurrence of E(2A) below E(1B) in linear polyenes is a paradigm for Coulomb correlations(67,5). In centrosymmetric systems, the even-parity ^1A_g states are two-photon allowed, related to Im $\chi^{(3)}(\omega;-\omega,\omega,-\omega)$, and hence probed in degenerate four-wave-mixing experiments. They also appear as resonances in THG spectra and are the sA virtual states discussed in Figure 2. Two-photon transition dipoles M(Y), Eq.(9), are shown in Figure 5 on a logarithmic scale, in units of E(1B), for PPP models of octatetraene and the indicated alternation δ. As previously noted, molecular PPP parameters account for E(1B) and E(2A) in polyenes and related ions, as well as π-π^* spectra of conjugated hydrocarbons. To reduce the N dependence, we use E(1B) as an internal standard and find(35) E(2A) around E(1B)/2 for the PA alternation $\delta = 0.07$, consistent with recent two-photon spectra(68) and older extrapolations(69). In contrast to the converged thresholds in Table II, however, the N = 8 spectra in Figure 5 are far from the polymer limit.

Increasing alternation leads in Figure 5 to a crossover of 2A and 1B. The lowest singlet excitation S_1 becomes one-photon allowed(70) at large δ and strong fluorescence is found in such polymers(71). Large alternation in PS is associated(20) with quite different t_+ for Si-Si bonds and t_- for two sp^3 hybrids of one Si. The bridgehead carbons in PPV lead to a topological alternation(72), which in Hückel models decomposes the π-system exactly into an extended alternating chain and orbitals localized on one ring. The nonconjugated S heteroatoms of the PT backbone in Figure 1 generate a charge-density-wave ground state with different site energies $\pm\varepsilon$ in Eq.(1) for the α and β carbons(73). When site energies based on semiempirical calculations are added to the PPP model with molecular parameters, E(2A) increases strongly and S_1 becomes 1B. The one and two-photon thresholds support the association(66) of polymer fluorescence with E(1B) < E(2A).

Quantum cell models account naturally for the 2A/1B crossover of centrosymmetric chains with electron-hole symmetry. Uniform t's in Hubbard chains rigorously(74,75) lead to vanishing E(2A) and finite E(1B) for any U > 0. The

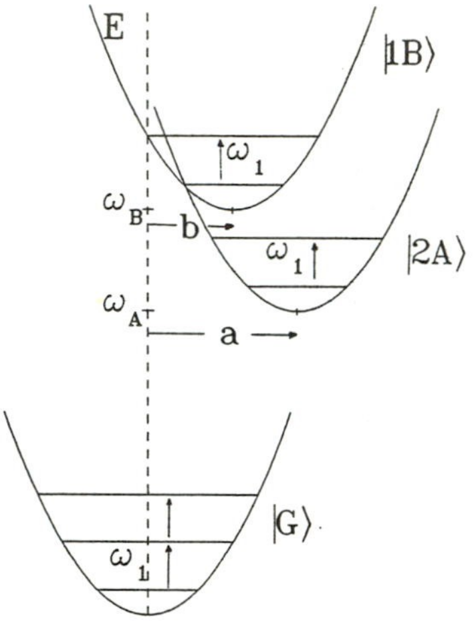

Figure 6. Schematic representation of three electronic states with excitation energies ω_A, ω_B, harmonic potentials with common frequency ω_1, and displacements a,b. Dipole-allowed transitions connect the odd-parity state 11B> to the even-parity ground state 1G> and excited states 12A>. Overlapping resonance occur for $\omega_A/2 \sim \omega_B/3$; from ref. 40.
(Reproduced with permission from reference 40. Copyright 1994 American Institute of Physics.)

symmetries are retained in $\delta = 0$ chains with arbitrary spin-independent $V(R_{pp'})$ in Eq.(1). Finite alternation generates a two-photon gap E(2A) whose position relative to E(1B) is a measure of correlations. The arithmetic mean of the band gap and band width is $2t_+ = 2t(1+\delta)$ in alternating chains, and its magnitude relative to U or $V(R_{pp'})$ controls(70) whether the correlations appear to be strong or weak. Since the conjugated polymers in Figure 1 are similar chemically and even the PS band width is ~10 eV, comparable correlations are expected(71) and the 2A/1B crossover is strongly linked to the alternation. Models provide the clearest understanding of excited-state crossovers and their connection to polymer fluorescence.

The strong two-photon feature around 1.5 E(1B) in Figure 5 is associated with two-electron excitation and, in the band limit, gives(35) a giant singularity at 2 E(1B). The strongest two-photon absorption(76) of PDA-PTS crystals peaks at 2.7 eV, or 1.35 E(1B), consistent with molecular PPP models. The corresponding Si parameters(20) place both E(2A) and the biexciton above E(1B), again consistent(62) with broad two-photon spectra and equally broad excited-state absorption and electroabsorption. Accurate determination of high-energy excitations is difficult either experimentally or theoretically and may be fundamentally limited by lifetimes in conjugated polymers. The sharp E_b threshold(63) for charge carriers in crystalline PDA's is some 0.2 eV below the biexciton, which thus overlaps a continuum.

VI. Vibronic and Conformational Contributions

Excited states of conjugated molecules are strongly coupled to backbone CC stretches, as shown by vibronic progression in linear and two-photon spectra or by intense resonance Raman scattering. Soliton or polaron formation in the SSH model describes excited-state relaxation following the addition of an electron or hole. Vibronic analysis is a separate field with many challenges even in small molecules. We touch here on two rather different aspects of vibronic contributions to quantum cell models. In principle, the distance dependences of t(R) or V(R) yield linear electron-phonon coupling constants such as $t'(R_e)$ in Eq.(1) and quadratic terms involving second derivatives evaluated at equilibrium positions. Since these microscopic parameters are defined for the delocalized electrons being modeled, a major challenge for their identification is to identify π-electron contributions accurately in vibrational spectra. This has been done in terms of reference force-fields(77,78) and Herzberg-Teller expansions. Rather different considerations arise in modeling vibronic contributions(79,40) to NLO coefficients. The Condon approximation usually suffices for the strong, dipole-allowed transitions associated with large NLO responses. Information about excited-state parameters becomes the limiting factor.

The β-carotene THG spectrum in Figure 3 illustrates a favorable case. Careful analysis of vibronic structure in the linear spectrum and of Raman excitation profiles yields(80) the displacement b along 9 normal modes in 1B. Single displacements b and a are shown in Figure 6 for harmonic potentials representing the 1B and 2A states. The strongest coupling is to C=C modes with ω_1 ~ 0.2 eV in both ground and excited state.

Analytical expressions are then available for all Franck-Condon factors, which modulate the electronic transition moment <X|μ|Y> over the vibrational levels. Closure simplifies the SOS expressions whenever the energy denominators $E(X) - n\hbar\omega$ are large compared to the vibrational spacings. Vibronic contributions then reduce to using Franck-Condon averages rather than 0-0 excitations for NLO coefficients. The actual displacements a, b and excitation energies ω_A, ω_B in Fig. 6 are needed for small energy denominators.

In β-carotene, the three-photon resonance to 1B and two-photon resonance to 2A overlap for radiation around the peak THG response in Figure 3. Overlapping resonance increase(40) the response by a factor of 3-5 near the peak and give the faster decrease towards high energy observed experimentally. The theoretical fit in Figure 6 has two displacements b and frequencies for C=C and C-C vibrations, as well as the transition dipole to 1B, taken from the linear spectrum. We estimated a and $\omega_A \sim 14{,}500$ cm^{-1} from two-photon polyene spectra(81), and an even lower value of 13,200 cm^{-1} has recently been reported(82) for β-carotene and supports the occurrence of overlapping resonances $\omega_A/2 \sim \omega_B/3$. Transition dipoles between 1B, 2A, and the even-parity state around 1.5 E(1B) were taken from exact PPP calculations on polyenes. NLO spectra incorporate excited-state and vibronic information gleaned from other sources. The fit in Figure 6 for the THG amplitude and a similarly good fit for the phase are based on related spectra, except for Lorentzian lifetime broadening. Additional Gaussian broadening gives superior fits to linear and THG spectra around the peak(41) and is reasonable for molecules dispersed in polystyrene. Such broadening smoothes out the vibronic features in Figure 6. The spectrum of overlapping resonances depends on the relative phases of the displacement a and b in the two excited states and such information may be accessible from high-resolution data.

Vibronic contributions to the static second hyperpolarizability(79) are related to resonance Raman cross sections. Indeed, vibrational data provides direct information about NLO coefficients(83). We will not pursue these applications of models. Conformational effects have been relatively neglected, although the flexibility of both backbones and side chains has been widely recognized in experimental studies(2). The thermochromic and solvatochromic trransitions of PS and PDA illustrate the coupling of the intense E(1B) transition to the backbone conformation. Segment models(84,85,62) account for many qualitative features by postulating defects that interrupt the conjugation. Detailed connection with electronic excitations is elusive, however. Hückel models reflect the connectivity rather than conformations, while the limitation of accurate correlated states to modest N hinders widespread application. Specific conformational effects can be treated as perturbations in $\Delta V(R_{pp'})$, the change in the potential from a reference geometry. For example, E(2A) and E(1B) of *all-trans* octatetraene change(87) far more on forming a *cis* single bond than a *cis* double bond, in agreement with direct PPP calculations. The perturbation analysis(88) is more instructive and economical, as it accounts for the change in terms of charge-correlation functions of the *all-trans* reference. Transition moments are also sensitive to molecular conformations and such degrees of freedom must be considered in comparing NLO coefficients.

VII. Concluding Remarks

We have touched on a variety of NLO calculations related to quantum cell models, Eq. (1), for extended systems. Model Hamiltonians are well suited for initial discussions of vibronic, conformational, lifetime, or other contributions to NLO spectra. The famous two or three-level models used to demonstrate the richness of NLO phenomena are even more idealized, with excited states and transition moments chosen to fit experiment. Quantum cell models provide the initial relation between such few-level schemes and the underlying electronic structure of conjugated molecules, polymers, or other extended systems with delocalized electrons. Subsequent studies of individual cases are needed to assess the strengths and limitations of quantum cell models. General issues such as the evolution from molecular to band states, excited-state crossovers, or size extensivity are most instructively discussed in terms of models.

The electronic excitations and transition moments appearing in SOS expressions for NLO coefficients emphasize quite different aspects of electronic structure than encountered in ground-state calculations. Correlations are far more important in excited states. Comparison of exact and SCF results for models with interacting electrons often uncovers qualitative differences in the ordering of even and odd-parity states or in signs of NLO coefficients. We have focused on PPP descriptions of THG and two-photon spectra of the conjugated polymers in Figure 1, and have found the Coulomb potential and other molecular parameters to provide an excellent starting point for neutral excitations of conjugated polymers. In a related context, models provide guidance for extending quantum-chemical methods for molecules and band-structure calculations of solids to include properly excited states and correlations. The excitation thresholds, photophysics, and NLO spectra of conjugated polymers bear striking resembles to large conjugated molecules, while charged excitations generated by chemical doping and vibronic structure indicate strong electron-phonon coupling. We expect quantum cell models in general, and PPP models in particular, to remain important approaches to the electronic states of extended systems with interacting electrons coupled to a lattice.

Acknowledgments

It is a pleasure to thank R.G. Kepler, S. Etemad, D.S. Galvão,G.W. Hayden and P.C.M McWilliams for many stimulating discussions and fruitful collaborations, and the National Science Foundation for support through DMR-9300163.

Literature Cited

1. Hanna, D.C.; Yuratich, M.A.; Cotter, D. *Nonlinear Optic of Free Atoms and Molecules,* Springer-Verlag, Berlin, 1979.
2. Shen, Y.R. *The Principles of Nonlinear Optics*, Wiley, New York, 1984.

3. *Molecular Nonlinear Optics: Materials, Physics, and Devices*, Zyss, J. ed.; Academic, New York, 1994.
4. Proc. Int. Conf. on the Science and Technology of Synthetic Metals: Tübingen, Germany; *Synth. Met.* **1991**,*41-43*; Göteborg, Sweden *Synth. Met.* **1993**, *55-57*.
5. Soos, Z.G.; Hayden, G.W. in: *Electroresponsive Molecular and Polymeric Systems*, Skotheim, T.A. ed. Marcel Dekker, New York, 1988; pp 197-265.
6. Baeriswyl, D.; Campbell, D.K.; Mazumdar, S. in: *Conducting Polymers*, Kiess, H. ed. Topics in Current Physics; Springer-Verlag, Heidelberg, 1992; pp 7-XX
7. *Handbook of Conducting Polymers*, T.A. Skotheim, ed., Marcel Dekker, New York, 1988;Vol. 2.
8. Soos, Z.G.; Klein, D.J. in: *Molecular Association,* Vol. 1, R. Foster, ed.; Academic, New York, 1975; pp. 1-109.
9. Pariser, R.; Parr, R.G. *J. Chem. Phys.* **1953**, *21*, 767.
10. Pople, J.A. *Trans. Faraday Soc.* **1953**, *42*, 1375.
11. Salem, L. *The Molecular Orbital Theory of Conjugated Systems*, Benjamin, New York, 1966.
12. Etemad, S.; Soos, Z.G. in *Spectroscopy of Advanced Materials*, R.J.H. Clark and R.E. Hester, eds.;Wiley, New York, 1991; pp 87-133.
13. Heilmann, O.J.; Lieb, E.H. *Trans. N.Y. Acad. Sci.* **1971**, *33*, 116.
14. Bondeson, S.R.; Soos, Z.G. *J. Chem. Phys.* **1979**, *71*, 3807.
15 Ohno, K. *Theor. Chim. Acta* **1964**, *2*, 219.
16. Schulten, K.; Ohmine, I.; Karplus, M. *J. Chem. Phys.* **1976**, *64*, 4422.
17. Roos, B.; Skancke, P.N. *Acta Chem. Scand.* **1967**, *21*, 233.
18. Chalvet, O; Hoaurau, J.; Jousson-Dubien, J; Rayez, J.C. *J. Chim. Phys.* **1972**, *69*, 630.
19. Labhart, H.; Wagniere, G. *Helv. Chim. Acta* **1967**, *46*, 1314.
20. Soos, Z.G.; Hayden, G.W. *Chem. Phys.* **1990**,*143*, pp 199-207.
21. Heeger, A.J.; Kivelson, S.; Schrieffer, J.R.; Su, W.P. Rev. *Mod. Phys.* **1988**, *60*, 781.
22. Soos, Z.G.; Hayden, G.W.; Girlando, A.; Painelli, A. *J. Chem. Phys.* **1994**, *100*, 7144.
23. Chemla, D.S.; Zyss, J. *Nonlinear Optical Properties of Organic Molecules and Crystals*, Vols. 1 and 2; Academic, New York, 1987.
24. Hermann, J.P.; Ducuing, J. *J. Appl. Phys.* **1974**, *45*, 5100.
25. Soos, Z.G.; Ramasesha, S. in: *Valence Bond Theory and Chemical Structure,* Klein, D.J.; Trinajstic, J. eds.: Elsevier, Amsterdam, 1990 pp. 81-109.
26. Takayama, H.; Lin-Liu, Y.R.; Maki, K. *Phys. Rev.* **1980**, *B21*, 2388.
27. Fesser, K.; Bishop, A.R.; Campbell, D.K. *Phys. Rev.* **1983**, *B27*, 4808.
28. Martin, C.H.; Freed, K.F. *J. Chem. Phys.* **1994**, *101*, 5929.
29. Orlandi, G.; Zerbetto, F; Zgierski, M.Z. *Chem. Rev.* **1991**, *91*, pp 867-891.
30. Hubbard, J. *Proc. Roy. Soc. Series A*, **1965**, *285*, 542.
31. Soos, Z.G.; Ramasesha, S. *J. Chem. Phys.* **1989**, *90*, 1067.
32. Ramasesha, S.; Soos, Z.G. *J. Chem. Phys.* **1993**, *98*, 4015.
33. Langhoff, P.W.; Epstein, S.T.; Karplus, M. *Rev. Mod. Phys.* **1972**, *44*, 602.

34. Soos, Z.G.; Hayden, G.W. *Phys. Rev.* **1989**, *B40*, 3081.
35. McWilliams, P.C.M.; Hayden, G.W.; Soos, Z.G. *Phys. Rev.* **1991**, *B43*, 9777.
36. DeMelo, C.P.; Silbey, R. *J. Chem. Phys.* **1988**, *88*, 2558,2567.
37. Soos, Z.G.; Hayden, G.W.; McWilliams, P.C.M. in, *Conjugated Polymeric Materials: Opportunities in Electronics , Optoelectronics, and Molecular Electronics*; Brédas, J.L.; Chance, R.R., Eds.; NATO ASI Series E182, Kluwer, Dordrecht, The Netherlands, 1990; pp 495-508.
38. Ramasesha, S.; Soos, Z.G. *Chem. Phys.* **1984**, *91*, 35.
39. Samuel, I.D.W.; Ledoux, I.; Dhenaut, C.; Zyss, J.; Fox, H.H.; Schrock, R.R.; Silbey, R.J. *Science* **1994**, *265*, 1070.
40. Soos, Z.G.; Mukhopadhyay, D. *J. Chem. Phys.* **1994**, *101*, 5515.
41. Van Beek, J.B.; Kajzar, F.; Albrecht, A.C. *J. Chem. Phys.* **1991**, *95*, 6400.
42. Van Beek, J.B.; Albrecht, A.C. *Chem. Phys. Lett.* **1991**, *187*, 269.
43. Van Beek, J.B.; Kajzar, F.; Albrecht, A.C. *Chem. Phys.* **1992**, *161*, 299.
44. Aramaki, S; Torruellas, W.; Zanoni, R.; Stegeman, G.I. *Opt. Commun.* **1991**, *85*, 527.
45. Hameka, H.F. *J. Chem. Phys.* **1977**, *67*, 2935.
46. Albert, I.D.L.; Das, P.K.; Ramasesha, S. *Chem. Phys. Lett.* **1990**, *168*, 454.
47. Shuai, Z.; Brèdas, J.L. *Phys. Rev.* **1991**, *B44,* 5962.
48. Mazumdar, S.; Soos, Z.G. *Phys. Rev.* **1981**, *B23*, 2810.
49. Yaron, D.; Silbey, R. *Phys. Rev.* **1992**, *B45*, 11655.
50. Spano, F.C.; Soos, Z.G. *J. Chem. Phys.* **1993**, *99*, 9265.
51. Agrawal, G.P.; Cojan, C.; Flytzanis, C. *Phys. Rev.* **1978**, *B17*, 776.
52. Heflin, J.R.; Wong, K.Y.; Zamani-Khamiri, O.; Garito, A.F. *Phys. Rev.* **1988**, *B38*, 1573.
53. Archibong, E.F.; Thakkar, A.J. *J. Chem. Phys.* **1994**, *100*, 7471.
54. Wong, K.Y.; Jen, A.K-Y.; Rao, V.P.; Drost, K.J. *J. Chem. Phys.* **1994**, *100*, 6818.
55. Stanton. J.F.; Bartlett, R.J. *J. Chem. Phys.* **1993**, *98*, 7029.
56. Takahashi, A.; Mukamel, S. *J. Chem. Phys.* **1994**, *100*, 2366.
57. Tavan, P.; Schulten, K. *J. Chem. Phys.* **1986**, *85*, 6602.
58. Mukherjee, D.; Pal, S. *Adv. Quantum Chem.* **1989**, *20*, 291.
59. Kundu, B.; Mukherjee, D. *Chem. Phys. Lett.* **1991**, *179*, 468.
60. Salter, E.A.; Sekino, H.; Bartlett, R.J. *J. Chem. Phys.* **1987**, *87*, 502.
61. Mukhopadhyay, D.; Hayden, G.W.; Soos, Z.G. *Phys. Rev. B*, submitted.
62. Kepler, R.G.; Soos, Z.G. in: *Relaxation in Polymers,* Kobayashi, T. ed.World Scientific, Singapore, 1994; pp. 100-133.
63. Weiser, G. *Phys. Rev.* **1992**, *B45*, 14076.
64. Abe, S.; Yu, J.; Su, W.P. *Phys. Rev.* **1992**, *B45*, 8264.
65. Abe, S. in: *Relaxation in Polymers,* Kobayashi, T. ed.World Scientific, Singapore, 1994; pp. 215-244.
66. Soos, Z.G.; Galvão, D.S; Etemad, S. *Adv. Mat.* **1994**, ***6***, 280.
67. Hudson, B.S.; Kohler, B.E., Schulten, K. in: *Excited States*, vol. 6; Lim, E.,ed; Academic, New York, 1982; Ch. 1.

68. Halvorson, C; A.J. Heeger, A.J. *Chem. Phys. Lett.* **1993**, ***216***, 488.
69. Kohler, B.E.; Spangler, C.; Westerfield, C. *J. Chem. Phys.* **1988**, *89*, 5422.
70. Soos, Z.G.; Ramasesha, S.; Galvão, D.S. *Phys. Rev. Lett.* **1993**, *71*, 1609.
71. Soos, Z.G.; Kepler, R.G.; Etemad, S. in *Organic Electroluminescence*, Bradley, D.D.C.; Tsutsui, T. eds.; Cambridge University Press, 1995; in press.
72. Soos, Z.G.; Etemad, S.; Galvão, D.S.; Ramasesha, S. *Chem. Phys. Lett.* **1992**, *194*, 341.
73. Soos, Z.G.; Galvão, D.S. *J. Phys. Chem.* **1994**, *98,* 1029.
74. Lieb, E.H.; Wu, F.Y. *Phys. Rev. Lett.* **1968**, *20*, 1445.
75. Ovchinnikov, A.A. *Soviet Phys. JETP* **1970**, *30*, 1100.
76. Lawrence,B.; Torruellas, W.E.; Cha, M.; Sundheimer, M.L.; Stegeman, G.I.; Meth, J.; Etemad, S.; Baker, G.L. *Phys. Rev. Lett.* **1994**, *73*, 597.
77. Girlando, A.; Painelli, A.; Soos, Z.G. *J. Phys. Chem.* **1993**, *98*, 7459.
78. Girlando, A.; Painelli, A.; Hayden, G.W.; Soos, Z.G. *Chem. Phys.* **1994**, *184*, 139.
79. Yaron, D.; Silbey, R.J. *J. Phys. Chem.* **1992**, *97*, 5607.
80. Mantini, A.R.; Marzocchi, M.P.; Smulevich, G. *J. Phys. Chem.* **1989**, *91*, 95.
81. Granville, M.F.; Holtom, G.R.; Kohler, B.E. *J. Phys. Chem.* **1980**, *72*, 4671.
82. Bondarev, S.L.; Knyukshto, V.N. *Chem. Phys. Lett.* **1994**, *225*, 346.
83. Castiglioni, C.; Gussoni, M.; Del Zoppo, M.; Zerbi, G. *Sol. State Commun.* **1992**, *82*, 13.
84. Miller, R.D.; Michl, J. *Chem. Rev.* **1989**, *89*, 1359.
85. Schweizer, K.S. in *Silicon-Based Polymer Science: A Comprehensive Resource,* Zeigler, J.M.; Fearon, F.W.G., eds; ACS Symposium Series 358; American Chemical Society, Washington, D.C. 1990; p. 379.
87. Kohler, B. in *Conjugated Polymers: The Novel Science and Technology of Conducting and Nonlinear Opritcally Active Materiala,* Brédas, J.L.; Silbey, R.J., eds.; Kluwer, Dordrecht, The Netherlands, 1991; pp.405-434.
88. Mukhopadhyay, D.; Akram, M.; Soos, Z.G. unpulished results.

RECEIVED June 14, 1995

Chapter 12

Semiempirical Quantum Cell Models for the Third-Order Nonlinear Optical Response of Conjugated Polymers

David Yaron

Department of Chemistry, Carnegie Mellon University, 4400 Fifth Avenue, Pittsburgh, PA 15213–3890

Semiempirical models are used to explore how electron-electron interactions may alter Huckel theory's simple physical picture of the origin of the nonlinear optical response. Huckel theory treats a conjugated polymer as a one-dimensional semiconductor and attributes the large hyperpolarizability to a migration process in which the first photon sees an insulator, and must create an electron-hole pair, while the second photon sees a conductor, moving the electron or hole within a one-dimensional band. The effects of limiting the excited-state charge transfer are explored using a model based on singles-configuration interaction theory and parameters appropriate for gas-phase polyenes. The effects of introducing Coulomb interactions between two different electron-hole pairs are also discussed. In both cases, Coulomb interactions have a quantitative effect, but Huckel theory's predictions remain qualitatively valid. The role of ground state correlation and interactions between polymer chains are also briefly discussed.

Conjugated polymers and other systems with delocalized electrons exhibit large third-order nonlinear optical responses (*1*). In this paper, semi-empirical quantum-cell models are used to gain a better understanding of the origin of the large nonlinear response. We first look carefully at how Huckel theory, the simplest model of delocalized electrons, connects electron delocalization to a large hyperpolarizability (*2,3*). We then discuss the effects electron-electron interactions may have on the simple qualitative picture presented by Huckel theory. From experimental (*4*) and theoretical (*5-7*) studies, electron correlation is known to have large qualitative effects on both the structure of the energy levels and the predicted hyperpolarizabilities of polyenes. But the effects of electron correlation in the long-chain, polymeric limit are not well understood. Electron-electron interactions are often needed to rationalize the resonant features of the nonlinear response (*8-10*), since without such interactions, the lowest two-photon

0097–6156/96/0628–0211$15.00/0

allowed state will be nearly degenerate with the lowest one-photon allowed state. However, in this work, we concentrate on the effects electron correlation may have on the magnitude of the nonresonant response, a quantity of importance in many nonlinear optical applications (*11*).

We begin by presenting the form of quantum cell Hamiltonians and discussing some general aspects of our use of semiempirical models. We then discuss the cancellation of unlinked clusters in the sum over states expression for the hyperpolarizability. This, together with the Huckel model, forms the basis for our approach to nonlinear optics. One possible deficiency of Huckel theory is that it may be overestimating the degree of charge transfer present in the low-lying excited states and we address this in the S-CI model presented below. Huckel theory also ignores the Coulomb interactions between electron-hole pairs, and we briefly describe our Coulomb scattering model designed to explore the effects of these interactions on the hyperpolarizability. We then briefly consider effects that may arise from ground state correlation and from interactions between chains.

Quantum Cell Hamiltonians

The models we use are based on semiempirical Hamiltonians for the delocalized π-electrons. These models start with the electronic Hamiltonian as written in a basis of spin orbitals (*12*),

$$H = \sum_{ij} [i|h_1|j]\, a_i^\dagger a_j + \sum_{ijkl} [ik|jl]\, a_i^\dagger a_j^\dagger a_l a_k \tag{1}$$

The indices i, j, k, and l label spin orbitals, ϕ_i, $a_i^\dagger$ creates an electron in the ith spin orbital, a_j destroys an electron in the jth spin orbital, $[i|h_1|j]$ is the one-electron Hamiltonian matrix element between ϕ_i^* and ϕ_j, and $[ik|jl]$ are the two-electron Hamiltonian matrix elements in chemist's notation:

$$[ik|jl] = \int dr_1 dr_2 \phi_i^\dagger(r_1)\,\phi_k(r_1)\,\frac{1}{r_{12}}\,\phi_j^\dagger(r_2)\,\phi_l(r_2) \tag{2}$$

We obtain a quantum cell model by assuming zero-differential-overlap (ZDO) between orbitals:

$$\int dr_1 \phi_i^\dagger(r_1)\,\phi_k(r_1) = \delta_{i,k}\ ;\ [ik|jl] = \Gamma_{i,j}\delta_{i,k}\delta_{j,l} \tag{3}$$

Within the ZDO approximation, each electron in eq. 2 can be assigned to a specific orbital. For a π-electron model with one p-orbital per carbon atom, the indices i and j label carbon atoms or sites, and each site has a quantum cell that can hold between 0 and 2 electrons.

The simplest quantum cell models are tight-binding models such as Huckel (*2,3*) or Su-Schreiffer-Heeger (SSH) theory (*13*) that ignore electron-electron repulsions entirely and include a one-electron matrix element or transfer integral, typically referred to as β or t, only between bonded atoms. The size of the transfer integral is related to the strength of the bond connecting the two atoms and a linear or exponential dependence of the transfer integral on bond length is typically assumed. Since the bond-length dependence of the transfer integral couples the electronic motion to the vibrational motion, a model that assumes a linear form, such as the SSH model, is often referred to as including linear electron-phonon coupling.

$\Gamma_{i,j}$ of eq. 3 gives the Coulomb repulsion energy between an electron on site i and an electron on site j. The on-site Coulomb repulsion energy, or Hubbard parameter $\Gamma_{i,i}$=U, is the Coulomb energy associated with placing two electrons on the same atom. A Hubbard model includes only on-site Coulomb repulsion, while an extended Hubbard model also includes interactions between sites. The Pariser-Parr-Pople model is an extended Hubbard model that includes interactions between all sites. In such a model, the on-site interaction energy, U, is typically set to the difference between the ionization potential and the electron affinity of an isolated carbon atom, IP–EA= 11.13eV (*5*). Since for both the IP and the EA of carbon an electron is either added to or removed from a p orbital, the difference results from the Coulomb energy associated with placing two electrons in the same orbital in the negative ion C^-. The Coulomb energy between different sites is typically assumed to be a function of the distance between sites, r, and is obtained by interpolating from U at r=0, to the Coulomb 1/r form at long distance. In this work, we use the Ohno interpolation formula (*14*):

$$\Gamma(r) = \frac{14.397\text{eV Å}}{\sqrt{\left[\frac{14.397\text{eV Å}}{U}\right]^2 + r^2}} \qquad (4)$$

The parameterization of the Coulomb energy discussed above ignores dielectric stabilization from adjacent chains which, as discussed below, may have large effects on the structure of the excited states.

Semiempirical Models Consisting of both a Hamiltonian and a Solution Method

Obtaining accurate solutions of even the simplest quantum cell model that includes Coulomb interactions is a difficult task. Leib and Wu (*15*) obtained an analytic solution of the Hubbard model for the special case of a one-dimensional chain of atoms in an all-bonds-equal geometry, and numerically exact solutions of the extended Hubbard model have been obtained by Soos (*6*) and Mazumdar (*10*) for systems with up to 12 carbons. But, in general, if we are interested in the long-chain polymeric limit, we are forced to work with approximate solution methods. The addition of an electron-electron repulsion term to the Hamiltonian, even in the simple form of an on-site Hubbard interaction, takes us from a relatively simple Huckel Hamiltonian to a Hamiltonian that is extremely difficult to solve. This reflects the fact that electron-electron interactions are responsible for many different qualitative effects and in attempting to solve a Hamiltonian that includes Coulomb interactions, we are attempting to include all of these effects at once. So if we view the Hamiltonian as defining a semiempirical model, it is difficult to obtain models with complexities that lie between that of the Huckel and extended Hubbard models. If instead, we view the semi-empirical model as consisting of both the Hamiltonian and the method used in its solution, we can, depending on our choice of solution method, construct intermediate models that include some but not all of the effects of electron-electron interactions. From a more formal perspective, such an approach is equivalent to constructing a semi-empirical

Hamiltonian by starting with a quantum-cell Hamiltonian and projecting this onto the space spanned by the trial solutions.

The models discussed below are first parameterized to the band gap and band width of the polymer and then used to predict the hyperpolarizability. If the solution method is viewed as part of the semi-empirical model, it makes sense to reparameterize the Hamiltonian for each method of solution. This is similar in spirit to the reparameterization of a one-electron Hamiltonian on addition of a two-electron term. For instance, to maintain agreement with experimental absorption spectra of polyenes, the difference in transfer integral for the single and double bonds, $|\beta_1-\beta_2|$, is set to about 0.9eV (*2*) in the Huckel Hamiltonian and lowered to about 0.35eV in the PPP Hamiltonian (*5*). This is easily rationalized since in Huckel theory, β_2 and β_1 must absorb effects due to electron-electron interactions and this is no longer necessary in the PPP model. Similarly, in our approach, changing the method of solution alters what is explicitly included in the model and what must be absorbed into the effective parameters.

Cancellation of Non-Size Extensive Terms

Our qualitative picture of the nonlinear response is rooted in Huckel theory and the following physical interpretation of the unlinked cluster theorem. The sum-over-states expression for the non-resonant third-order hyperpolarizability is (*1*):

$$\gamma_{xxxx} = \sum_{ABC} \frac{\langle GS|x|A\rangle\langle A|x|B\rangle\langle B|x|C\rangle\langle C|x|GS\rangle}{E_A E_B E_C} - \sum_{AC} \frac{\langle GS|x|A\rangle\langle A|x|GS\rangle\langle GS|x|C\rangle\langle C|x|GS\rangle}{E_A^2 E_C} \quad (5)$$

where $|GS\rangle$ is the ground electronic state, $|A\rangle$, $|B\rangle$ and $|C\rangle$ are excited electronic states, x is the dipole operator and E_A is the energy of state $|A\rangle$ relative to the ground state. Eq. 5 is obtained by using fourth-order perturbation theory to describe the effects of an applied electric field on the energy. In fourth-order perturbation theory, unlinked clusters appear in the sum-over-states expression (*16,12*). These unlinked clusters are terms in the first summation of eq. 5 that are exactly cancelled by terms in the second summation.

In an independent electron model, such as Huckel theory, the hyperpolarizability may be written as a sum over molecular orbitals, rather than the many-electron states of eq. 5. The summation over molecular orbitals has many easily identified unlinked clusters which may be explicitly removed by using, for example, the unlinked cluster theorem of diagrammatic perturbation theory. Andre et. al. (*17*) used such an approach to obtain an efficient expression for the hyperpolarizability within a noninteracting electron model. But in going beyond Huckel theory, it is convenient to work with a summation over many-electron states, and it is useful to consider how unlinked clusters appear within this context. By first understanding how unlinked clusters appear in a summation over the many-electron states of the Huckel model, we can perhaps understand how they will appear in the summation over the many-electrons states of a correlated model.

The terms in eq. 5 will be represented schematically as:

$$|GS\rangle \rightarrow |A\rangle \rightarrow |B\rangle \rightarrow |C\rangle \rightarrow |GS\rangle \; - \; |GS\rangle \rightarrow |A\rangle \rightarrow |GS\rangle \rightarrow |C\rangle \rightarrow |GS\rangle$$

The first summation can be viewed as a sum over all possible virtual two-photon absorption processes. (Since we are off resonance, the molecule does not actually absorb energy from the electric field, nevertheless we will use the term absorption as a convenient language for the sum over virtual states in eq. 5. Alternatively, the molecule can be viewed as absorbing the photon for a very short time, as set by the time-energy uncertainty principle, with the uncertainty in energy being the distance off resonance (*18*).) In the first summation of eq. 5, an unlinked cluster results when each of the two photons, $|GS\rangle \rightarrow |A\rangle$ and $|A\rangle \rightarrow |B\rangle$, creates an excitation and these two excitations do not interact (19-21,7)

$$|GS\rangle \rightarrow |1 \text{ excitation}\rangle \rightarrow |2 \text{ non-interacting excitations}\rangle \rightarrow |1 \text{ excitation}\rangle \rightarrow |GS\rangle$$
$$- \; |GS\rangle \rightarrow |1 \text{ excitation}\rangle \rightarrow |GS\rangle \rightarrow |1 \text{ excitation}\rangle \rightarrow |GS\rangle = 0$$

The type of the excitation depends on the model, for instance, in Huckel theory the excitations are electron-hole pairs. But independent of model, we expect that in the limit of a long polymer chain, it will be possible to create two excitations that do not interact. Non-interacting means that the presence of the first excitation has no effect on either the energy or transition moment involved in the creation of the second excitation. More precisely, for state $|B\rangle$ to contain the excitations present in $|A\rangle$ and $|C\rangle$ and for these excitations to be non-interacting, the energy of the state containing both excitations must be the sum of the energies for the states containing either excitation individually, $E_B=E_A+E_C$; the transition moment for creating the excitation in $|C\rangle$ on top of that in $|A\rangle$ must be the same as that for creating the excitation in $|C\rangle$ on the ground state, $\langle B|x|A\rangle=\langle C|x|GS\rangle$; and similarly, $\langle B|x|C\rangle=\langle A|x|GS\rangle$.

The origin of the above cancellation can be understood by considering a gas of N non-interacting molecules. If we restrict the summations in eq. 5 such that in each term, the excited states, $|A\rangle$, $|B\rangle$, and $|C\rangle$ are those of one specific molecule, then the summation changes into a sum over individual molecules and we get the expected result; namely, that the hyperpolarizability of the gas is N times that of a single molecule. If instead, we sum over the excited states of the gas as a whole, we will include states $|B\rangle$ in which two different molecules are excited. Since the gas molecules are noninteracting, these states contain two noninteracting excitations, and as discussed above, their contribution to the first summation of eq. 5 will be identically cancelled by terms in the second summation, specifically, those terms in which $|A\rangle$ and $|C\rangle$ are the excited states of two different molecules. As we turn on interactions between gas molecules, states containing two different excited molecules will begin to contribute to the hyperpolarizability. A polymer is analogous to the interacting gas, since in the long chain limit, it is possible to create two excitations that interact either weakly or strongly.

Understanding the cancellation between the two summations of eq. 5 is especially important in work on polymers, since in the polymeric limit, the cancellation between the two terms becomes infinitely large (19-21,7). In the infinite chain limit, the hyperpolarizability should scale as the number of unit cells in the polymer ($\gamma \propto N$). However, both the first and second summations in eq. 5 scale as the square of the length of the polymer, N^2, and it is only the difference that has the correct linear dependence on chain length. Thus in the infinite chain limit, eq. 5 gives the hyperpolarizability per unit cell γ/N, a finite quantity, as the difference between two infinite quantities. Due to this large cancellation, the origin of the nonlinear response can not be understood by looking at either summation individually. A two-photon state, $|B\rangle$, that contributes a large amount to the first summation of eq. 5 can not necessarily be identified as an important contributor to the nonlinear response, since it may be contributing primarily to the infinity that is to be cancelled by the second summation. As discussed above, this will occur when state $|B\rangle$ contains two non-interacting excitations and the origin of the N^2 dependence of the first term relates to there being N^2 different ways to create two noninteracting excitations on a long chain. (For example, in a noninteracting gas of N molecules, the number of states containing two different excited molecules scales as N^2.) The difficulty is that in a correlated model of a polymer, it is not easy to identify a state as containing two noninteracting or weakly interacting excitations. Overcoming this difficulty is the motivation for our scattering approach, described below.

Knowing which processes do not contribute to the hyperpolarizability may allow us to infer the characteristics of a material that lead to a large nonlinear optical response. That the creation of two noninteracting excitations does not contribute to the hyperpolarizability can be taken to mean that a nonlinear response results only when the first photon changes the material in a manner that has an effect on the absorption of the second photon. It seems reasonable to extend this and say that for a material to have a large nonlinear optical response, the first photon must have a large qualitative effect on the nature of the material. In the next section, we identify two qualitatively different ways for the second photon to see the effects of the first photon.

The Huckel Model

Within the Huckel model, a conjugated polymer is a one-dimensional semiconductor (*2,3*). For instance, in the Huckel model of polyacetylene, the alternation between single and double bonds, or Peierls distortion (*13*), leads to the formation of two bands of molecular orbitals, a filled valence band consisting of π bonding orbitals and an empty conduction band consisting of π^* anti-bonding orbitals. The distance from the bottom of the valence band to the top of the conduction band is given by the sum of the transfer integrals between the single and double bonds, $2|\beta_1+\beta_2|$, and the band gap is given by the difference, $2|\beta_1-\beta_2|$. The lowest-energy excited state is obtained by promoting an electron from the highest occupied molecular orbital (HOMO), creating a hole, to the lowest unoccupied molecular orbital (LUMO), creating an electron. One of the

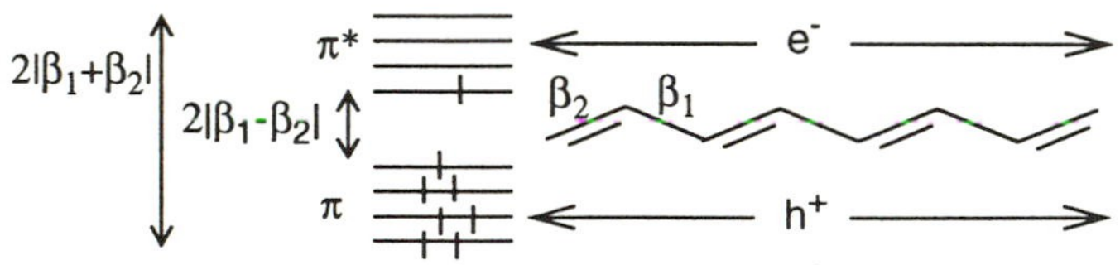

FIGURE 1. Schematic representation of the energy levels obtained from a Huckel model of a polyacetylene-like (2 sites per unit cell) polymer.

fundamental approximations of Huckel theory relates to the large degree of charge transfer present in the low-lying excited states. Since the orbitals occupied by the electron and hole are delocalized over the entire polymer, the probability of finding the electron and hole at remote positions on the chain, corresponding to long-range charge

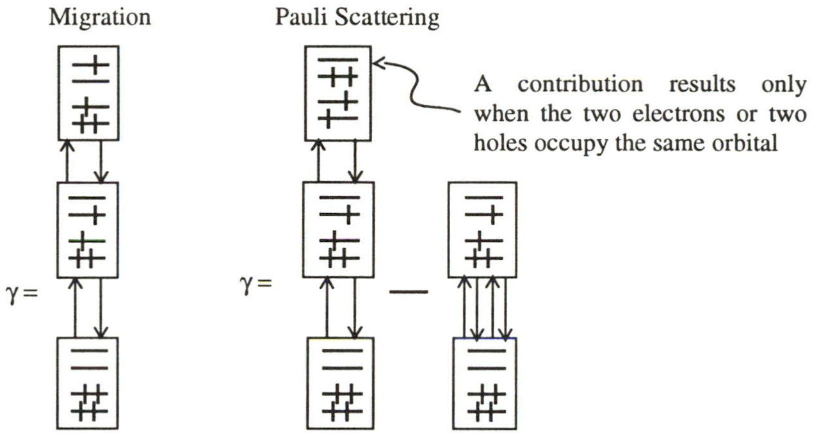

FIGURE 2. Schematic representation of terms in eq. 5 that contribute to the hyperpolarizability within a Huckel model. The difference between migration and scattering relates to the number of electron-hole pairs in the two-photon excited states.

transfer, is equal to the probability of finding the electron and hole near one another, corresponding to short-range charge transfer.

The polyacetylene-like polymer of Figure 1 provides a general model system of a conjugated polymer as a chain of monomers connected by transfer integrals, β_1. Each monomer consists of two sites connected by β_2. The energy needed to create charges is given by the band gap, $E_g=2|\beta_1-\beta_2|$, and once these charges are created, they are delocalized in a one-dimensional metallic band. The band width of both the conduction and valence bands is given by $2|\beta_1|$. When $\beta_1=0$, the model consists of a row of non-interacting ethylene-like molecules, which means the band width of the conduction and valence band is zero and the electrons and holes are immobile. As we increase $|\beta_1|$, the conduction and valence bands widen and the electron and hole become increasingly delocalized. Below, we will consider the dependence of the hyperpolarizability on the band gap and band width.

The absorption of the first photon in eq. 5 will move electrons between bands and create an electron-hole pair. The second photon can then do one of two things, giving rise to the two processes shown in Figure 2 (*21*). By moving electrons within a band, the second photon can modify the electron-hole pair created by the first photon. We will refer to this type of process, in which the second photon modifies the excitation created by the first photon, as a *migration* process. Alternatively, by moving electrons between bands, the second photon can create an additional electron-hole pair. The creation of a second electron-hole pair will contribute to the hyperpolarizability only when it interacts with the electron-hole pair created by the first photon; otherwise, the process involves the creation of two non-interacting excitations, giving rise to an unlinked cluster and making no contribution to the nonlinear response. We will refer to this second process as a *scattering* process, since the second photon creates an excitation that interacts with, or scatters with, that created by the first photon. Within Huckel theory, the electron-hole pairs interact only through Pauli exclusion, that is,

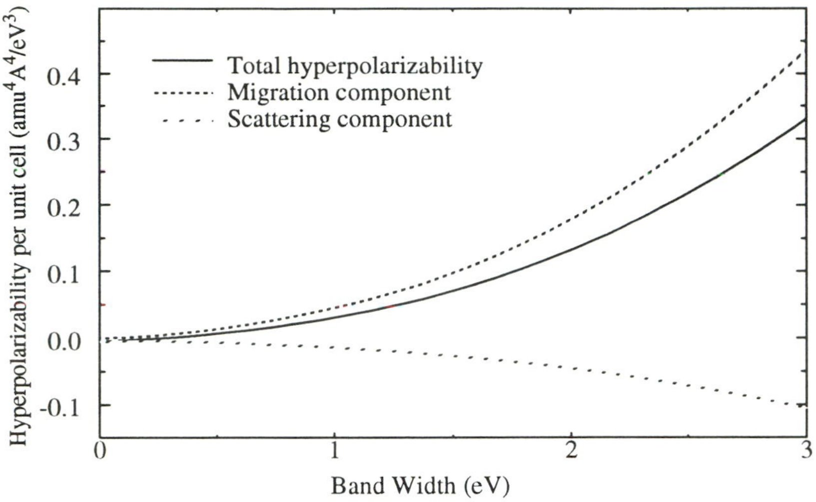

FIGURE 3. Migration and scattering contribution to the hyperpolarizability of Huckel theory, plotted versus the band width of the conduction and valence band, for a chain with 142 carbons (71 unit cells). The band gap is fixed at 1.8eV. Migration dominates for band widths greater than about 0.2eV.

they interact only when the electron-hole pair created by the second photon shares an orbital with the electron-hole pair created by the first photon. We note that Agrawal, Cojan and Flytzanis (*2*) used the band formalism of Genkin and Mednis (*22*) to obtain the nonlinear optical response of polyacetylene within Huckel theory, and the two terms of the band theory expression correspond to the migration and scattering contributions discussed here, not to the first and second summations of eq. 5 (*21*).

Figure 3 shows the migration and scattering contributions to the hyperpolarizability of a polyacetylene-like polymer. The results are shown as a function of the width of the valence and conduction band, $2|\beta_1|$, holding the band gap fixed at 1.8eV. When the band width is zero, the second photon can not move the electron and hole within a band, and the migration process can not occur. Zero band width corresponds to $\beta_1=0$, for which the polymer of Figure 1 becomes a row of two-site molecules, and the hyperpolarizability is N times that of a single two-site molecule. In a Huckel model of a two site system, there is only one π bonding and one π^* antibonding orbital, the second photon can not move the electron or hole within a band and migration is not allowed. As the band width is increased, migration becomes increasingly important and for band widths greater than about 0.2eV, migration becomes the dominant nonlinear process. The Huckel parameters typically used for polyacetylene yield a band width of about 4eV (*2,3*), placing it well into the regime where migration dominates the response.

The migration and scattering components of the hyperpolarizability have opposite signs and this can be rationalized as follows. A positive nonresonant hyperpolarizability results when the second photon sees a material that is more polarizable than that seen by the first photon. This is the case in migration, where the first photon creates an electron-hole pair, essentially changing the material from a insulator into a conductor, and the second photon operates on the electron or hole. In Huckel theory, the scattering component is negative because the electron-hole pair created by the first photon suppresses, through Pauli exclusion, the creation of the second electron-hole pair.

Discussion of the Huckel Model

Huckel theory provides an intuitively appealing picture of the origin of the hyperpolarizability. As discussed at the end of the *Cancellation of Non-Size Extensive Terms* section, a large nonlinear response results when the first photon changes the material in a manner that makes the absorption of the second photon very different from the absorption of the first photon. In the one-dimensional semi-conductor picture presented by Huckel theory, a large nonlinear response results because the electron-hole pair created by the first photon changes the material from an insulator into a conductor. This allows the second photon to do something very different than the first photon; namely, it can participate in a migration process and move either the electron or hole within the conduction or valence band. The migration process of Huckel theory involves long-range charge transfer between monomers, and to the extent that long-range charge transfer dominates the hyperpolarizability, Huckel theory may capture the essential aspects of the nonlinear response. While it seems likely that Huckel theory does not provide an accurate description of the structure of the monomer, especially within our simple two-site model, if the hyperpolarizability is dominated by the transfer of charge between monomers, the detailed structure of the monomers may not be of central importance. The crucial question appears to be whether or not Huckel theory provides a reasonable description of the long-range charge transfer process. By parameterizing the model to the band gap, we have built in the energy required to perform the initial charge separation process. The issue is then whether or not the motion of these charges, once created, is adequately modelled by the simple band structure of Huckel theory. The following models are designed to address this issue.

Coulomb Interactions and the Degree of Charge Transfer (S-CI model)

Within Huckel theory, absorption of a photon generates charges that enter a one-dimensional metallic band structure and are delocalized independently along the polymer chain. As discussed above, this corresponds to an essentially infinite degree of charge transfer in the low-energy excited states. However, Coulomb interactions may limit the degree of charge transfer in the low-energy excited states. For a model to describe this effect, it must be possible to parameterize the model to include the cost of separating the electron and hole to form a charge-separated configuration. This specific consequence of electron-electron interactions can be included by using singles-configuration interaction (S-CI) theory to solve a extended Hubbard model (*21*,*23*,*24*,*8*). Since the excited states of S-CI theory are constrained to contain exactly one electron and one hole, this model can be used to model the migration process but not the scattering process, which involves the creation of two electron-hole pairs. Using S-CI theory to solve the PPP Hamiltonian of polyacetylene leads to the energy levels shown in Figure 4 (*21*). Also shown are the wavefunctions, which in periodic boundary conditions may be written:

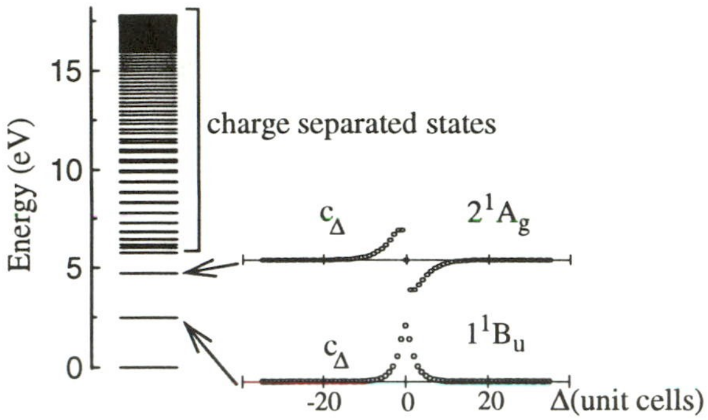

FIGURE 4. Energy levels and wavefunctions from a S-CI calculation on a chain of polyacetylene with 71 unit cells, using the PPP Hamiltonian with gas-phase Ohno parameters (21). Periodic boundary conditions were used and only K=0 states are shown. The 1^1B_u and 2^1A_g states contains bound electron-hole pairs, or excitons. States above about 6eV contain free electron-hole pairs and would form a continuum on an infinitely long polymer.

$$|\text{excited state}\rangle = \sum_{\Delta} e^{iKn} c_\Delta a^\dagger_{n+\Delta} b^\dagger_n |\text{Hartree-Fock ground state}\rangle \tag{6}$$

where K is the wavevector, and $a^\dagger_n$ and $b^\dagger_n$ are the electron and hole creation operators for Wannier functions centered on the nth unit cell of the polymer. Figure 4 shows c_Δ, the probably amplitude for charge transfer over a distance of Δ unit cells, for the two lowest-energy excited states. The degree of charge transfer in the lowest energy state is limited to about 4 unit cells. This limited separation between the electron and hole can be interpreted as the formation of a bound electron-hole pair, or exciton. Since we are using the gas-phase PPP parameters (*5*), which ignore dielectric effects from surrounding polymer chains, we are probably underestimating the degree of charge transfer. Nevertheless, the S-CI hyperpolarizability is only about a factor of two smaller than that obtained from a Huckel model parameterized to yield the same band gap and band width as that obtained in the S-CI model (21). (Since S-CI theory does not include scattering, we compare the S-CI results to the migration component of the Huckel hyperpolarizability.) This qualitative agreement suggests that even when charge transfer is limited to nearby unit cells, Huckel theory still captures the essentials of the nonlinear response.

Coulomb Scattering (SD-EOM Model)

In the scattering process of Huckel theory, the electron-hole pairs interact only through Pauli exclusion, whereby the presence of the electron-hole pair created by the first photon suppresses the formation of a second electron-hole pair. The resulting Pauli scattering makes a negative contribution to the hyperpolarizability that is offset by the large positive contribution from migration. But by ignoring Coulomb interactions between electron-hole pairs, Huckel models ignore what may be an important contributor to the nonlinear response, Coulomb scattering. The low-lying 2^1A_g state of polyenes(*4,5,25*) can be taken as evidence for strong Coulomb interactions between e-h pairs, since although this state has significant character from double electron-hole pair configurations, it has an energy below that of the single electron-hole pair 1^1B_u state (*25*). To model Coulomb scattering requires the inclusion of double electron-hole pair configurations and to obtain a valid polymeric limit for the hyperpolarizability, this must be done in a size-consistent manner. While S-CI theory is size-consistent, the inclusion of double excitations within a configuration interaction formalism (SD-CI) is not (*26,12*). Instead, we use an equation-of-motion method (*27*) with single and double excitations (SD-EOM) (*28*). Like SD-CI theory, SD-EOM theory allows us to describe excited states with up to two electron-hole pairs, but unlike SD-CI theory, it ignores ground state correlation. In SD-EOM theory, the ground state is constrained to remain the Hartree-Fock ground state, and it is only due to this constraint that it is size-consistent.

To make SD-EOM calculations numerically feasible in the long chain limit, we use a contracted "scattering" basis set (*29*). In constructing the basis, we restrict the one-photon states, $|A\rangle$ and $|C\rangle$ of eq. 5, to the 1^1B_u state of the S-CI model. (Higher-order configuration interaction calculations find that the 1^1B_u state of polyenes is composed primarily of single electron-hole pair configurations (*25*), and both S-CI and exact solutions of the PPP model find that this state carries essentially all of the one-photon intensity (*21,6*).) The two photon states, $|B\rangle$, are obtained from an SD-EOM calculation with a complete set of single electron-hole pair configurations, but with a contracted set of double electron-hole pair configurations. Since the 1^1B_u states carries most of the one-photon intensity and this state contains a bound electron-hole pair, a 1^1B_u exciton, we expect that in the long chain limit, the primary effect of the second photon will be the creation of another 1^1B_u exciton. States containing two noninteracting excitons give rise to unlinked clusters that are cancelled by the second summation of eq. 5 and by including basis functions consisting of all possible separations between two excitons, we can let the calculation decide how strongly interacting the excitons are and whether or not they contribute to the hyperpolarizability. When the excitons get close together, the interactions will be strong enough to alter the form of the excitons. This distance defines the size of the scattering region, within which we include a full set of double electron-hole pair configurations, consistent with some limit on the maximum allowed separation between electrons and holes. This truncated basis is defined by two parameters, the maximum allowed separation between electrons and holes, and the size of the scattering region. Our results are converged with respect to these parameters and the SD-EOM hyperpolarizability, calculated using the PPP Hamiltonian of polyacetylene, is about 40% larger than that obtained with a S-CI model (*29*). Since the SD-EOM model adds Coulomb scattering between excitons to the S-CI model, this qualitative agreement suggests that Coulomb scattering has an effect but does not dominate the nonresonant nonlinear optical response. The inclusion of double excitations through a SD-EOM formalism does, however, have large qualitative effects on the energy levels and thus on the resonant structure of the nonlinear response (*28*). For instance, SD-EOM theory finds a low-lying 2^1A_g, although, due to the lack of ground state correlation, the calculated energy is probably still overestimated. Further studies on the effects of Coulomb scattering on the nonlinear optical response are ongoing.

Ground State Correlation

Up to this point, we have been discussing models that may be parameterized to include the Coulomb energy associated with charge separation in the excited states. This requires an electronic structure theory that includes the "static" correlation associated with mixing between a few important electronic configurations. For example, the S-CI 1^1B_u state in Figure 4 is a linear combination of a few configurations corresponding to different electron-hole pair separations. This "static" correlation can be viewed as determining the structure of the excitation created by absorption of a photon. Such models ignore the dynamic correlation associated with, for instance, the breakdown of the Hartree-Fock approximation for the ground state. Breakdown in the Hartree-Fock approximation mixes excited configurations into the ground state, and this mixing can be viewed as the presence of virtual excitations in the correlated ground state. The size-consistent description of these virtual excitations is a challenging problem, and it is unclear how these "virtual" excitations modify the form of the "actual" excitations created on absorption of a photon. By using perturbation theory and *ab initio* Hamiltonians, Suhai *(30)* and Liegener *(31)* demonstrated that dynamic correlation has large effects of both the band gap and exciton binding energy of polyacetylene.

Solid State Effects

It seems likely that interactions between polymer chains plays an important role in materials based on conjugated polymers. Beyond the local field effect, through which the polarization on adjacent chains modifies an externally applied electric field, interactions between chains may also have large effects on the electronic structure of the excited states *(32-34)*. For instance, in the S-CI calculations of Figure 4, the energy difference between the 1^1B_u state and the start of the charge-separated states, the exciton binding energy, is about 3.5eV. While this may be a reasonable estimate for an isolated polymer chain, dielectric effects from adjacent chains will stabilize the charge separated state and lower the exciton binding energy *(32)*. Based on Hartree-Fock calculations of a point charge surrounded by polyacetylene chains, we estimate the solvation energy of a single point charge to be about 1.5eV. The solvation energy associated with the charge separated states is then about 3eV. Using the electron-electron interaction potential of eq. 4, the Hubbard parameter U is the difference in energy between a neutral configuration, which has one electron per site, and a charge-separated configuration, which has one empty and one doubly occupied site separated by a large distance. If we solvate the charge-separated state by 3eV, we lower the effective Hubbard parameter from 11eV to 8eV, which has a relatively minor effect on the exciton binding energy (32). This screening of the electron-electron interaction potential of eq. 4 implicitly assumes that the dielectric response of the adjacent chain is sufficiently fast that it follows the charge fluctuations on the excited chain. This assumes, for example, that the polarization of adjacent chains follows the motion of the electron-hole pair in the 1^1B_u state of S-CI theory. If instead, we assume the 1^1B_u state is non-polar and therefore not well solvated by the surrounding chains, then the 3eV solvation energy of the charge separated states is to be subtracted from the exciton binding energy, which in Figure 4 is about 3.5eV. The numbers reported here are rough estimates, especially since our estimated exciton binding energy ignores dynamic correlation

(*30,31*), but it is clear that the largest dielectric effects are obtained when the polarization of the adjacent chains stabilizes the charge separated states, but is too slow to stabilize states containing neutral excitations, such as the 1^1B_u exciton state of the S-CI model. If this is the case, and the time-scale of the dielectric response plays an important role in determining the structure of the excited states, it seems unlikely that dielectric effects can be modelled by screening the electron-electron interaction potential of eq. 5 (*8,24*) and a more sophisticated treatment will be required.

Concluding Remarks

The Huckel, one-dimensional semiconductor model of conjugated polymers attributes the large nonlinear response to a migration process in which the first photon sees an insulator, and must create an electron-hole pair, and the second photon sees a conductor, moving the electron or hole within a one-dimensional band. We have considered some specific ways in which Coulomb interactions may alter this simple picture. We use a simple two-site quantum cell Hamiltonian for the monomers, since if it is charge transfer between monomers that gives rise to the large nonlinear response, then the structure of the monomers themselves is probably not of central importance. By using S-CI theory to solve this quantum cell model, we allow Coulomb interactions to limit the degree of charge separation present in the excited states. We find that limiting the degree of charge transfer, while keeping the band gap and band width constant, lowers the migration contribution to the hyperpolarizability by about a factor of two but does not fundamentally change the predictions of Huckel theory. We have also considered the effects of Coulomb scattering on the response and again find basic agreement with the predictions of Huckel theory. This is the case even though these studies were done using unscreened, gas phase parameters that probably overestimate the strength of Coulomb interactions. While this work suggests that Huckel theory captures the essential aspects of the nonresonant third-order nonlinear optical response, further work that includes higher-order effects from Coulomb interactions, and that includes interactions between polymer chains is clearly needed.

Acknowledgements

We thank the Petroleum Research Fund for partial support of this work.

Literature Cited

1. *Nonlinear Optical Properties of Organic Molecules and Crystals*; Chemla D.; Zyss, J., Eds.; Academic Press, Inc. New York, 1987.
2. Agrawal, G.; Cojan, C.; Flytzanis. C. *Phys. Rev. B* **1978**, *17*, 776.
3. Beratan, D.; Onuchic, J.; Perry, J. *J. Phys. Chem.* **1987**, *91*, 2696.
4. Kohler, B.; Spangler, C.; Westerfield C. *J. Chem. Phys.* **1988**, *89*, 5422.
5. Schulten, K.; Ohmine, I.; Karplus M. *J. Chem. Phys.* **1976**, *64*, 4422.
6. Soos, Z.; Ramasesha, S. *J. Chem. Phys.* **1989**, *90*, 1067.
7. Shuai, Z.; Bredas, J.-L. *Phys. Rev. B* **1991**, *44*, 5962.
8. Hasegawa, T.; Iwasa, Y.; Sunamura, H.; Koda, T.; Tokura, Y.; Tachibana, H.; Matsumoto, M.; Abe, S. *Phys. Rev. Lett.* **1992**, *69*, 668.

9. Soos, Z.; Kepler, R. *Phys. Rev. B* **1991**, *43*, 11908.
10. Dixit, S.; Guo, D.; Mazumdar, S. *Phys. Rev. B* **1991**, *43*, 6781.
11. Nalwa, H.S. *Adv. Mater.* **1993**, *5*, 341.
12. Szabo, A.; Ostlund, N. *Modern Quantum Chemisty* McGraw Hill, 1989.
13. Heeger A.; Kivelson, S.; Schrieffer, J.; Su, W.-P. *Rev. Mod. Phys.* **1988**, *60*, 781.
14. Ohno, K. *Theor. Chim. Acta* **1964**, *2*, 219.
15. Lieb, E; Wu, F. *Phys. Rev. Lett.* **1968**, *20*, 1445.
16. Brandow, B. *Rev. Mod. Phys.* **1967**, *39*, 772.
17. Andre, J.M.; Barbier, C.; Bodart, V.; Delhalle, J. in reference (*1*), pp. 137-158.
18. Lee, S.; Heller, E. *J. Chem. Phys.* **1979**, *71*, 4777.
19. Soos, Z.; Hayden, G.; McWilliams, P. in *Conjugated Polymeric Materials, Opportunities in Electronics, Optoelectronics and Molecular Electronics*, Kluwer Academic Publishers, Netherlands, **1990**, pp. 495–508.
20. Ishihara, H.;Cho, K. *Phys. Rev. B* **1990**, *42*, 1724.
21. Yaron, D.; Silbey, R. *Phys. Rev. B* **1992**, *45*, 11655.
22. Genkin V.; Mednis P. *Sov. Phys.-JETP*, **1968**, *27*, 609.
23. Yaron D.; Silbey R. in *Electrical, optical and magnetic properties of organic solid state materials*, Chiang, L.; Garito, A.; Sandman, D. Eds.; Materials Research Society Symposia Proceedings, 1991.
24. Abe S.; Yu J.; Su W.-P. *Phys. Rev. B* **1992**, *45*, 8264.
25. Goldbeck, R.; Switkes, E. *J. Chem. Phys.* **1985**, *89*, 2585.
26. Heflin J.; Wong, K.; Zamani-Khamiri, O.; Garito, A. *Phys. Rev. B* **1988**, *38*, 1573.
27. Rowe, D. *Rev. Mod. Phys.* **1968**, *40*, 153.
28. Yaron, D. *Mol. Cryst. Liq. Cryst.* **1994**, *256*, 631.
29. Yaron, D. *Polym. Preprints, Am. Chem. Soc.* **1994**, *35*, 327.
30. Suhai, S. *Int. J. Quant. Chem.* **1986**, *29*, 469.
31. Liegener, C.-M. *J. Chem. Phys.* **1988**, *88*, 6999.
32. Lubchenko,V.; Yaron D. *Polymer Preprints, Am. Chem. Soc* **1995**, *36*, (in press).
33. Soos, Z.; Hayden, G.; McWilliams, P.; Etemad S. *J. Chem. Phys.* **1990**, *93*, 7439.
34. Guo, D.; Mazumdar S. *J. Chem. Phys.* **1992**, *97*, 2170.

RECEIVED July 11, 1995

Chapter 13

Multiphoton Resonant Nonlinear Optical Processes in Organic Molecules

Paras N. Prasad and Guang S. He

Photonics Research Laboratory, Department of Chemistry, State University of New York, Buffalo, NY 14260–3000

Based on a general and phenomenological theoretical description of nonlinear optical responses in media, the roles and application of multi-photon resonances in various nonlinear optical processes are discussed. The latest research progress in two specific application aspects, i.e. multi-photon absorption based optical limiting and two-photon pumped upconverted lasing are presented. Some issues and opportunities for theoretical studies on optical nonlinearity of organic molecular systems are discussed.

Nonlinear optical effects in organic systems have received a great deal of attention[1]. The interest has been two-fold: (i) a fundamental understanding of the processes and structure-property relationship and (ii) the important roles of these processes in the technology of photonics. The development of the technology of photonics which involves photons to transmit, process and store information is crucially dependent on the availability of materials which possess the required optical nonlinearity and at the same time meet other ancillary requirements. Organic structures are particularly attractive because they provide the flexibility to tailor their structures at the molecular level to optimize their nonlinear optical response. Strong nonlinear optical effects in organic structures have already been established[1]. However, in order to take advantage of the full potential of structural tailoring ability to optimize specific optically nonlinear functional response, there is still a need of an improved theoretical understanding of the structure-nonlinear optical properties[2]. This deficiency in the theoretical understanding is even more so for the nonlinear optical processes under resonance conditions.

A major thrust of the current investigations has been focused on nonresonant nonlinear optical processes i.e. processes which are observed at the optical frequencies at which the system does not exhibit one or multiphoton resonances[1,2]. From a theoretical point-of-view, computation of nonresonant nonlinear optical coefficients (where one considers only virtual excitations) is much simpler. From technological perspectives,

0097–6156/96/0628–0225$15.00/0

nonresonant conditions are desirable for applications such as frequency mixing (harmonic,parametric oscillation, sum and difference frequency generation), frequency shifting (stimulated Raman, Brillouin, Kerr scattering), optical beam induced refractive index change (self-focusing, self-modulation, phase-conjugation), optical switching and bistable devices, and so on. Under a fully non-resonant condition there is no attenuation due to one-, two-, or multi-photon absorption of the nonlinear optical media at frequencies of the interacting optical waves. Therefore, a higher effectiveness of the desirable process can be obtained. As there is no photon-resonance absorption, the thermal effects can be avoided which in some cases are undesirable and could greatly reduce the response speed of optical switching or bistable devices. However, on other hand, the nonlinear response of a given medium can be considerably enhanced when frequencies (or their linear combinations) of the interacting optical waves approach to one-, two-, or multi-photon absorption frequencies, or to Raman-mode frequencies[3]. Typical examples are two-photon resonance enhanced third harmonic generation and four wave frequency mixing (FWFM), coherent anti-Stokes spectroscopy (CARS), Raman-induced Kerr effect (REKE), two-photon or Raman resonance enhanced refractive index changes useful for optical phase-conjugation and bistability techniques. In these particular cases, a practical compromise between the undesirable one-, two- or multi-photon absorption induced attenuation of interacting beams and the desirable resonance enhancement of the signal beam could be achieved by controlling, for example, the concentration of the solute or dopant, or by using a small detuning from the center of a resonance band. One may find that in these cases the overall excitation of population at any real excited level via one- or two-photon absorption can be still neglected. For this reason we may term this kind of nonlinear processes as quasi-resonant processes. Naturally, there are also some other nonlinear optical processes, in which the real two- or multi-photon absorption are dominant and the population excited to upper real energy levels cannot be neglected any longer. Typical examples include: two-photon absorption (TPA) induced fluorescence emission, TPA based optical power limiting, multi-photon induced ionization, dissociation and photoconductivity. In general, resonance-related nonlinear optical effects are of considerable interest. First, they offer challenging opportunities for theoretical computation. Second, they also provide insight into the dynamics of excited states as they probe real excitations[1,3]. Finally, resonant nonlinear optical processes can also be useful for a number of applications such as TPA based nonlinear spectroscopy, two-photon pumped lasing and TPA based optical power limiting.

In this article, we briefly review the resonant nonlinear optical processes by using a phenomenological description. Then we discuss experimental probes to study these nonlinear optical processes. Then two specific applications of the resonant nonlinear optical processes are presented: (i) up-conversion lasing and (ii) optical power limiting. Finally we review the status of microscopic theory in relation to predicting resonant nonlinear optical properties and conclude with opportunities which exist for theory and computational work in this field.

Theoretical Background

Nonlinear optical processes occur under the action of an electric or optical field when the polarization of a medium is no longer expressed only by the linear term in the electric field. In such a case the polarization vector **P** of the medium is expressed as a power series in the electric field vector **E** as[1,3]

$$P(E) = \chi^{(1)}E + \chi^{(2)}EE + \chi^{(3)}EEE + \chi^{(4)}EEEE + \chi^{(5)}EEEEE + \ldots \quad (1)$$

Where $\chi^{(1)}$ is the linear susceptibility of the medium, $\chi^{(2)}$, $\chi^{(3)}$ and $\chi^{(n)}$ are the second-, third- and n-th order nonlinear susceptibilities, respectively. In general, $\chi^{(1)}$, $\chi^{(2)}$... and $\chi^{(n)}$ are tensor coefficients depending on symmetry characteristics of a given optical medium. It can be theoretically proved that under dipole approximation, all even-order terms of nonlinear susceptibilities ($\chi^{(2)}$, $\chi^{(4)}$ ...) should be zero for any kind of isotropic or centro-symmetric media[3]. For simplicity of qualitative discussions, by neglecting the tensor property of the different susceptibilities and the vector property of the electric field E, the generalized refractive index of the medium can be expressed as

$$n = n_0 + n_1E + n_2EE + n_3EEE + n_4EEEE + \ldots \quad (2)$$

In this expression n_0 is the linear refractive index determined by $\chi^{(1)}$; n_1 , n_2... are nonlinear refractive index coefficients determined by $\chi^{(2)}$, $\chi^{(3)}$ ..., respectively. The second-order susceptibility $\chi^{(2)}$ describes second harmonic generation (SHG) and optical parametric oscillation (OPO); n_1 describes linear electro-optical modulation (Pockels effect)[3]. The third order nonlinear susceptibility $\chi^{(3)}$ describes third harmonic generation (THG) and general four wave mixing (FWM); n_2 describes the intensity dependent refractive index change. Near a resonance, the generalized refractive index becomes complex, i.e. n_0, n_1, n_2...are also complex. In such a case, the imaginary part of n_0 represents the linear absorption properties of the medium, and the real part of n_0 determines the frequency dispersion of the real refractive index. The imaginary parts of the higher order terms describe nonlinear absorption (or gain) properties. For isotropic or centro-symmetric media, Eq. (2) can be simplified as

$$n = n_0 + n_2'I + n_4'I^2 + \ldots \quad (3)$$

where $I \propto E^2$ represents the intensity of the optical field. In resonant interaction for this kind of media, the real part of n_2' (or $\chi^{(3)}$) describes an induced refractive index change proportional to I, and the imaginary part of n_2' (or $\chi^{(3)}$) describes two-photon absorption (TPA), as well as (Stokes) Raman gain effect or reverse (anti-Stokes) Raman attenuation effect. Similarly, the imaginary part of n_4' (or $\chi^{(5)}$) describes a three-photon absorption process.

At a molecular level, in a way similar to Eq. (1), the optical field induced dipole moment vector **p** in the molecule can be expressed as[1]

$$p = \alpha E + \beta EE + \gamma EEE + \delta EEEE + \zeta EEEEE + \ldots \quad (4)$$

where α is the linear polarizability , and β, γ, δ, ζ... are different orders of hyperpolarizabilities of a molecule. In general, α, β, γ ... are tensor coefficients. In principle, if we know the quantized energy-level structures and the related complete eigenfunctions for a given molecular system, the α, β, γ ... can be theoretically determined[2]. Furthermore, if we also know the macroscopic symmetric property of an assembly consisting of a great number of molecules, and assume weak Van der Waals' interaction, the different order susceptibilities, $\chi^{(1)}$, $\chi^{(2)}$... $\chi^{(n)}$ can also be determined[1].

Experimental Methods to Probe Multiphoton Resonances

There are a number of experimental techniques which can be used to investigate multiphoton resonances. In our laboratory we have used the following techniques.

(i) Nonlinear transmission: Here one studies the optical input-output relation, as will be discussed later for optical power limiting function. Then from the theoretical modeling of the quantitative dependence of the output intensity on the input intensity one can determine whether one is dealing with two-photon or higher order absorption, and the corresponding nonlinear absorption coefficient can be directly measured[4]. The disadvantage of this simple method is that it does not provide one with the information as to whether the nonlinear absorption is a direct multiphoton absorption or a sequential stepwise multiphoton (more than one step) absorption.

(ii) Up-converted emission:There are some systems which show frequency up-converted fluorescence emission based on two- or multi-photon excitation. The spectral properties of two- or multiphoton excitation related absorption and upconverted emission of the samples can be easily investigated if the excitation source is wavelength tunable. One advantage of this technique is the high sensitivity for detecting the upconverted emission signal by using very sophisticated photon-detection systems. This technique can be also employed to identify the mechanisms that cause the observed up-converted emission. For example, most TPA induced fluorescence emission measurements clearly demonstrated a quadratic dependence of its fluorescence on the input excitation intensity[5].

(iii) Transient absorption: This is a two beam pump-probe experiment where the absorption (or transmission) of a weak probe pulse in the presence of a strong pump pulse is monitored. By conducting a time-resolved experiment in which the time delay between the pump and the probe pulses is varied, one can get direct information on whether the absorption is a sequential stepwise multiphoton process or a direct multiphoton process[6].

(iv) Four-wave mixing: In this method, two pulses cross at an angle to produce an intensity modulation due to interference. The intensity modulation can lead to refractive index modulation by several different possible mechanisms[1]: (a) the intensity dependent refractive index change such as n_2' I term of equation (3) above, which is a purely electronic effect and occurs both under nonresonant and resonant conditions; (b) change in the linear refractive index due to population redistribution along the excited levels, which is highly sensitive to intensity and wavelength tuning change of the incident excitation beams; and (c) thermal change of refractive index due to radiationless transition processes, which usually possesses a much longer relaxation time.

The resultant refractive index modulation, which forms a grating is then probed by the diffraction of a probe pulse. In the time resolved study, the probe pulse is delayed with respect to the pump pulses and, therefore, one can monitor the build-up and the decay of the refractive index grating. Because different grating mechanisms are characterized by different intensity dependence and temporal responses, therefore, some useful conclusions could be abstracted by using this technique[7].

Applications of Multiphoton Resonances

Multiphoton resonance has been widely used for nonlinear optics, laser technology and opto-electronics. The typical examples can be summarized as follows.

(1) multiphoton resonance enhanced optical frequency mixing techniques including optical harmonic generation, sum or difference frequency generation, optical parametric oscillation and amplification;

(2) multiphoton-resonance based nonlinear spectroscopic techniques, such as Doppler-free two-photon absorption (TPA) spectroscopy, upconverted emission spectroscopy, multiphoton ionization spectroscopy etc.;

(3) two-photon resonance enhanced refractive index change, such as TPA enhanced or Raman-resonance enhanced refractive index change;

(4) multiphoton resonance excited dissociation, ionization, energy transfer as well as isotopic separation of molecular systems;

(5) multiphoton absorption based optical power limiting that can be used for protection of human eye and optical sensors;

(6) multiphoton pumped frequency upconverted emission and lasing devices.

Our research group is currently focusing on two specific applications utilizing organic molecular systems: multiphoton absorption based optical limiting and frequency upconversion lasing.

Multi-Photon Absorption Based Optical Limiting

Optical limiting effects and devices are becoming more interesting in the area of nonlinear optics and opto-electronics because of their special application potential. There are several different mechanisms which can lead to optical limiting behavior, such as reverse saturable absorption (RSA), two-photon absorption (TPA), nonlinear refraction (including all types of beam induced refractive index changes), and optically induced scattering [8]. A number of research studies of optical limiting effects related to TPA processes have been reported, most of them have focused on semiconductor materials [8]. According to the basic theoretical consideration, if the beam has a Gaussian transverse distribution in the medium, the TPA induced decrease of transmissivity can be expressed as [8,9]

$$T(I_0)=I(L)/I_0=[\ln(1+I_0L\beta)]/I_0L\beta \tag{5}$$

where I_0 is the incident intensity, L the thickness of a given sample, and β is the TPA coefficient of a given medium. Furthermore, the TPA coefficient β (in units of cm/GW) of the sample is determined by

$$\beta=\sigma_2N_0=\sigma_2N_Ad_0\times10^{-3} \tag{6}$$

Here, N_0, σ_2, and d_0 are the molecular density (in units of $1/cm^3$), molecular TPA coefficient or cross-section (in units of cm^4/GW), and concentration (in units of M/L) of the dopant in a solid matrix, respectively. Finally, N_A is the Avogadro number. For a known experimental relationship between $T(I_0)$ and I_0, the value of β or σ_2 coefficient (or cross-section) can be determined. In some reference papers the molecular TPA cross-section was also defined as

$$\sigma_2' = h\nu\sigma_2 \tag{7}$$

where hν is the energy of an incident photon, σ_2' is in units of cm^4-sec.

It is well known that many organic compound systems exhibit a strong two-photon absorption (TPA) property. Recently we have surveyed quite a number of existing or newly synthesized organic compounds for TPA study[4]. The compound 2,5-benzothiazole 3,4-didecyloxy thiophene (or BBTDOT hereafter), possessed both a large molecular TPA coefficient and very high solubility in common organic solvents. For this reason BBTDOT can be further used to dope into solid matrices for TPA purpose. Preliminary experimental results on optical limiting and stabilization behavior were obtained in BBTDOT doped solid matrix rods[10]. The incident beam was provide by an ultrashort-pulse laser source with a wavelength of 602 nm, pulsewidth of 0.5 ps, spectral width of 60 cm^{-1}, and repetition rate of 30 Hz. The transmissivity as a function of the input intensity is shown in Fig. 1 for two BBTDOT doped samples:(a) a 2.4 cm-long epoxy rod with a dopant concentration of 0.09 M/L and (b) a 1.1 cm-long composite glass rod with a dopant concentration of 0.1 M/L. It shows a clear optical limiting effect. At an intensity level of 500 MW/cm^2, the transmissivity of the samples decreased to less than 50% of its initial value. In Fig. 1 the points represented the measured data, and the lines were the best fitted curves by using Eq. (5). Based on these nonlinear absorption measurements the molecular TPA cross-section of BBTDOT in the epoxy and composite glass rods were estimated as $\sigma_2'=1.4\times10^{-47}$ cm^4-sec and $\sigma_2'=3\times10^{-47}$ cm^4-sec, respectively.

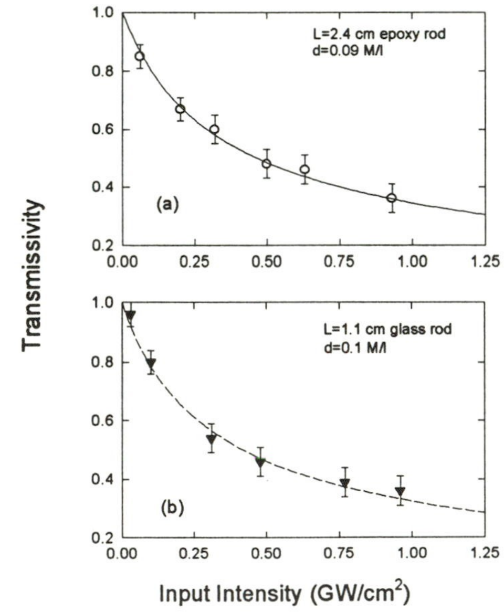

Figure 1 Transmissivity as a function of input intensity for a BBTDOT doped L=2.4 cm-long epoxy rod (a) and L=1.1 cm-long composite glass rod (b).

We recently synthesized and studied a new dye, trans-4-[p-(N-ethyl-N--hydroxyethylamino)styryl]-N-methylpyridinium tetraphenylborate, abbreviated as ASPT hereafter. This dye manifested a superior one-photon pumped lasing performance[11]. Our recent investigation showed that ASPT possessed much greater TPA cross-section in comparing with Rhodamine and other commercial dye materials[12].

For studying the two-photon absorption induced power limiting effect at 1.06 μm for this material, the incident near IR laser beam was provided by a Q-switched Nd:YAG pulsed laser source with a wavelength of 1.06 μm, pulsewidth of ~8 ns, spectral width of ~1 cm^{-1}, and repetition rate of 3 Hz. A 2 cm-long ASPT doped epoxy (EPO-TEX301) rod with a concentration of d_0 = 0.004 M/L was used. The measured transmitted intensity as a function of the incident intensity is shown in Fig. 2[12]. Here the solid line is the theoretical curve predicted by Eq. (5) by using a best fit parameter of β = 6 cm/GW, and the dashed line is given by assuming that there is no TPA (β=0). In Fig. 2 one can see that there is a clear TPA induced optical limiting behavior at the incident intensity levels of 50-

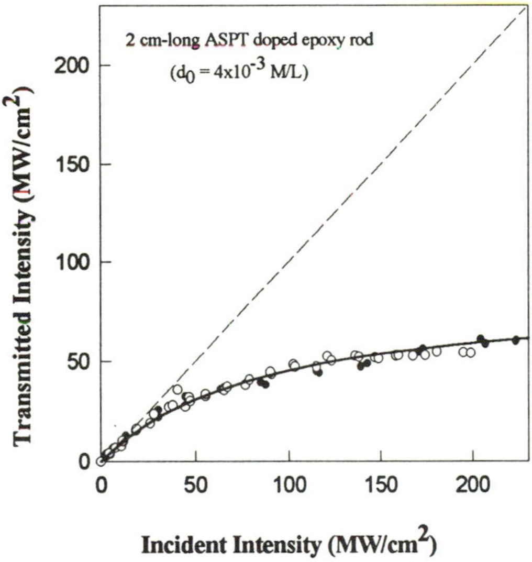

Figure 2 Transmitted intensity as a function of the 1.06 μm input intensity for a 2 cm-long ASPT-doped epoxy rod.

250 MW/cm^2. Based on the known value of β = 6 cm/GW and Eqs.(6) and (7), the values of molecular TPA cross-section of ASPT in epoxy can be estimated as $\sigma_2' = 4.7 \times 10^{-46}$ cm^4-sec. This measured value of σ_2' for dye ASPT in the EPO-TEX301 matrix is greater than

the corresponding values of Rhodamine dyes by 2~3 orders of magnitude [13,14].

In addition, three-photon absorption has been also observed in several solid and liquid materials as well as some gas systems[15-19]. In principle, three- or multi-photon absorption mechanisms can also be employed for optical-limiting performance. For instance, we have found that BBTDOT could demonstrate relatively strong upconverted blue fluorescence and optical power limiting behavior based on three-photon absorption (3PA) pumped with Q-switched 1.06 μm laser pulses.

According to the basic theoretical consideration of three-photon absorption, the intensity change of an excitation beam along the propagation (z axis) direction can be written as

$$dI(z)/dz = -\gamma I^3(z) \quad (8)$$

where γ is the three-photon absorption coefficient of the given sample medium. The solution of Eq. (8) can be simply obtained as[20]

$$I(z) = I_0/\sqrt{(1+2\gamma z I_0^2)} \quad (9)$$

here I_o is the incident intensity of the excitation beam and z is the propagation distance within the sample medium. From Eq. (9) one can see that the three-photon absorption coefficient γ can be experimentally determined by measuring the transmitted intensity as a function of the incident intensity for a given medium with a known thickness of the sample. Fig. 3 shows the measured transmitted 1.06-μm laser beam intensity versus the incident intensity from a 10 cm-long liquid cell filled with BBTDOT solution in TFH of 0.18 M/L concentration[20]. For comparison, the hollow circles represent the data obtain from a 10 cm-long pure THF solvent sample. The transmitted intensity dependence on the incident intensity clearly shows an optical power limiting behavior. The solid line in Fig. 3 is the theoretically fitted

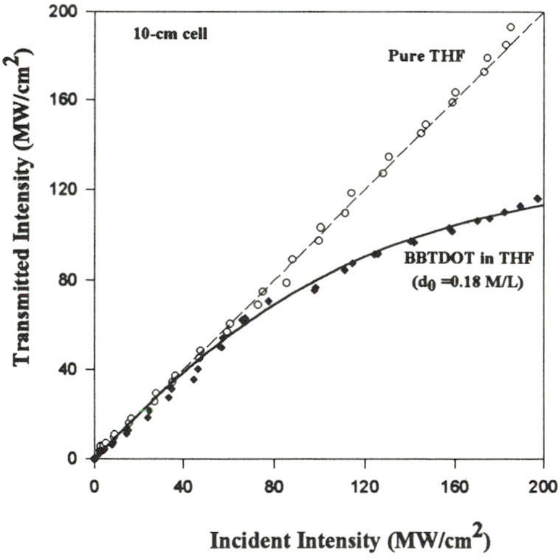

Figure 3 Transmitted intensity as a function of the incident intensity of the 1.06 μm beam for a solution sample (solid squares) and pure THF sample (hollow circles).

curve given by Eq. (9) by using the best fit value of $\gamma = 2.7\times10^{-18}$ cm^3/W^2. It is known that for a given solution sample the γ value is related to the solute concentration d_0 (in units of M/L) by

$$\gamma = \sigma_3 N_A d_0 \times 10^{-3} \tag{10}$$

where σ_3 is the molecular three-photon absorption coefficient or cross-section in units of cm^6/W^2 and N_A the Avogadro number. In some reference papers the molecular three-photon absorption cross-section was defined as

$$\sigma_3' = \sigma_3 (h\nu)^2 \tag{11}$$

where $h\nu$ is the photon energy of the excitation beam, and the σ'_3 is in units of cm^6-sec^2. Based on the known γ value of the measured solution sample, the values of $\sigma_3 = 2.5\times10^{-38}$ cm^6/W^2 and $\sigma'_3 = 8.8\times10^{-76}$ cm^6-sec^2 can be obtained for BBTDOT molecule system in THF.

Two-Photon Pumped Lasing in Novel Dye Doped Bulk Matrixes

Frequency upconversion lasing is an important area of research and is becoming more interesting and promising in recent years. Compared to other coherent frequency upconversion techniques, such as optical harmonic generation or sum frequency mixing based on second- or third-order nonlinear optical processes, the main advantages of upconversion lasing techniques are i) elimination of phase-matching requirement, ii) feasibility of using semiconductor lasers as pump sources, and iii) capability of adopting waveguide and fiber configurations. To date, two major technical approaches have been used to achieve frequency upconversion lasing. One is based on direct two-photon (or multi-photon) excitation of a gain medium (two-photon pumped); the other is based on sequential stepwise multi-photon excitation (stepwise multi-photon pumped). Since 1970's, several reference papers reported experimental results of two-photon pumped (TPP) lasing behavior in organic dye solutions[21-24]. Recently, TPP upconversion stimulated emission was reported in a DCM dye doped PMMA channel waveguide configuration[25].

In order to achieve TPP lasing the gain medium should have a larger TPA coefficient and a higher fluorescence yield for an appropriate pump wavelength. As we mentioned above, the dye ASPT exhibits high molecular TPA cross-section and excellent solubility, therefore, it can be used for two-photon pumped lasing. Pumped with a nanosecond 1.06 μm pulsed laser beam, we have achieved, to the best of our knowledge, the first two-photon pumped cavity lasing in ASPT-doped bulk matrix (polymer, solgel glass, and VYCOR glass) rods.

In our experimental set-up, the input pump IR laser beam was provided by a Q-switched Nd:YAG pulsed laser source with a wavelength of 1.06 μm, pulsewidth of ~10 ns, spectral width of ~1 cm^{-1}, angular divergence ~1.3 mrad, and a variable repetition rate of 1~10 Hz. A 7 mm-long ASPT doped poly-HEMA rod ($d_0 \approx 8\times10^{-3}$ M/L) was used for lasing observation[12]. Once the pump intensity increased to a certain threshold level, simultaneous forward and backward highly directional superradiation could be observed from the rod samples. In order to achieve cavity lasing, two parallel plane dielectric-coating mirrors were employed to form a cavity. Fig. 4 shows the emission spectra of the same 7 mm-long ASPT doped rod sample under different excitation conditions: (a) 532 nm one-photon excited fluorescence emission, (b) 1.06 μm two-photon excited fluorescence emission, and (c) 1.06 μm pumped cavity lasing. One can find that cavity lasing occurred at the central region of the TPA induced fluorescence band, but the lasing band width (~8 nm) was much narrower than the ordinary one-photon excited fluorescence band width (~60 nm) due to lasing threshold requirement.

The lasing output energy versus the pump input energy is shown in Fig. 5 for the rod cavity lasing. In Fig. 5, the solid line is the best fitted curve based on the square law that should be followed for a two-photon excitation process.

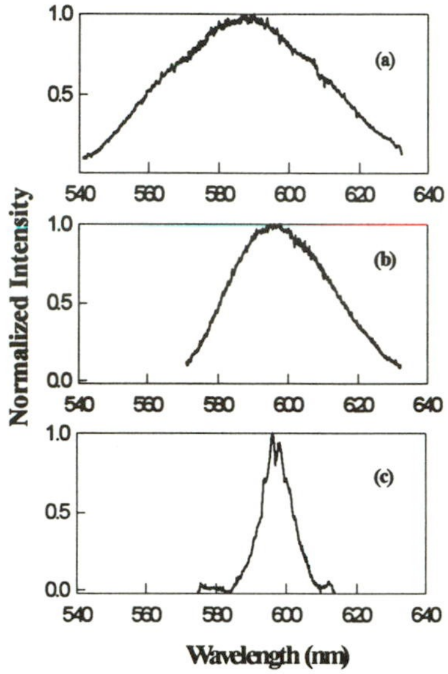

Figure 4 (a) one-photon fluorescence, (b) two-photon fluorescence, and (c) two-photon pumped lasing.

Computational Methods: Issues and Opportunities

Quantum mechanical approach to compute the two-photon absorption cross-section of molecular systems have been investigated by many researchers[26-28]. Most of the past approaches have used the sum-over-states approach which is a perturbative approach involving summation of mixing with many states. In the sum-over-states approach the third-order molecular nonlinear coefficient γ can be expressed as[27,28]

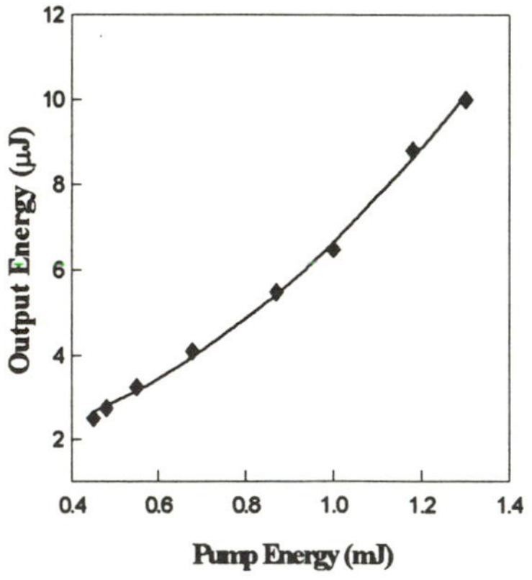

Figure 5 600 nm lasing output as a function of 1.06 μm pump energy.

$$\gamma_{abcd}(-\omega;\ \omega,\ \omega,\ -\omega) = \left(\frac{2\pi}{h}\right)^3 K(-\omega;\ \omega,\ \omega,\ -\omega)e^4$$

$$\times\left\{\sum_p\left[\sum_{i,j,k}{}' \frac{<o|r_a|k><k|r_b|j><j|r_c|i><i|r_d|o>}{(\omega_{ko}-\omega)(\omega_{jo}-2\omega)(\omega_{io}-\omega)}\right]-\sum_p\left[\sum_{j,k}{}' \frac{<o|r_a|j><j|r_b|o><o|r_c|k><k|r_d|o>}{(\omega_{jo}-\omega)(\omega_{jo}-\omega)(\omega_{ko}+\omega)}\right]\right\} \quad (12)$$

Near a two-photon resonance, the energy (frequency) is taken as complex

$$\omega_{jo} \rightarrow \omega_{jo} + i\Gamma \quad (13)$$

Then the imaginary part of γ is related to the two-photon absorption cross-section. One uses a semi-empirical approach for this method.

The second approach used to calculate nonlinearities of organic structures is the derivative approach[2]. In this method one calculates the energy (or the induced dipole moment) in the presence of the field. Then various derivatives of the energy provides the various nonlinear optical coefficient. We have extensively used time-dependent coupled perturbed Hartree-Fock method (TDCPHF) at the ab-initio level to obtain analytically the various derivatives as a function of the frequency of the optical field [27-29]. This method, therefore, provides frequency dependent nonlinear coefficients. However, this method in the past has been used only for computation of nonlinear coefficients under non-resonant conditions. The method can be extended to compute also the imaginary components of the nonlinear coefficients and, therefore, the various multiphoton absorption coefficients by making the frequency a complex quantity. The method would require input of an appropriate damping parameter. Also, the TDCPHF method as used in the past has not involved electron-correlation, which may play an even more important role in the calculation of resonant nonlinear optical coefficients.

In order to design new molecular structures with enhanced two-photon (or multiphoton absorption), at a desired wavelength, the issues to be addressed by computational chemists are the prediction of the position, dispersion and the strength of two-photon and higher order resonances. Such prediction can guide efforts of synthetic chemists in tailoring chemical structures to enhance the multiphoton absorption and to shift energy in appropriate direction by chemical modifications. We, therefore, see a tremendous opportunity for computation chemistry in understanding the relationship between chemical structure and the multiphoton resonance probability.

Acknowledgements

The review presented here was supported by the Air Force Office of Scientific Research and the Polymer Branch of Wright Laboratory under contract F49620-9. We thank Bruce Reinhardt of the Polymer Branch of Wright Laboratory for helpful discussion and providing us with the BBTDOT sample which was synthesized in his laboratory.

Literature Cited

1. Prasad, P. N.; Williams, D. J. Introduction to Nonlinear Optical Effects in Molecules and Polymers, Wiley, New York, 1991.
2. Prasad, P. N.; Karna, S. P. Int. J. Quantum Chem. **1994**, 28, 395.
3. Shen, Y. R. The Principles of Nonlinear Optics, Wiley, New York; 1984.
4. He, G. S.; Xu, G. C.; Prasad, P. N.; Reinhardt, B. A.; Bhatt, J. C.; McKellar, R.; Dillard, A. G. Opt. Lett. **1995**, 20, 435.
5. He, G. S.; Zieba, J.; Bradshaw, J. J.; Kazmierczac, M. R.; Prasad, P. N. Opt. Commun. **1993**, 104, 102.
6. Pang, Y.; Samoc, M.; Prasad, P. N. J. Chem. Phys. **1991**, 94, 5282.
7. Zhao, M. T.; Cui, Y.; Samoc, M.; Prasad, P. N.; Unroe, M. R.; Reinhardt, B. A. J. Chem. Phys. **1991**, 3, 864.
8. Tutt, L. W.; Boggess, T. F. Prog. Quantum Electron. **1993**, 17, 299.
9. Boggess, T. F.; Bohnert, K.; Mansour, K.; Moss, S. C.; Boyd, I. W.; Smirl, A. L. IEEE J. Quantum Electron. **1986**, QE-22, 360.
10. He, G. S.; Gvishi, R.; Prasad, P. N.; Reihardt, B. A. Opt. Commun, **1995**, 117, 133.
11. He, G. S.; Zhao, C. F. Park, C.; Prasad, P. N. Opt. Commun. **1994**, 111, 82.
12. He, G. S.; Bhawalkar, J. D.; Zhao, C. F.; Prasad, P. N. Appl. Phys. Lett. **1995**, 67, 2433.
13. Brunner, W.; Duerr, H.; Klose, E.; Paul, H. Kvantovaya Electron. (Moscow), **1975**, 2, 832.
14. Qui, P.; Penzkofer, A. Applied Physics B, **1989**, 48, 115.
15. Selden, A. C. Nature, **1971**, 229, 210.
16. Agostini, P.; Bensoussan, P.; Boulassier, J. C. Opt. Commun. **1972**, 5, 293.
17. A. Penzkofer, A.; Falkenstein, W. Opt. Commun. **1976**, 16, 247.
18. Kramer, M. A.; Boyd, R. W. Phys. Rev. **1981**, B23, 986.
19. Brost, G.; Braeunlich, P.; Kelly, P. Phys. Rev. **1984**, B30, 4675.
20. He, G. S.; Bhawalkar, J. D.; Prasad, P. N.; Reinhardt, B. A. Opt. Lett. **1995**, 20, 1524.
21. Rapp, W.; Gronau, B. Chem. Phys. Lett. **1971**, 8, 529.
22. Topp, M. R.; Rentzepis, P. M. Phys. Rev. **1971**, A3, 358.
23. Rubinov, A. N.; Richardson, M.C.; Sala, K.; Alcock, A. J. Appl. Phys. Lett. **1975**, 27, 358.
24. Kwok, A. S.; Serpenguzel, A.; Hsieh, W. F.; Chang, R. K.; Gillespie, J. B. Opt. Lett. **1992**, 17, 1435.
25. Mukherjee, A. Appl. Phys. Lett. **1993**, 62, 3423.
26. Lin, S. H.; Fujmura, Y.; Nausser, H. J.; Schlag, E. W. Multiphoton Spectroscopy of Molecules, Academic Press, Orlando, Florida, 1984.
27. Birge, R. R.; Pierce, B. M. J. Chem. Phys. **1979**, 70, 165.
28. Pierce, B. M. J. Chem. Phys. **1989**, 91, 791.
29. Karna, S. P.; Prasad, P. N.; Dupuis, M. J. Chem. Phys. **1991**, 94, 1171.

RECEIVED November 29, 1995

INDEXES

Author Index

Affiliation Index

Subject Index